NATIONAL CURRICULUM EDITION

UNDERSTANDING MATHEMATICS

Second Edition

C. J. Cox & D. Bell

JOHN MURRAY

Acknowledgements

The authors are grateful for advice received from the following: Dr Kath Hart, Brian Bolt (Exeter University), Andrew Rothery (Worcester College of Education), Alec Penfold, Jenny Byrom (Huish Episcopi School), Jacqueline Gilday (Wells Blue School), Hazel Bevan (Millfield School), John Wishlade (Uffculme School), John Collins, Marion Fletcher and Mr R. A. Batts (Texas Instruments).

The authors thank their publishers for their help, their families for their tolerance, and all the many teachers and pupils who have helped in the testing and revising of the course.

Illustrations by Tony Langham and Technical Art Services.

The following examination boards have kindly given permission for the use of past examination questions. The authors accept full responsibility for any errors in these questions or in the answers to the same.

Midland Examining Group (MEG)   **Southern Examining Group** (SEG)

University of London Council (ULEAC)   **Northern Board** (NEAB)

Welsh Joint Education Committee (WJEC)

Pupils' Book 5   ISBN 0-7195-5032-7
Teachers' Resource Book 5   ISBN 0-7195-5033-5

© C. J. Cox and D. Bell 1987, 1993

First published in 1987
by John Murray (Publishers) Ltd
50 Albemarle Street, London W1X 4BD

Reprinted 1987, 1989

Second edition 1993

Typeset by Blackpool Typesetting Services Ltd,
Blackpool

Printed and bound in Great Britain at the
University Press, Cambridge

A catalogue entry for this title can be obtained
from the British Library

ISBN 0-7195-5032-7

About this book

This is the second book in the widely used two-book **Understanding Mathematics** course for Key Stage 4 leading to GCSE mathematics at National Curriculum Levels 5 to 10.

Although best used after the first four books in the series, Understanding Mathematics Book 5 may be used as a separate revision book for the UK National Curriculum Key Stage 4 and for overseas GCE and local equivalent (e.g. CXC) examinations.

This **Pupils' Book** is divided into six main sections:

- **Exercises:** banks of graded questions (see below). These start on page 1.
- **Papers:** mixed topic questions for revision and homework (pages 170–200).
- **Reference notes:** notes and worked examples, coded to the chapters (pages 201–282).
- **Information for aural tests** (pages 283–286)
- **Answers** (pages 287–314)
- **Glossary/Index:** defining terms, and telling you where to find notes and examples (from page 315).

The exercises are divided in four sections:

- Introductory questions up to Level 8.
- Reinforcement ('starred questions').
- Development questions, up to Level 10. Those at Levels 9 and 10 are marked H.
- Boxed questions, to extend the most able.

This structure helps you to learn at your own pace and to build up your confidence.

The **Teachers' Resource Book** has teaching notes and demonstration examples; transparency masters; answers, including diagrams; aural tests; practical worksheet masters; assessment tests; computer teaching programs; attainment target tests; and photocopiable reference notes.

Notes on the National Curriculum edition

Book 5 has undergone major restructuring.

- The order has been changed to align with National Curriculum Levels 5 to 10. To cater for both the central and upper bands, questions and notes at Levels 9 and 10 have been coded H, so users know which topics are not essential for the central tier GCSE.
- Questions from recent GCSE and specimen Key Stage 4 papers have replaced many of the GCE/CSE questions in the first edition.
- Much new material has been added, particularly in algebra (sequences; iteration) and data (sampling; dispersion; critical-path analysis).
- The topic of sets has been deleted, except for the use of Venn diagrams.
- The reference section has been re-arranged and individual topics made easier to find.
- The glossary has been extended and adapted to serve a dual purpose as an index.

Contents

Topics given in brackets are at National Curriculum Levels 9 and 10. Chapters 30 to 33 cover major topics that only appear at Level 10.

Number

Algebra

Shape

Data

Level 10

Reference notes contents

1   Knowing about numbers

Kinds of numbers	Standard form
Estimation	Inequality
Approximation	(Rationality)

● You need to know . . .

Basic arithmetic 1 to 6 (page 201)
Computation 2 (page 204)
Basic algebra 6 (page 220)

In all exercises, questions marked with * are reinforcement at Level 8 or below, while questions marked H are above Level 8. All boxed questions may be above Level 8.

1 $A = \{15, 16, 17, 18, 19, 20\}$

From set A, state:
(a) two prime numbers   (b) a square number   (c) a factor of 100
(d) a multiple of 9   (e) a triangular number.

2   What number, when squared, gives the same answer as its square root?

3 n is an integer such that $3 < n \leqslant 5$. Which numbers could n be?

4   Using a calculator, find:
(a) 1.65×14.2 correct to 1 decimal place
(b) $1.65 \div 14.2$ correct to 2 significant figures.

5   Figure 1:1 shows a riverside flood-warning post.
(a) At what level is the water now?
(b) If the water rises $1\frac{1}{2}$ feet, what level will it be at?
(c) If the water then fell $2\frac{1}{2}$ feet, what would be the reading?

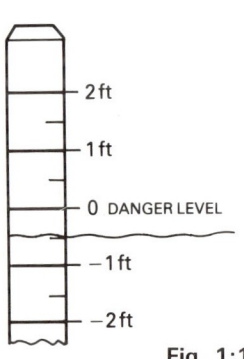

Fig. 1:1

6   State for 15, 20 and 25:
(a) their highest common factor
(b) their lowest common multiple.

7   Write these in order, from least to greatest:
3.14, 0.651, 0.009, 3.142, 0.6501, 0.6512

8   Find for 4 and 6:
(a) their sum   (b) their difference
(c) their product expressed as a product of primes
(d) their two possible quotients expressed both as common and decimal fractions.

9 Tower Bridge was completed in 1894, when it opened 6000 times. It was operated 24 hours a day by 119 men. In 1991 it opened 200 times and had a staff of 15.

(Mark Waters)

(a) Which numbers do you think are exact?

(b) What is the maximum and minimum possible number of openings in 1894 and 1991 if the 6000 is correct to the nearest ten and the 200 is correct to the nearest five?

10 Write as an ordinary number:

(a) 6^2 (b) 2^5 (c) $4^{\frac{1}{2}}$ (d) $8^{\frac{1}{3}}$ (e) 3.4×10^2 (f) 6.38×10^{-2}.

11 Write in standard form:
(a) 98.5   (b) 0.46
(c) the answer to $3 \times 10^2 + 2.5 \times 10^{-1}$
(d) the answer to $3 \times 10^{-1} \times 2 \times 10^3$
(e) the answer to $5 \times 10^3 \times 7 \times 10^{-2}$.

12 Write:
(a) $\frac{2}{5}$ as a decimal fraction   (b) $\frac{3}{5}$ as a percentage   (c) 0.07 as a common fraction
(d) 15% as a common fraction as simply as possible.

*13 List all six factors of 12.

*14 From your answer to question 13 list:
(a) the prime numbers   (b) the odd square numbers   (c) the factors of 20
(d) the multiples of 4   (e) the triangular numbers   (f) the rectangular numbers.

*15 Write in figures:
(a) one thousand   (b) one million   (c) two hundred thousand and fifty-six.

*16 Write your answers to question 15 in standard form.

*17 A calculator shows 0.000 176 as 1.76   −04. How would it show:
(a) 0.000 28   (b) 0.0365   (c) 0.842?

*18 A 15 cm bar of metal shrinks by 0.1 mm when the temperature drops by 1 °C. What is its new length in:
(a) mm   (b) cm?

*19 Simplify:
(a) $2 \times 2 \times 5$ (b) $2 \times 3 \times 3 \times 7$.

*20 Express as a product of prime factors, as in question 19:
(a) 18   (b) 39   (c) 42.

*21 Write the numbers at points A to E in Figure 1:2:
(a) as fractions or mixed numbers   (b) as decimal fractions.

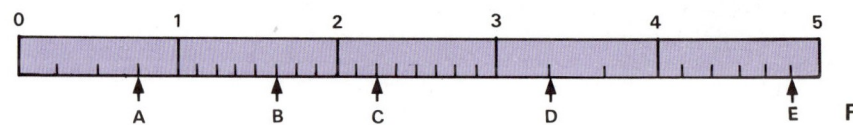

Fig. 1:2

*22 Copy Figure 1:3. Show with arrows, as in Figure 1:2, the points:
A, 2.3   B, 3.7   C, −1.3   D, −1.8   F, −2.5   F, −0.6

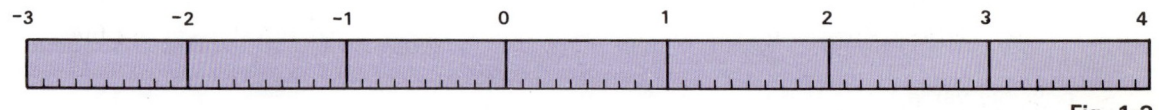

Fig. 1:3

*23 Suggest the most likely next two terms in the following sequences.
(a) 17, 14, 11, 8   (b) 1, 4, 9, 16   (c) 3, 6, 12, 24   (d) 6, 12, 20, 30
(e) $\frac{1}{2}, \frac{1}{4}, \frac{1}{8}, \frac{1}{16}$ (f) 10, 1, 0.1   (g) 11, 8, 5, 2   (h) 0.6, 0.4, 0.2   (i) 1, 8, 27, 64
(j) 26, 17, 10, 5, 2.

*24 The tables in Emma's classroom are arranged in four equal rows. The tables in John's classroom are arranged in six equal rows. Both classrooms have the same number of tables. What is the smallest possible number of tables in each room?

*25 Ninety-six packets of butter are to be packed in a box, four packets deep. Find four ways to arrange the packets, then choose one of them as a 'best way', saying why you chose it.

*26 Ten children and two adults are to travel 200 miles on the school minibus which costs 21p per mile to hire. The adults are to pay twice as much as the children. What charges should be made?

*27 Arrange in order of size, largest first:
15%, 4/5, 0.6, 4/7, 1/3, 2/3.

*28 On January 11, the temperature at midnight was −9 °C. On July 2, the temperature at midday was 24 °C. What is the difference between these two temperatures?

*29 (a) The number 0.07 can be written as $\frac{7}{c}$. Find c.

(b) The number 0.07 can be written as 7×10^n. Find n.

*30 Which numbers from 5 to 12 cannot be written as the sum of two prime numbers?

*31 If a is an even number, b is an odd number, and c can be either, state whether the answers to the following are odd, even, or either.
(a) $3 \times a$ (b) $3 \times b$ (c) $a + b$ (d) $a + b + c$ (e) $a - b$ (f) a^2
(g) $a^2 + b^2$ (h) $a^2 + 2b^2$.

H32 You must make sure you can explain what a rational and an irrational number is, and learn some examples. Use the glossary. Also see chapter 5, questions 18 to 21.

Are the following rational or irrational?

(a) $\sqrt{4}$ (b) $\sqrt{14}$ (c) $0.\dot{3}$ (d) π (e) 0.162 (f) $\frac{1}{4}$ (g) $\dfrac{\sqrt{3}}{\sqrt{2}}$

(h) 2^{-1} (i) $(\sqrt{3})^4$

(j) $\sqrt{3}.\sqrt{12}$

(k) $\pi\sqrt{7}$ (l) $\sqrt{3} + 2$

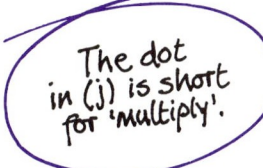
The dot in (j) is short for 'multiply'.

H33 $\dfrac{m}{n}$ is a rational number, where m and n are two different irrational numbers. Suggest a possible value for m and n.

H34 $\dfrac{a \times b}{c}$ is approximately equal to 56. What is an approximate value for $\dfrac{10a \times 20b}{100c}$?

H35 State the reciprocal of:
(a) 2 (b) 7, correct to 3 significant figures (c) $\frac{3}{4}$ (d) 4/5.

H36 Simplify:
(a) 2^{-1} (b) 3^{-2} (c) $4^{\frac{1}{2}}$ (d) $4^{-\frac{1}{2}}$ (e) $8^{\frac{2}{3}}$ (f) $8^{-\frac{1}{3}}$ (g) $100^{1\frac{1}{2}}$.

---

37 Write the nth term for each sequence:
(a) 2, 4, 6, 8, 10, . . . (b) 1, 3, 5, 7, 9, . . . (c) 1, 4, 9, 16, . . .
(d) 1, 8, 27, 64, 125, . . . (e) 3, 7, 11, 15, 19, . . . (f) 1, 0.1, 0.01, 0.001, . . .
(g) 2, 6, 12, 20, 30, . . . (h) 6, 12, 20, 30, 42, . . .

38 In this question, i is a positive integer.
(a) What can you say about $2i$?
(b) What can you say about $2i + 1$?
(c) Prove the answers to question 31 are always true.
(d) Investigate question 31 and this question if the numbers a, b and c could be negative.

39 A picture frame needs two 15.4 cm lengths and two 12.6 cm lengths of moulding. How should I cut two 2-metre lengths of moulding to make the maximum number of frames?

40 Find a multiple of 7 that leaves remainder 1 when divided by 2, 3, 4, 5 or 6.

41 The area of Scotland is about $7.88 \times 10^4 \text{km}^2$ and the area of England is about $1.30 \times 10^5 \text{km}^2$. The population of Scotland is about 5.2×10^6 and the population of England is about 4.65×10^7.

Giving answers in full (not in standard form) state:

(a) how much larger in area England is than Scotland

(b) how many times as many people live in England as in Scotland

(c) the density of population in each country, in people per km^2 (Comment on your answer.)

(d) the largest and smallest possible area for England according to the given information

(e) the difference it would make to your answer to part (d) if England's area had been given as $1.3 \times 10^5 \text{km}^2$.

42 Find a four-digit number that reverses when multiplied by 4. Investigate further.

43 Find a number n that leaves a remainder of 1 less than the divisor when it is divided by 2, 3, 4, 5, 6, 7, 8, 9 or 10 (e.g. $n \div 7$ leaves remainder 6).

2 Using numbers

The four rules
Directed numbers
BODMAS

● **You need to know . . .**

Computation 1 to 4 (page 204)

1 Show full working to calculate:
(a) 306×48 (b) $1248 \div 6$ (c) $435 \div 15$.

2 The chart shows the maximum and minimum temperatures on five days.

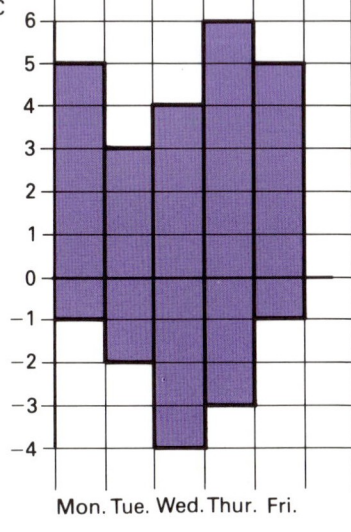

(a) What is the difference between the maximum and minimum temperatures each day? Write your answers like this: Sunday $8 - -4 = 12\,°C$.

(b) How much colder was Wednesday's minimum then Friday's?

(c) The answer to part (b) could be expressed as
$-1 - -4 = -1 + 4 = 3$.
Calculate $-3 - -7$.

3 Write in order from lowest to highest:
$-7\,°C$, $3\,°C$, $-4\,°C$, $8\,°C$, $0\,°C$, $-6\,°C$.

Mon. Tue. Wed. Thur. Fri.   **Fig. 2:1**

4 Calculate:
(a) $4 - 8$ (b) $5 - -2$ (c) $-6 + -3$ (d) $-9 - -2$ (e) $-3 + 9$.

5 Work out $s = ut + \frac{1}{2}at^2$ when $t = 0.5$, $a = -3$ and $u = 8$.

6 (a) $\frac{3}{4} + 1\frac{2}{5}$ (b) $4\frac{3}{5} - 1\frac{3}{4}$ (c) $\frac{2}{5} \times 3\frac{1}{3}$ (d) $2\frac{1}{5} \div 3\frac{2}{3}$.

7 What should 0.2 be multiplied by to give 1.68?

8 The average wage at a car factory is £231.80 per 38 hour week. Estimate roughly, showing clearly your method:

(a) the average annual wage (b) the hourly rate.

9 Roy plans to put $\frac{1}{3}$ of his garden to lawn, $\frac{2}{3}$ of what is left he will use for fruit, and the remainder will be for vegetables.

   (a) What fraction of his garden will be for vegetables?

   (b) His garden has an area of 240 square yards. About how many square **metres** does he plan to plant with fruit?

*10 Show full working to calculate:
   (a) 160×27 (b) 402×89 (c) $1236 \div 4$ (d) $2184 \div 26$.

*11 State the difference between these temperatures:
   (a) $7\,°C$ and $-1\,°C$ (b) $-5\,°C$ and $9\,°C$ (c) $-3\,°C$ and $-8\,°C$
   (d) $32\,°C$ and $-12\,°C$.

*12 Write the eight temperatures given in question 11 in order from lowest to highest.

*13 Find v when $v^2 = u^2 + 2as$ and $u = 12$, $a = -4$ and $s = 5\frac{1}{2}$.

*14 (a) $\frac{1}{4} + \frac{3}{5}$ (b) $\frac{5}{7} - \frac{2}{3}$ (c) $\frac{7}{9} \times \frac{3}{7}$ (d) $\frac{1}{8} \div \frac{1}{3}$.

*15 (a) $1\frac{1}{2} + \frac{3}{8}$ (b) $2\frac{2}{3} - 1\frac{1}{2}$ (c) $\frac{1}{3}$ of $2\frac{2}{5}$ (d) $2\frac{3}{4} \div 6\frac{7}{8}$.

*16 Ahmed borrows two-thirds of the cost of his new £54 000 house. How much does he borrow?

*17 An organism grows by cell division. Every cell divides into two cells every 15 minutes. How many cells will develop from one cell during 2 hours?

*18 My video tape records for 195 minutes. Can I record the whole of a concert which starts at 5:25 and finishes at 8:45? If so, how many minutes are spare? If not, how many minutes short is the tape?

*19 Jodi travels 18 miles in 20 minutes. What is her average speed in miles per hour?

20 A teacher gives each pupil 20 marks to start with. A right answer scores $+2$ marks and a wrong answer scores -3 marks. What will be the final score of a pupil with five correct answers and seven wrong answers?

21 If $\dfrac{x}{10} = \dfrac{5}{6}$, find x.

22 Find:
   (a) $(3 + 5) \div 8$ (b) $3 + (5 \div 8)$ (c) $\dfrac{70\,000 \times 60\,000}{800}$.

23 The formula $C = \dfrac{5(F - 32)}{9}$ converts degrees Fahrenheit (F) to degrees Celsius (C). What is the Celsius equivalent of:

   (a) $0\,°F$ (b) $-10\,°F$?

24 (a) $(\frac{1}{3} - \frac{1}{4}) \div (\frac{1}{2} - \frac{1}{3})$ (b) $\dfrac{6}{\frac{4}{5} - \frac{2}{3}}$

25 In Figure 2:2, ABC is an equilateral triangle. M and N are midpoints. What fraction of the triangle is area X?

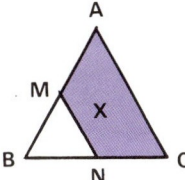

Fig. 2:2

26 British Airways carried about 14.6 million passengers one year, using a fleet of 148 aircraft which flew a total of 138 million miles on 177 routes. For their major long routes they used the 340 ton Boeing 747 jumbo jets. Each jet can carry 407 passengers.

Let us follow a jumbo flight from London to Sydney, Australia. At 2050 on Saturday the plane, carrying 121 tons of fuel and 407 passengers, sets off on the 3627 miles to its first refuelling stop at Muscat. It is on its 10 378 th take-off, and has flown for 45 369 hours. The ideal altitude is 37 000 feet, but our journey starts at 27 000 feet until it reaches Austria, when it climbs to 33 000 feet.

The passengers settle down. Those going to Sydney have paid £3796 return 1st class, or £1822 club class, or £938 economy class.

On landing at Muscat at 0345 Sunday (London time) the crew is changed. The first crew will wait 24 hours, then take over in their turn. This pattern continues for 21 days, when they will be given 6 days' leave in London.

The second stop, Kuala Lumpur, is reached at 1110 London time, then on to Singapore at 1250. Finally, 10 946 miles from its departure point in London, the jumbo lands at Sydney at 2152 on Sunday, London time.

(British Airways)

(a) On average, how many passengers were carried each month that year?

(b) On average, how many miles were flown per aircraft that year?

(c) If the average weight of a passenger and luggage is 80 kg, what is the total weight of the Boeing 747, its passengers and fuel on take-off? (You may take 1 tonne as the same as 1 ton.)

(d) How long does the plane take to reach Muscat?

(e) What is the average speed of the plane between London and Muscat?

(f) Mount Everest is five and a half miles high. One mile is 5280 feet. About how many feet higher than Mount Everest is the plane flying once it has passed Austria?

(g) What is the cost per mile for a club-class passenger travelling from London to Sydney and back again?

(h) How long does the Boeing take (ignoring time on the ground) between
    (i) Muscat and Kuala Lumpur   (ii) Kuala Lumpur and Singapore
    (iii) Singapore and Sydney   (iv) London and Sydney?

(i) What is the average speed of the plane over the whole journey?

---

27 In a sheep-dog trial a dog managed to divide the flock into two equal groups, three equal groups, four equal groups, five equal groups, six equal groups and seven equal groups . . . well, almost. You see, every time the same small sheep escaped from the flock, and was never put into a group. What is the smallest possible number of sheep in the flock?

28 A factory produces canned fish of three kinds, caviar, salmon and tuna, each in a different size can. Boxes are needed of one size only, for transportation. Find the size of the smallest possible box that will be completely filled by each type of can, for cans of the following sizes:

Caviar: 6 cm diameter, 4 cm high
Salmon: 8 cm diameter, 6 cm high
Tuna: 12 cm diameter, 8 cm high.

Draw a plan and elevation of the box filled with each kind of can, and say how many of each are in the box.

29 Split the numbers 28, 35, 42, 44, 48, 61, 63, 77, 84 and 88 into three sets such that the total of the first set is 180, the total of the second is 190 and the total of the third is 200.

3 Percentages

% of an amount   % change
One amount as a % of another

● **You need to know . . .**

Percentages 1 to 8 (page 206)

1 Find 9% of £150.

2 Alison scores 55 out of 80 in French and 48 out of 60 in German. By changing each to a percentage mark, find which subject she did best in.

3 My car insurance costs £140 less 60% no-claims discount. How much do I actually pay?

4 Arlene invests £1800 in some shares. Three months later she sells them for £4800. What is her percentage profit?

5 Write as a simplified common fraction and as a decimal fraction:
   (a) 75%   (b) 16%   (c) 24%   (d) 7%   (e) $33\frac{1}{3}$%.

6 (a) A bookseller orders 50 copies of a book with a list price of £18.50. She pays the publisher £560.60 for the books. What percentage profit on her cost price will she make if she sells all the books at the list price?

   (b) In fact only 33 copies are sold at £18.50. If the rest are sold on special offer at 40% off list price what will now be the bookseller's final profit?

   (c) More bad news: the last eight books remain unsold and are put in the Bargain Basement. When these have been sold the shopkeeper has made 35% over all. What was the price of the book in the Bargain Basement?

7 In an advertisement we are told that a germ population increases to the square of its previous population every hour. A young girl calculates the population by chanting; 'Two times two equals four; four times four equals sixteen; sixteen times sixteen equals . . .'

   (a) List the populations at the end of each of the first six hours (4, 16, etc.). Use standard form where sensible.

   (b) What is the percentage increase over the first three hours?

*8 A pullover priced at £60 is reduced in a sale by 20%. What is the sale price?

*9 A car which cost £2400 was sold with a loss of 5% of the cost price. What was the selling price?

*10 A bottle of squash cost 76p five years ago. Since then its price has increased by 50%. What does it now cost?

*11   In a sale 15% is taken off the price of a skirt. The original price was £46.40. Calculate the sale price.

*12   The wholesale price of a television is £260. The retailer sells it for £312. What is the retailer's percentage profit?

*13   What overall percentage profit is made when a car bought for £1260 is sold, after a respray costing £465, for £2760?

*14   A bicycle bought for £168 was later sold for £126. Find:
(a) the loss   (b) the loss as a percentage of the buying price.

*15   A greengrocer buys lettuce at £1.80 a dozen. For how much must he sell each lettuce to make a profit of 40% on his cost price?

*16   To buy a flat costing £22 500 Tony has to pay a 10% deposit.
(a) How much is the deposit?
(b) How much would he still have left to pay?

*17   A bus fare was increased from 40p to 50p. By what percentage was it increased?

*18   A motorbike has a list price of £2400. It can either be paid for immediately, in which case a 10% discount is given, or bought on credit terms of 20% deposit and 24 payments of £90.

What is the total amount paid in both cases?

*19

Canford Workers Offered 8% Pay Rise

(a) Indira works at the Canford factory and is paid £150 per week. If the pay rise is accepted, how much will Indira be paid per week?

(b) The workers at the factory had a ballot and 1200 workers voted. There was a majority of 3 to 2 in favour of accepting the pay rise. How many workers voted to accept the pay rise?

(c) Altogether the factory has 1500 workers. What percentage of them voted to accept the pay rise?

(MEG)

*20   Mr Gambol bought £1000 worth of shares in Tryall Ltd. In the first year they increased by 8%, in the second year they decreased by 10%, and in the third year they increased by 2%. What was the value of the shares at the end of each of the three years?

21   The population of the UK in 1985 was 56 million. Fifteen per cent of these were aged 65 or over, an increase of 2 million on the 1961 figure, when the population was 55 million.

(a) How many people aged 65 or over were there in the UK in 1985?

(b) What percentage of the population were aged 65 or over in 1961?

22 A computer system sells for £5748 including 17.5% VAT. James Ltd can claim back the VAT they paid. How much should they claim?

23 Freda has an 8% pay rise, giving her earnings of £800 per month. What was her pay before the rise, correct to the nearest pound?

24 In an MOT testing station 10% of the cars tested one year had faulty brakes and of these 25% had faulty shock-absorbers. What percentage of the cars tested had both faults?

25 Between 1969 and 1992, a house price rose by 1400%. In 1992 it was valued at £90 000. What was its value in 1969?

26 In a general election candidates lose their deposit if they do not obtain at least 5% of the total votes cast in the constituency for which they are seeking election.

(a) In a certain constituency three candidates, A, B and C, had put up for election.

Out of the 42 560 votes cast, 21 523 people voted for candidate A, 18 862 people voted for candidate B, and the rest voted for candidate C.

Decide whether, and by how many votes, candidate C lost or saved her deposit.

(b) In another constituency there were just two candidates, R and S. The winner, who was candidate R, received 16 017 votes. By letting x be the number of people who voted for candidate S, or otherwise, calculate the least number of votes candidate S would have to obtain in order not to lose his deposit.

(NEAB)

27 Once a year a scientist measured the mass of a certain piece of decaying radioactive element. His results are shown.

Time (years)	0	1	2	3
Mass (kg)	20	18	16.2	14.58

(a) Calculate the annual percentage decrease of the mass.

(b) Calculate the mass after 10 years.

(c) Estimate the half-life of the element (i.e. the time it takes to lose half its mass).

(SEG)

28 An engineer is using the formula $s = \frac{1}{2}gt^2$. The value of g she uses has a maximum error of 2% and her value of t has a maximum error of 3%. What is the maximum percentage error in calculating s?

(ULEAC)

29 Road gradients are given as a percentage. What is it a percentage of? What do you think about this use of percentages?

30 Find out how the interest is worked out on a credit card. Write about your findings, including a worked example. Program a calculator or a computer to do the calculation.

4 Ratio

Direct and inverse proportion

Rate

Proportional division

• **You need to know . . .**

Ratio 1 to 6 (page 206)
Variation 1 (page 240)

1 State whether the following are examples of direct or inverse proportion.

(a) The speed of a car and the distance travelled per minute.

(b) The circumference of a circle and its diameter.

(c) The number of people travelling on a coach outing and the cost per person if they share the cost of hiring the coach equally.

(d) The extension of a spring and the load hung on it.

2 Which graph in Figure 4:1 could show that the price of houses has risen less in the last twelve months than in the previous twelve months?

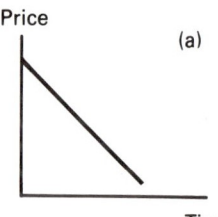
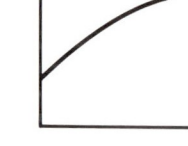
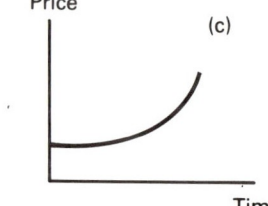
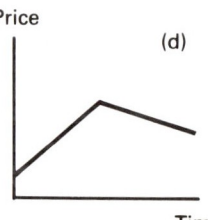

Fig. 4:1

3 Simplify the following ratios:
(a) $16:8$ (b) 4 cm to 2 m (change them both to cm first)   (c) 15 cm to 10 mm
(d) $3\frac{1}{4}:2\frac{1}{2}$.

4 Express in the form $n:1$
(a) $12:4$ (b) $12:5$ (c) $6:8$.

5 The ratio of boys to girls in a youth club of 25 members is $2:3$. How many members are girls?

6 Tanya and Sandra share their pool winnings in the ratio $5:4$. Tanya receives £300. How much does Sandra receive?

7 A house plan is drawn to a scale of $1:100$. The front wall is 10 cm long on the plan. How long is it really?

8 The scale of a map is $1:25\,000$. Find the distance in km represented by 20 cm on the map.

13

*9   To make orange paint, red and yellow are mixed in the ratio 2 : 3. If I have 12 litres of yellow paint and 12 litres of red paint, how much orange paint can I make?

*10   When a 100 franc note is worth about £10, about how much is a 25 franc note worth?

*11   A coin is made of three parts copper and five parts nickel.
(a) What fraction of the coin is copper?
(b) What percentage of the coin is copper?
(c) The coin weighs 10.4 g. What weight of copper is in it?

*12   Which of the toothpaste packs shown in Figure 4:2 is the best buy?

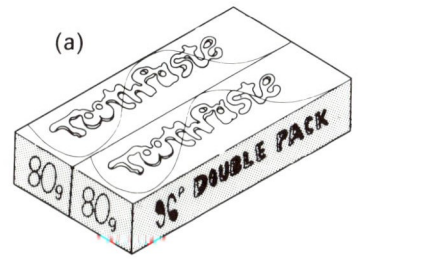
(a)

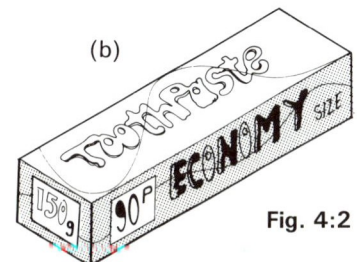
(b)

Fig. 4:2

*13   Taking 8 km ≈ 5 miles, find the equivalent of 60 m.p.h. in km/h.

*14   A model aeroplane is to a scale of 1 to 50. The real plane is 50 metres long. How many cm long is the model?

*15   Figure 4:3 shows a 3-speed gear box. Wheels A have 60 teeth, wheels B have 40 teeth and wheels C have 20 teeth.

The gear ratio is the number of turns made by shaft X to the number of turns made by shaft Y, either in the form $n : 1$ or $1 : n$. State three gear ratios obtainable as shaft Y is moved to the right.

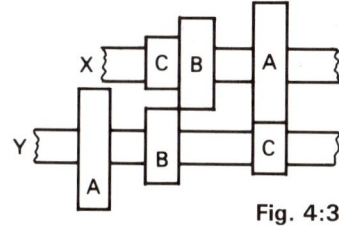
Fig. 4:3

*16   A builder reckons he will take 12 days to build a garage if he works 8 hours a day. Due to illness he has to complete the job in 10 days. How many hours a day must he now work?

17   State a formula to find the cost C in pounds of x articles given that the cost of y of the same articles is z pounds.

18   At 0900 one day a six-foot pole casts a nine-foot shadow. At the same time a tree casts a fifteen-foot shadow. How tall is the tree?

19   A packing machine packs a box of chocolates in 12 seconds, at a rate of 150 chocolates a minute. A newer machine can deliver 250 chocolates per minute. How long will the new machine take to pack a box?

20 Why is the method shown in Figure 4:4 a good way of dividing a given distance into equal divisions?

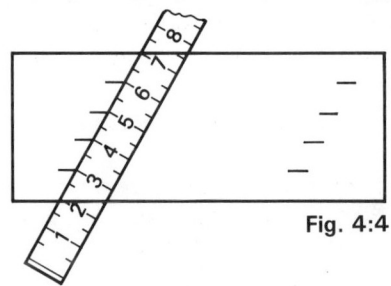

Fig. 4:4

21 Duracolor sells paint in cans of three sizes, 500 ml, 1 litre and $2\frac{1}{2}$ litres. The empty cans cost 15p, 20p and 25p respectively. The paint itself costs £1.75 a litre to produce and the firm allows 30p per litre of paint produced to cover production expenses and overheads. The profit on each can is to be 40% of its total production cost. Work out the price at which Duracolor should sell its paint to the retailers.

22 In a factory the ratio of the number of articles produced per day to the number of employees is 100:1. New production methods give a 15% cut in the number of employees with a 20% rise in output. What is the new ratio of articles produced to the number of employees, giving your answer in the form $n:1$?

23 The coefficient of friction (μ) can be worked out from the ratio $F:R$, where F is the force required to move a block and R is the weight of the block plus weights put on it. This is represented in Figure 4:5.

The table gives some experimental results. Plot the points, draw a straight line of best fit, and give a value for μ for the block being tested.

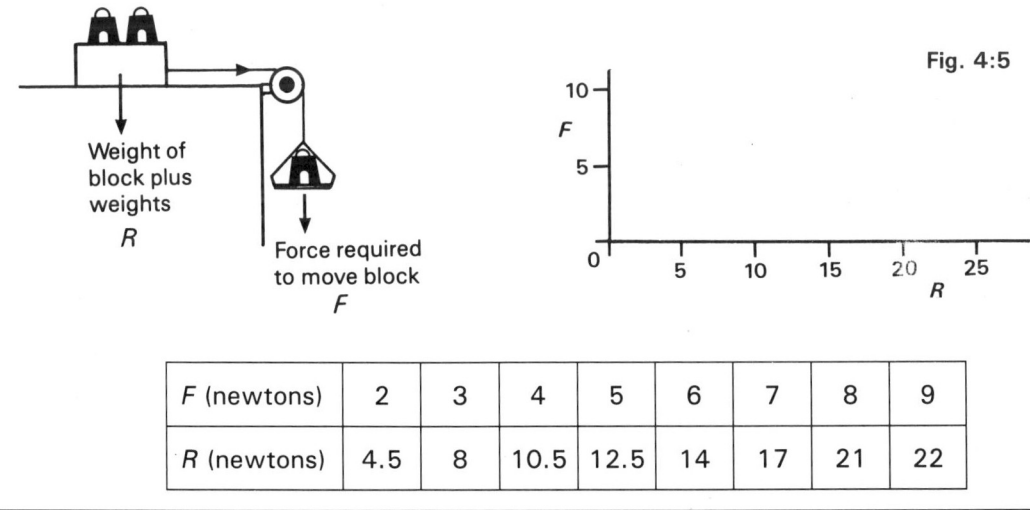

Fig. 4:5

F (newtons)	2	3	4	5	6	7	8	9
R (newtons)	4.5	8	10.5	12.5	14	17	21	22

5 Using a calculator

Efficient use   (Recurrence and rationality)
Trial and improvement   (Iteration)

● You need to know . . .

Calculators 1 to 7 (page 211)

1 Does your calculator follow the BODMAS rule? See the reference notes *Calculators 1* if you are not sure.

2 Does your calculator have an accumulating memory and/or a store memory? Explain the differences between them. See the reference notes *Calculators 3* if you are not sure about this.

3 By rounding each number to one-figure accuracy, give an approximate answer to:
   (a) $16.1 + 8.9 + 15.1 + 231 + 7.6 + 8.14$
   (b) $301 + 17.9 + 45.5 + 54.6 + 56.3 + 17.8$
   (c) $21p + 9p + 86p + 47p + 35p + 10p + 79p$
   (d) $8.75\,m + 16\,m + 4.17\,m + 121\,m + 9.9\,m.$

4 Use your calculator to work out accurate answers to question 3.

5 By approximating each number to one non-zero figure accuracy, state a rough answer to each of the calculations in question 6.

6 Calculate exactly:
   (a) 3.6×4.7 (b) $8.1 - 9.6$ (c) $4.15 \div 1.6$ (d) $4 \times 3.6 + 1$

   (e) $3 + 4 \times 5$ (f) $\dfrac{8 \times 6}{7.5}$ (g) $\dfrac{7}{4.6 \times 2.1}$ (h) $\dfrac{3}{9.1 \times 7.3}$ (i) $4.6 \div \frac{3}{4}$

   (j) $\dfrac{2.9 \times 3.1}{4.5 \times 7.6}$ (k) $\dfrac{2.3 + 4.7}{1.6 + 24}$ (l) $\dfrac{4.7}{2.6 + 1.3}$

7 Find correct to 3 significant figures where appropriate:
   (a) x if $x^2 = 20$ (b) $\sqrt{x}$ if $x = 12$ (c) the reciprocal of 7
   (d) the reciprocal of the reciprocal of 7.

8 Using trial and improvement, not any calculator root or power keys, find correct to 3 significant figures:
   (a) $\sqrt{2}$ (b) $\sqrt[3]{6}$ (c) d when $d^2 - d = 10$.

*9 (a) £14.60 + £8.30 + £16.10 − £3.50.
   (b) Find the cost of 145 breeze-blocks at 38p each.
   (c) How many 19p stamps can I buy for £5?

*10 By approximating each number to one non-zero figure accuracy, state a rough answer to each of the calculations in question 11.

*11 Calculate exactly:

(a) $3.8 \times (7.6 \times 5.3)$ (b) $(3.8 + 7.6) \times 5.3$ (c) $\dfrac{4}{7.1 \times 2.8}$ (d) $\dfrac{3.1}{1.6 + 1.7}$

(e) $\dfrac{8.9 \times 360}{4.1 \times 19}$

*12 Using trial and improvement, showing clearly your working, find correct to 3 significant figures:

(a) $\sqrt{20}$ (b) $\sqrt[3]{12}$ (c) n when $n^2 - 2n = 1$.

*13 (a) Sum 111, 333, 555, 777 and 999.

(b) Now replace six of the figures in part (a) by zeros so that the total is 1111. (Note: numbers like 007 are allowed.)

14 (a) Take 0.165 away from -3.5.
(b) Take 3.5 away from -0.165.
(c) Multiply -4.6 by -3.8.
(d) Divide 0.15 by -0.04.

15 Find correct to 3 significant figures:

(a) 1.6^5 (b) $\sqrt[3]{20}$ (c) $\sqrt[3]{0.6}$

(d) $\sqrt[5]{0.14}$ (e) n when $n^5 = 0.002\,43$

16 Figure 5:1 is Amanda's method to convert a decimal fraction to a common fraction. Investigate it.

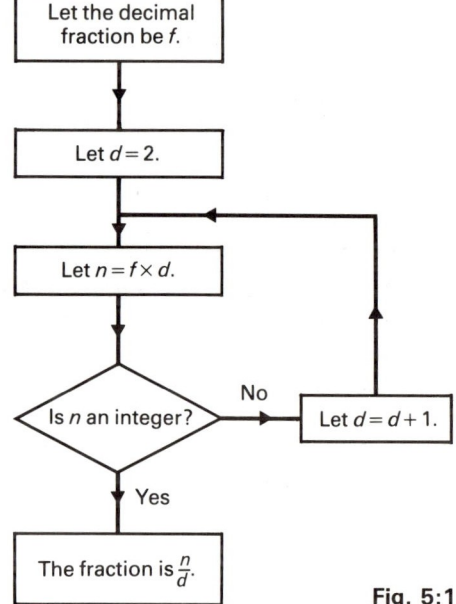

Let the decimal fraction be f.

Let $d = 2$.

Let $n = f \times d$.

Is n an integer? — No → Let $d = d + 1$.

Yes

The fraction is $\dfrac{n}{d}$.

Fig. 5:1

17 A **perfect number** is one whose factors, including 1 but not the number itself, have a sum equal to the number.

Show that although the formula $P = 2^{n-1}(2^n - 1)$ gives some perfect numbers it is not perfectly correct!

18

$\dfrac{1}{17}$ 0.058823 | 529411 | 764705 | 88235

$\times 17$ rem. 9 | $\times 17$ rem. 13 | $\times 17$ rem. 15 |

So $0.\dot{0}58\ 823\ 529\ 411\ 764\ \dot{7}$

Fig. 5:2

Jen has left these notes scribbled on a piece of paper under a calculator. What was she working out, and how?

19 Express as a recurring decimal the reciprocal of:
(a) 23   (b) 29.

20 **How to express a recurring decimal as a common fraction**

    Example $0.\dot{1}\dot{2}$

$$\text{Let } x = 0.121212\ldots$$
$$\text{Then } 100x = 12.121212\ldots$$
$$\text{Subtracting gives } 99x = 12, \text{ so } 0.\dot{1}\dot{2} = \tfrac{12}{99} = \tfrac{4}{33}$$

This shows that a recurring decimal is always rational

    Express as a simplified fraction:

    (a) $0.\dot{1}$ (b) $0.0\dot{9}$ (c) $0.\dot{0}76\,92\dot{3}$ (d) $0.\dot{4}28\,57\dot{1}$ (e) $0.2\dot{7}$

21 Write a simple rule to convert any recurring decimal into a common fraction.

22 Have you ever wondered how your calculator works out a square root? One method is to use iteration with the formula:

$$u_{n+1} = \frac{(u_n^2 + n)}{2u_n}$$

To use it, estimate a value for the answer, call this u_n, then substitute to find u_{n+1}. Now use this value as u_n, and so on, until the answer stops changing at the required number of significant figures. When you are doing this, make full use of your memory, bracket and square keys.

Use the given formula to calculate the following square roots correct to 7 s.f. Compare your answers with that given by using your calculator square root key.

(a) $\sqrt{7}$ (b) $\sqrt{13}$ (c) $\sqrt{27}$.

23 Investigate the iterative formula $u_{n+1} = \sqrt{\dfrac{n}{u_n}}$.

24 An alternative rule for square roots by iteration uses the fact that if $\dfrac{x}{y}$ is an approximation for $\sqrt{n}$ then $\dfrac{x + ny}{x + y}$ is a better one. Investigate.

25 **Step 1**   Let x be any multiple of 3.

    Step 2   Let y be the sum of the cube of the digits of x.

    Step 3   Let $x = y$.

    Step 4   Repeat steps 2 to 4.

    Investigate; rewrite as a flow chart; write as a computer program; etc.

6 Measure

SI system   Density

Time   Reading dials, scales and tables

Speed

● You need to know . . .

Measure 1 and 2 (page 214)

1 Many people still use imperial measures for everyday things. Suggest something that might be measured using:

(a) inches   (b) feet and inches   (c) pints   (d) gallons   (e) stones
(f) pounds weight.

2 How many:
(a) g make 1 kg   (b) kg make 1 tonne   (c) cm^3 make 1 litre
(d) m^2 make 1 hectare   (e) cm^2 make $1 \, m^2$ (f) cm^3 make $1 \, m^3$?

3 Figure 6:1 is a chart to convert between ounces and grams. Use the chart to find:
(a) how many ounces equal 200 g   (b) how many grams equal 12 ounces.

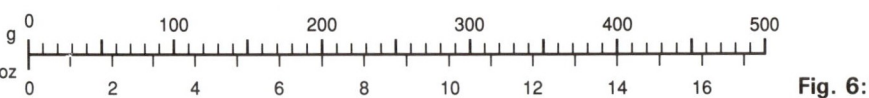

Fig. 6:1

4 In Figure 6:2, which of sections (i) to (viii) link with the following?

(a) about 16 km   (b) about $\frac{7}{8}$ inch   (c) about 2400   (d) about $\frac{1}{2}$ lb
(e) about 1 litre   (f) about 100 g   (g) about 2 m high   (h) about $\frac{3}{4}$ hectare.

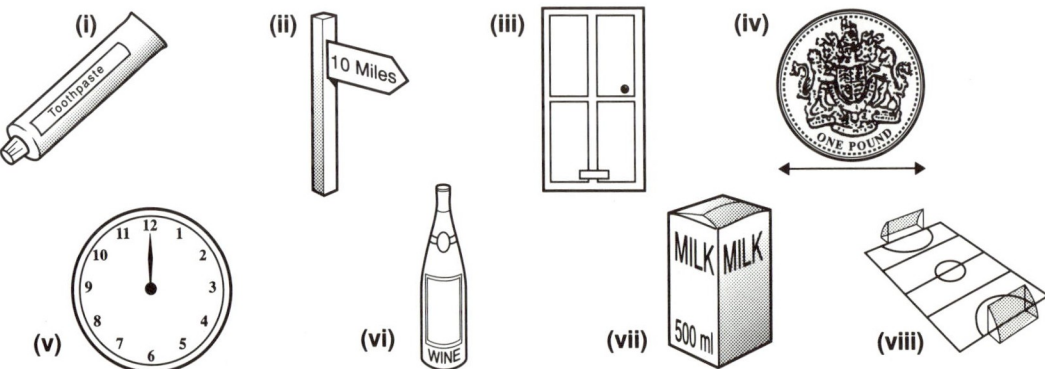

Fig. 6:2

5 Convert:
(a) 15 mm to m (b) 12.01 m to km (c) 16.5 kg to g (d) 5 m 5 cm to m
(e) 5 kg 7 g to kg (f) 500 ml to cl (g) 785.5 cm² to m².

6 Pattentown is 15 miles from Clarkesville, correct to the nearest mile.

(a) What is the greatest and least possible distance they are apart?

(b) John cycles from Clarkesville to Pattentown in 75 minutes. What is his average speed in m.p.h?

(c) How long should Donna allow for a journey from Clarkesville to Pattentown if she hopes to average 40 m.p.h?

7

Speed	Thinking		Braking		Total	
m.p.h.	Time	Distance	Time	Distance	Time	Distance
20	0.7 s	20 ft	0.7 s	20 ft	1.4 s	40 ft
40	0.7 s	40 ft	1.4 s	80 ft	2.1 s	120 ft
60	0.7 s	60 ft	2.1 s	180 ft	2.8 s	240 ft

This table is drawn up from information given in the Highway Code.

(a) Read from it:
   (i) the total stopping distance at 40 m.p.h.
   (ii) the thinking time at any speed
   (iii) the braking time at 60 m.p.h.

(b) In the thinking column why are the times the same but the distances different?

(c) Use the following conversion details to draw up an SI version of the table, giving reasonable approximations suitable for everyday use.

20 m.p.h. ≈ 30 km/h.
20 ft ≈ 6 metres.

8 Sian has to manufacture a rod of diameter 14.5 mm ± 0.1 mm. What is the maximum and minimum acceptable diameter?

9

Substance	Petrol	Pure water	Gold
Density (g/cm³)	0.8	1.0	19.3

(a) Bill has let some water get into his car petrol tank. Will the water be above or below the petrol?

(b) Juanita puts 50 litres of petrol into her car petrol tank. How much heavier is the car now?

(c) A bar of pure gold weighs 10 kg. It has a square cross-section of area 16 cm². How long is the bar, to the nearest mm?

10 Is it possible for two bars of different metal to have the same weight and the same volume? Explain.

11 Air has a density of 1.3 kg/m³. Estimate the weight of the air in your classroom, in both kg and lb.

12 A temperature can be measured as °F (Fahrenheit) or °C (Celsius). An approximate relationship between *F* and *C* is given by the following rule:
'To find *F*, add 15 to *C* and double your answer.'

(a) Write this relationship as a formula.

(b) Use this relationship to find an approximate value of *F* when *C* = −20.

(c) Give a rule for finding an approximate value of *C* when the value of *F* is known.

(d) The exact relationship between *F* and *C* is given by $F = \dfrac{9C}{5} + 32$.

Find the exact value of *F* when *C* is −20.

(e) The temperature is given as 18 °C correct to the nearest degree.
  (i) Calculate the range of possible values of the temperature in degrees Fahrenheit.
  (ii) A weather forecaster predicts a temperature will be 18 °C correct to the nearest degree. She gives the equivalent temperature in degrees Fahrenheit, also correct to the nearest degree. What values might she give?

(MEG)

*13 Figure 6:3 is drawn full size. Find the approximate area of the leaf in cm², stating how near to the true area you think your answer is.

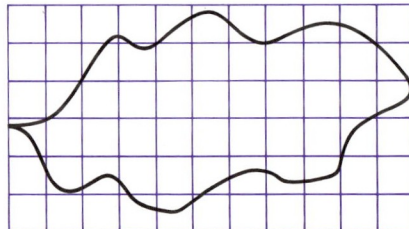

Fig. 6:3

*14 (a) How long should I allow for a 100-mile journey, 60 miles of which is on a motorway, 10 miles in a busy town, and 30 miles on an A-road?

(b) Elizabeth travels 1500 miles in 2½ hours. What is her average speed?

(c) A spaceship is travelling at 25 000 m.p.h. How many miles per second is this?

(d) About how long will it take the spaceship to travel 80 000 000 miles from Earth to Mars?

*15 Stock cubes of side 2 cm are packed in a 40 cm by 6 cm by 2 cm box. What other sizes of box could have been used to hold the same number of the same cubes? Which size would you choose if you were the manufacturer?

(*Water Authorities Association*)

16

Blackdown Reservoir

Area of gathering ground	4120 hectares
Capacity of reservoir	15 100 million litres
Surface area of full reservoir	161 hectares
Average rainfall	148.4 mm per year

(a) A hectare is the area of a square of side 100 metres. How many square metres is the gathering ground?

(b) One litre is a volume of 1000 cm^3.
$1 m^3 = 100 cm \times 100 cm \times 100 cm = 1$ million cm^3
What is the volume of water in the full reservoir, in m^3?

(c) The average depth of water multiplied by the surface area gives the volume of water in the reservoir. Calculate the average depth of water in the full reservoir in metres correct to the nearest centimetre.

(d) What volume of water falls on the gathering ground per year in litres? (Hints: Multiply your answer to (a) by the average rainfall in metres. Use the information in part (b) to change your m^3 volume to litres.)

17 Figure 6:4 shows a volt/ohm-meter dial. Note that the dial is read clockwise for volts but anticlockwise for ohms. The needle reading is different, depending on the position of the switch. As drawn in Figure 6:4 the dial shows kΩ from 0 to 1 kΩ, so the needle reads 0.35 kΩ. What is the reading when the needle is at the position shown and the switch is set to:

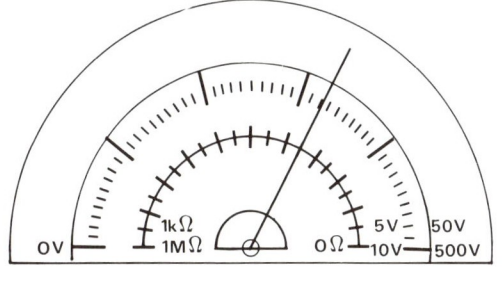

(a) 5 V   (b) 50 V   (c) 10 V
(d) 500 V   (e) 1 MΩ?

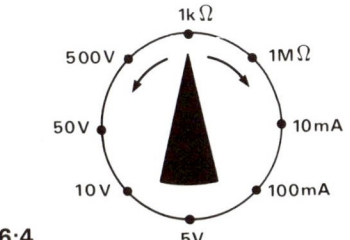

Fig. 6:4

H18 The plan of a room measures 25 mm by 15 mm on a 200 to 1 scale drawing. If my measuring is accurate to within 1 mm, what is the maximum and minimum floor area of the room in m^2?

H19 A book weighed on scales which are accurate to within 10 grams weighs 0.780 kg.

(a) What is the maximum possible reading on the scales when five copies of the same book are weighed together?

(b) The weight of 50 of the books is found by multiplying 0.78 kg by 50. What is the maximum possible error in the answer?

---

20 (a) Two trains travel on parallel tracks towards each other at 60 m.p.h. and 90 m.p.h. respectively. At twelve o'clock they usually pass two points A and B respectively, 80 miles apart. Find where and when the trains pass each other.

(b) One day the slower train was late and passed the express train at a point 6 miles nearer to A than the usual passing point. Assuming the express train to be punctual, and both trains to be travelling at the usual speeds, find how many minutes later than usual the slower train was that day.

(WJEC)

21 Wallpaper comes in rolls 10 metres long and 50 cm wide. Design a chart to help customers decide how many rolls to buy for various sizes of rooms, for heights between 2 m and 3.05 m at 15 cm intervals, and for perimeters from 12 m to 20 m at 1 m intervals. Doors and windows should be ignored, as pattern matching and wastage will more than compensate for these.

---

7 Finance

Household budgeting   Earning, saving and spending
Mortgage and insurance   Foreign exchange, including Europe

● You need to know . . .

Percentages 1 to 8 (page 206)
Practical graphs 3 (page 217)

1   A family visiting Germany and France changes £500 into German marks at the rate of DM2.90 to the £1. They spend DM676 before changing the remaining marks into French francs at the rate of 3.35 francs to the mark. Calculate the number of marks and the number of francs they bought.

2   Miriam has to pay her electricity board a quarterly charge of £10.20 plus 7.15 pence for every unit used. Her meter reading at the beginning of the quarter was 14 742 and at the end of the quarter was 15 774. How much will Miriam have to pay?

3   **Telephone charges (1992)**

Fees: 4.20p per unit for the first 2381 units,
        3.99p per unit for the next 3759 units up to 6140 units,
        3.86p for each unit over 6140 units.

All fees are subject to an additional charge of 17.5% VAT.

Time allowed in seconds for each unit fee			
Type of call and charge band	CHEAP All weekend Mon–Fri 6 pm–8 am	STANDARD Monday to Friday 8 am–9 am 1 pm–6 pm	PEAK Mon–Fri 9 am–1 pm
Local	220.00	80.00	57.50
National a	80.80	36.15	27.00
National b	50.35	32.00	19.20
Mobile	11.40	7.61	7.61

Find the cost correct to the nearest penny, inclusive of VAT, of the following calls. Remember that the charges rise in steps, that is you pay the same for a CHEAP Local call of 221 seconds as for one of 439 seconds. Assume the fees are charged at the 0 to 2381 units rate.

(a)   6 min from 8:15 a.m. on Monday at Local rate
(b)   6 min from 8:15 a.m. on Tuesday at Mobile rate
(c)   18 min from 1215 on Friday at National a rate
(d)   40 min from 1225 on Sunday at National b rate.

4   A television listed at £345 was bought with a £40 deposit and 12 monthly payments of £30.

   (a) Find the total cost of the monthly payments.

   (b) Find the total cost of buying the television.

   (c) If cash is paid, a discount of 20% on the list price is given. Find the cash price.

   (d) How much less is the cash price than the total cost of the instalment plan?

   (e) If the same television can be rented for £1.90 per week, how long will it be before the total rent paid becomes greater than the list price?

5   Niri owns a hundred £5 preference shares in British Metals. Each share cost her £12.50 and they bring in 8% of their face value (£5) simple interest each year. She sells them after receiving five years' interest, receiving £11.75 per share.

   (a) The interest is taxed at 25%. What net interest does she receive over the five years?

   (b) Including her interest, how much is her overall profit or loss on her speculation?

6   (a) Read about compound interest in the reference notes: *Percentages 8* (page 208).

   (b) Calculate the amount resulting from:
       (i) £1200 invested for 2 years at 5% p.a.
       (ii) £1150 invested for 3 years at 8% p.a.
       (iii) £400 invested for 5 years at 4% p.a.

7   Andrea borrows £20 000 for five years at 12% per annum compound interest. If she makes no repayments of the loan or interest until the end of the five years, how much would she then have to pay to repay the loan and compound interest?

8   A car bought for £12 000 depreciated by 20% in its first year, by 12% in the second, and by 5% in the third. What was its value at the end of each of the three years?

9   A speculator reckons that the price of gold will increase by an average of 7% p.a. over the next five years. What does she expect a gold coin worth £50 today to be worth in five years' time?

10   A building society will advance a mortgage loan equal to $2\frac{1}{2}$ times a person's annual salary, up to a limit of £40 000. How much can be advanced if a person earns £12 000 a year?

11   A young couple want to buy a flat for £42 000. A building society will give them a 90% mortgage. How much must they find themselves?

12   Calculate the commission an estate agent will charge on an £84 000 house according to the following scale of charges: 2% on the first £20 000 and 1% on the remainder.

13 The following table shows part of an insurance company's promotion leaflet. The amount payable on the death of the insured person depends on the sum assured, and grows each year for 15 years. If the insured person is still alive after 15 years, he/she will receive the amount shown (the maturity value).

15 year endowment assurance						
Premium	£10		£20		£30	
Age next birthday	Sum assured £	15 year maturity value £	Sum assured £	15 year maturity value £	Sum assured £	15 year maturity value £
Male Female						
15–29 15–33	1827	3595	3763	7405	5697	11 211
31 35	1824	3589	3757	7393	5688	11 193
33 37	1821	3583	3751	7381	5679	11 176
41 45	1792	3526	3690	7261	5587	10 995
51 55	1699	3343	3498	6884	5297	10 424

(a) For £10 a month:
   (i) How much could *you* expect to receive in 15 years?
   (ii) How much would you have paid during those 15 years?

(b) For £30 a month, for a woman aged 36:
   (i) How much would she expect to receive in 15 years?
   (ii) How much would she have paid during those 15 years?

(c) How much less insurance cover does a 50-year-old man receive than a 30-year-old man if both pay £20 a month?

(d) Why does a 50-year-old man receive less insurance cover than a 30-year-old man when he pays the same amount?

(e) How much more insurance cover does a 32-year-old woman receive for £30 a month than a 32-year-old man? Why is there this difference?

14 Mr Reynolds earns £23 500 per year gross and has tax allowances of £5165. He pays tax at 20% on the first £2000 of his taxable income, then 25% tax on the remainder. How much tax does he pay?

15 Glenys is a salesperson. She has a basic salary of £12 000 per year plus commission on her sales as follows:
The first £20 000   10%
The next £20 000 $12\frac{1}{2}$%
Sales over £40 000 $17\frac{1}{2}$% on the remainder

Find her total gross salary for years in which her sales total:
(a) £18 000   (b) £40 000   (c) £124 000.

16 Jock Fairhurst is a lorry driver. He is paid a flat rate of £140 a week, and in addition receives:
£12 a day subsistence allowance
9p per mile mileage bonus
50p per tonne tonnage bonus.

One week his records show:

	Monday	Tuesday	Wednesday	Thursday	Friday
Miles	289	175	348	246	191
Tonnes	25	15	30	20	15

(a) What is his subsistence allowance for the week?

(b) What is his mileage bonus for the week?

(c) What is his tonnage bonus for the week?

(d) What is his gross pay for the week?

17 Julie and Ron insure their home for £90 000 and its contents for £30 000. The rate for buildings is 16p per £100 and that for contents is 27p per £100. Find the total premium.

*18 Study the finance company's advertisement in Figure 7:1.

23.1% APR ANY PURPOSE LOANS FOR HOMEBUYERS/OWNERS WITH FREE LIFE INSURANCE, OPTIONAL SICKNESS/REDUNDANCY COVER AVAILABLE

BORROW	EQUIVALENT WEEKLY REPAYMENTS TO NEAREST PENNY OVER				
	36 mths £	60 mths £	90 mths £	120 mths £	180 mths £
£2000	18.00	12.93	10.58	9.55	n/a
£3000	27.01	19.39	15.88	14.33	n/a
£4000	36.01	25.86	21.17	19.11	n/a
£5000	45.01	32.32	26.46	23.89	21.87
£6000	54.01	38.79	31.75	28.66	26.25

e.g. £2500 × 36 mths at £97.52 per mth = £3510.72. Big rebates of interest for early settlement. Loans are secured on property for low cost.

Fig. 7:1

(a) What is a finance company?

(b) What interest rate ('Annual Percentage Rate') is quoted by the company?

(c) If you borrow £4000 over 10 years, how much will you have to pay back:
     (i) each week   (ii) altogether?

(d) Building societies were charging about 13% interest at the time of the advertisement. Why do you think the finance company charges so much more?

(e) If the interest payable per year is calculated on the amount owing at the start of the year, how much of the first year's payments repay part of the £4000 loan?

*19 A sewing-machine salesperson is paid a basic salary of £875 a month plus a commission of £35 for each machine she sells.

(a) In May she sells 35 machines. How much pay does she receive?

(b) If she needs at least £18 000 a year how many sewing machines must she aim to sell a month on average?

(c) She receives a 6% rise in her basic salary. Now how many machines must she aim to sell per month?

*20 (a) A petrol-pump attendant is paid at the rate of £3.40 per hour for a basic working week of 38 hours. Deductions from earnings are £20.50. Calculate the weekly net income.

(b) On overtime she is paid time and a half, and deductions are 30% of the gross overtime pay. Calculate the hourly overtime rate before deductions and the total net income for a week in which she works seven hours overtime.

---

21 The Western Building Society pays compound interest on the amount in your account on the 28th day of each month. The monthly rate of interest is one-twelfth of the yearly rate. If you pay in £50 on the 1st of each month in a year when the interest rate is 8% per annum, how much will you have in your account at the end of six months?

22 The estimated cost of building a house is £84 000, of which 55% is for wages and the rest is for materials. Before work starts wages are increased by 7% and the materials rise by an average of 8%. Calculate the increased cost of building the house.

23 Study the Mortgage Saver Account details in Figure 7:2. Part of the bank's leaflet (not shown) gave a detailed example for a 'typical saver'. Write such an example to help customers understand the implications of the scheme.

Your Mortgage Saver Account gives you guaranteed access to a mortgage from us. Under this guarantee, which remains open for 6 months after you have completed your savings, you may apply for a mortgage:

- up to 95% of the value of the house you want to buy (or the purchase price if that is lower)
- up to 3 times your gross annual income if you are the main income earner and $1\frac{1}{2}$ times the second income earner
- between £12 000 and £40 000.

The amount you eventually borrow will depend on the property you choose and your ability to meet repayments.

You may apply for a mortgage greater than £40 000 (up to a maximum of £150 000) but in these circumstances different criteria will apply.

The following table illustrates the different ways in which the total savings could be accumulated.

Monthly savings £	over 18 ms	over 24 ms	over 36 ms
50	900	1 200	1 800
75	1 350	1 800	2 700
100	1 800	2 400	3 600
200	3 600	4 800	7 200
500	9 000	12 000	18 000

To keep the examples as simple as possible the interest and bonus on the Mortgage Saver Account, paid to you each half year, is not included in these figures.

Fig. 7:2

8 Journey graphs

Distance/time   **(Area under a graph (trapezoidal rule))**
(Speed/time)

● You need to know . . .

Practical graphs 1 and 2 (page 216)
Algebraic graphs 7 (page 238)

1 Figure 8:1 represents the journeys made by a bus and a car starting at Sheffield, travelling to Leicester and returning to Sheffield.

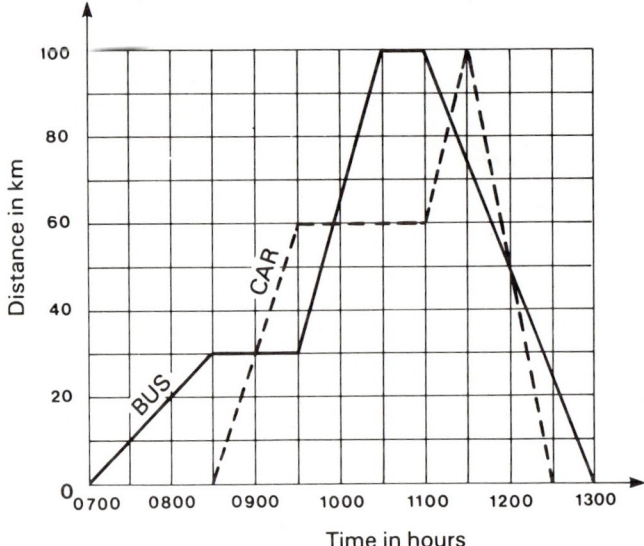

Fig. 8:1

(a) How much longer, including stops, did it take the bus to complete the journey from Sheffield to Leicester than it did the car?

(b) At approximately what time, between 0900 and 1000, did the bus overtake the car?

(c) What was the greatest speed attained by the car during the entire journey?

(d) What was the average speed attained by the car during the entire journey?   (MEG)

2 A passenger train leaves Motown at 1000 hours, averaging 40 m.p.h. on its journey to Nadir, 60 miles away. After a 30-minute halt it returns at 60 m.p.h. on average.

At 1100 a goods train leaves Nadir for Motown, travels at 40 m.p.h. for 30 minutes, then completes the journey at an average speed of 20 m.p.h.

Draw a graph to illustrate the journeys, taking the time axis from 1000 to 1300. Hence find when the trains pass and how far they then are from Motown.

*3 Describe the three journeys shown in Figure 8:2 (a), (b) and (c).

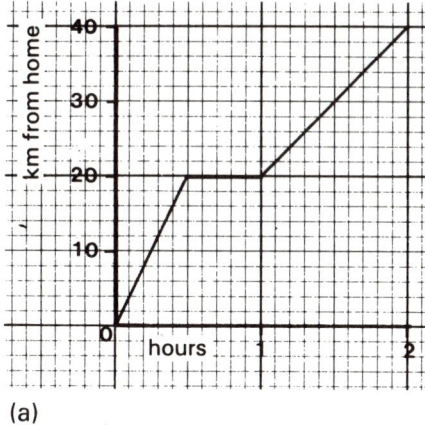

(a)

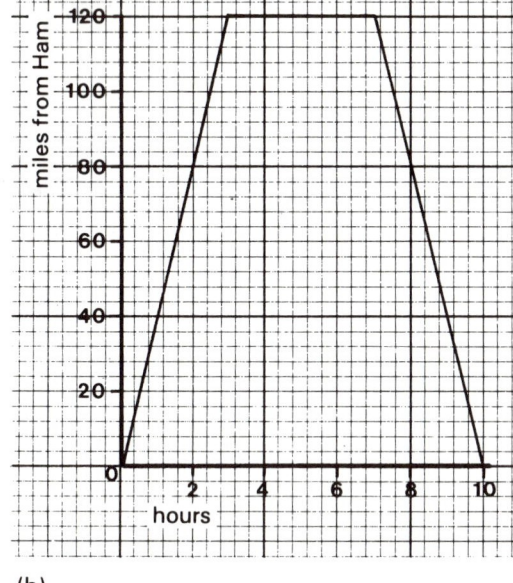

(b)

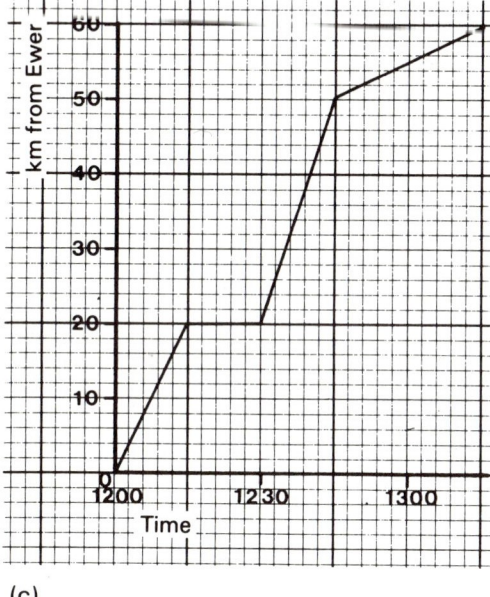

(c)

Fig. 8:2

*4 State the speeds for each part of each journey in Figure 8:2. Then, by finding the total distance and the total time, find the average speed for the whole journey.

*5 Figure 8:3 is a distance/time graph for a car journey. Tangents to the curve give the speed at various moments. One such tangent is shown 5 minutes from the start of the journey. What is the speed of the car at this moment?

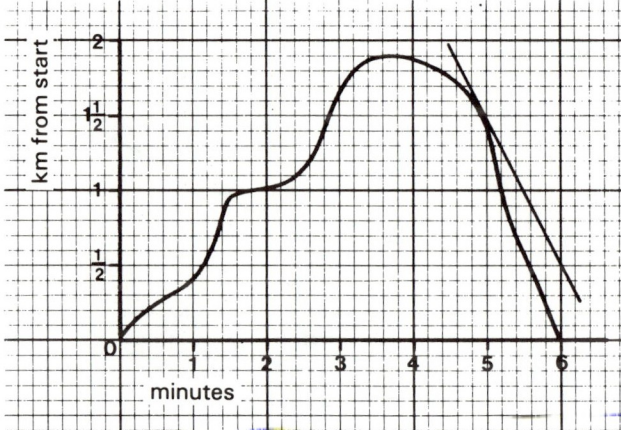

Fig. 8:3

H6 Figure 8:4 shows a speed/time graph. From A to B a constant speed of 10 m/s is shown and during the next second, from B to C, there is an acceleration of 10 m/s² (10 metres per second per second). Describe what happens between C and F.

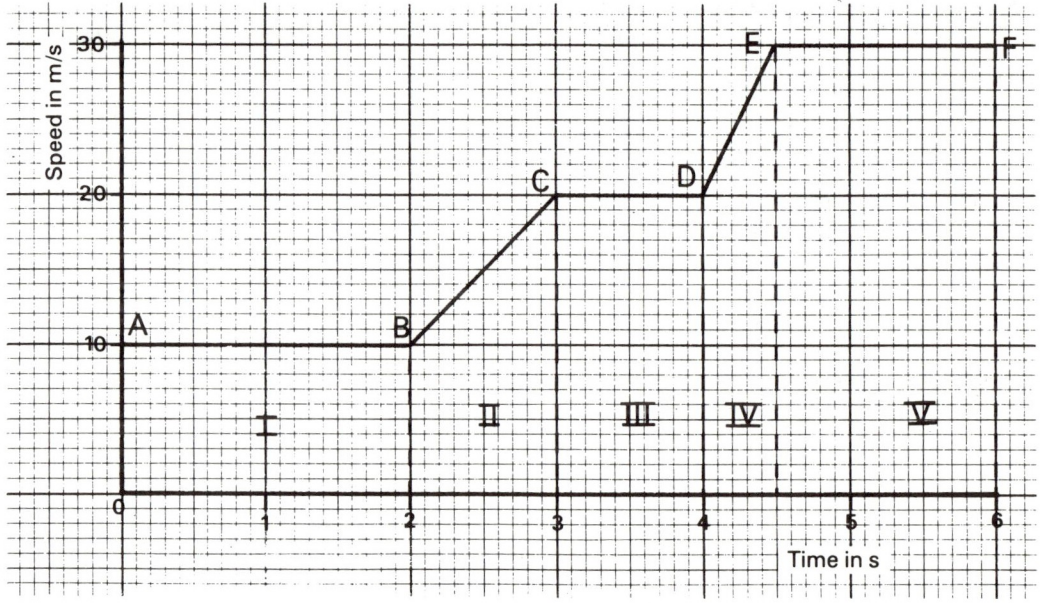

Fig. 8:4

H7 Because speed multiplied by time gives distance, the area under a speed/time graph gives the distance travelled.

In Figure 8:4, area I represents a distance of $10 \times 2 = 20$ metres. Area II represents a distance of $\frac{1}{2}(10 + 20) \times 1 = 15$ metres (area of trapezium = half the sum of the parallel sides times the distance between them).

Find areas III, IV and V, and hence the total distance covered in the six seconds.

H8 Figure 8:5 is the speed/time graph for a particle which moves with speed v m/s at time t seconds.

(a) Calculate the acceleration of the particle when $t = 10$.

(b) Calculate the total distance travelled in the 12 seconds, and hence the average speed of the particle for the whole journey.

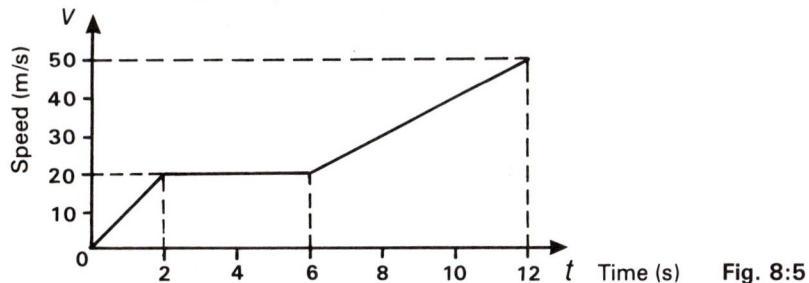

Fig. 8:5

H9 (a) The speed of a car is observed at 10-second intervals over 1 minute. The readings are given in the table.

Time (s)	0	10	20	30	40	50	60
Speed (m/s)	0	5	14	26	34	38	40

Plot the points on axes, remembering that the time axis should be the horizontal one. Join the points with a smooth curve.

(b) By drawing two tangents to your curve, estimate the acceleration of the car:
(i) after 30 seconds   (ii) when its speed is 35 m/s.

(c) The area under the curve cannot be found exactly, but by joining the plotted points with straight lines a reasonable approximation can be obtained from the resulting triangles and trapeziums.
Use this method (the 'trapezoidal method') to calculate the distance covered by the car during these 60 seconds.

---

10 An aeroplane flying at a constant height of 320 m drops a load fitted with a parachute. During the time intervals stated the motion of the load is as follows:

0 s–2 s   free fall, negligible resistance, acceleration due to gravity of 10 m/s^2
2 s–6 s   parachute opens, vertical speed decreases uniformly to 2 m/s
after 6 s   the load falls with a constant speed of 2 m/s.

(a) Sketch a speed/time graph for the first 8 s. (Speed, vertical, 0–20 m/s. Time, horizontal, 0–8 s.)

(b) Determine the distance fallen in the first 6 s.

(c) Determine the total time taken for the load to reach the ground.   (SEG)

11 The speed of a train at intervals of 1 minute during a journey of 8 minutes are 0, 11, 21, 28, 30, 30, 26, 15, 0 m.p.h. Draw the speed/time graph and from it estimate the total distance travelled.

Conversion   **Growth and decay**
Flow   **Interpretation**

● **You need to know . . .**

Practical graphs 3 (page 217)

1 You will meet graphs in nearly all areas of school work and everyday life. Figures 9:1 to 9:9 are just a few examples for you to think about and discuss.

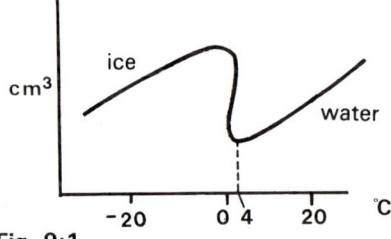

Fig. 9:1

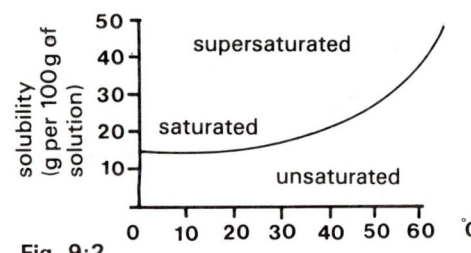

Fig. 9:2

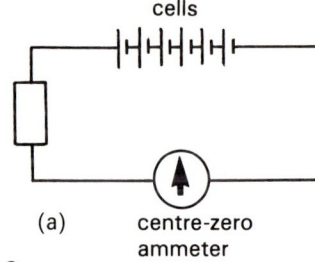

Fig. 9:3

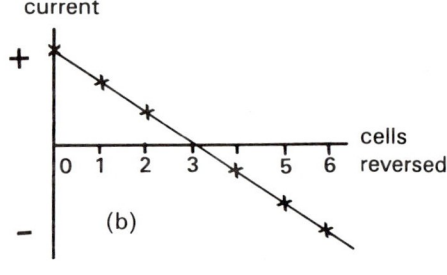

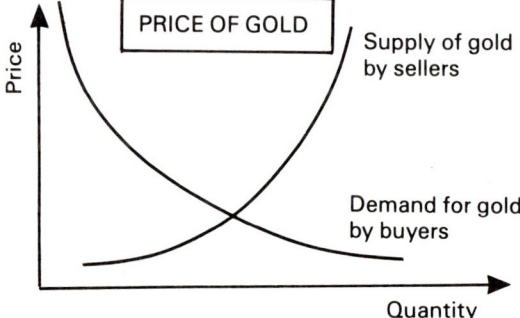

Fig. 9:4

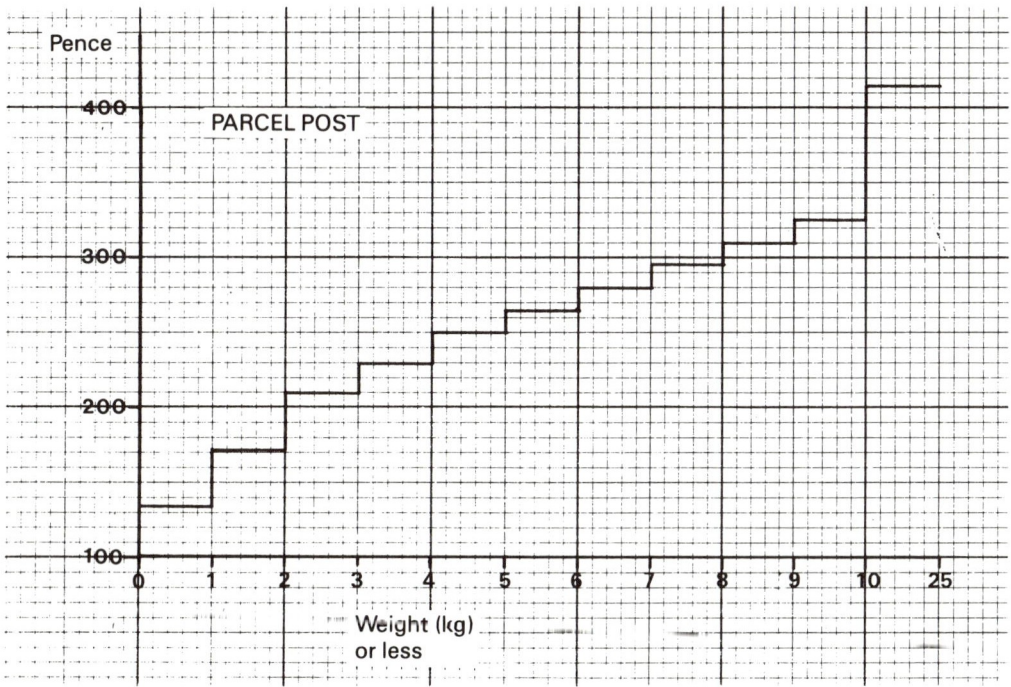

Fig. 9:5

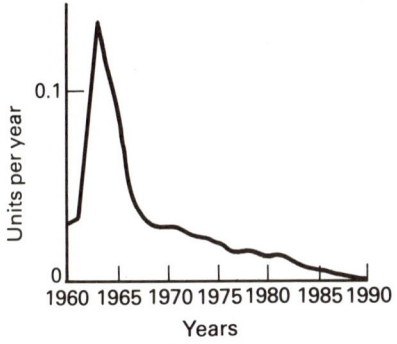

Fig. 9:6   Growth and decay of radiation exposure from nuclear weapons tests. Note: The average annual radiation dose from natural sources is about 2.2 units. The 1986 Chernobyl accident resulted in an average extra dose of 0.03 units in London during that year. A week's holiday in Cornwall will give you a dose of 0.1 units, due to the radioactive rocks. There is no evidence of an effect on human health of doses of under 100 units a year. Source: Office for Official Publications of the European Communities.

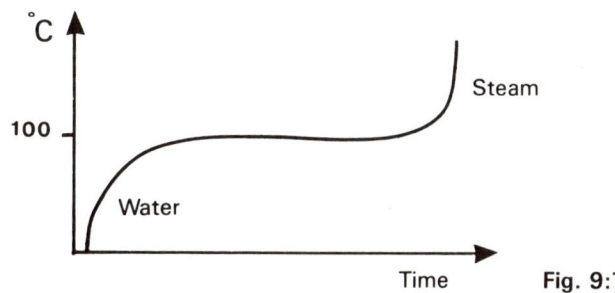

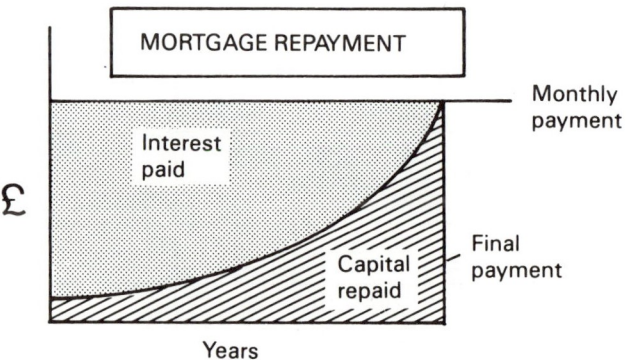

MORTGAGE REPAYMENT

Monthly payment

£

Interest paid

Capital repaid

Final payment

Years

Fig. 9:8

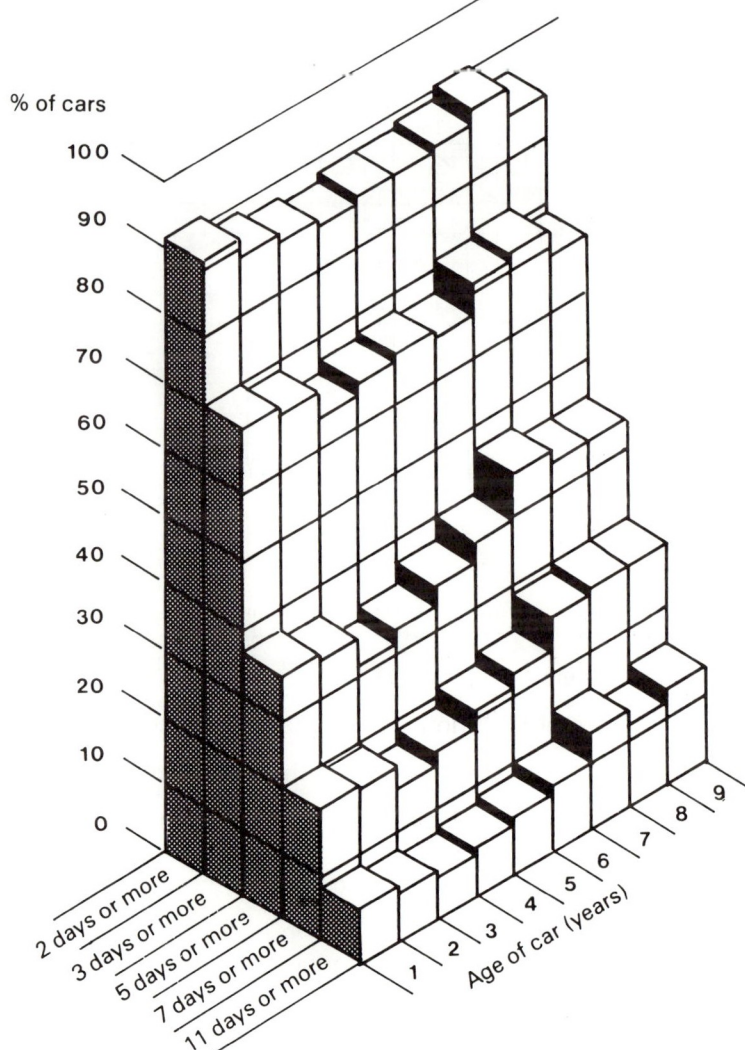

% of cars

Fig. 9:9

2 A percentage conversion graph provides a quick way to read off percentage amounts without calculation.

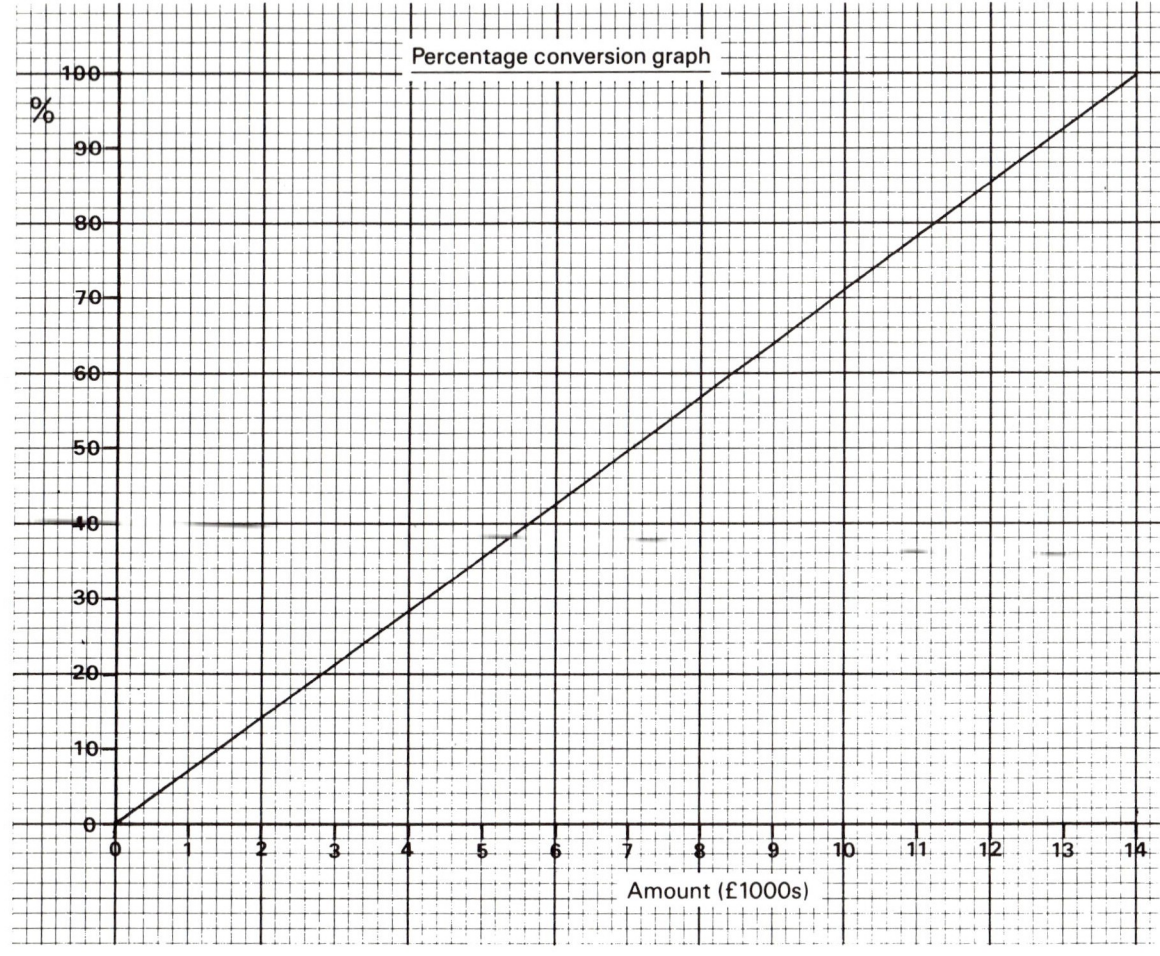

Fig. 9:10

Figure 9:10 shows Sharon's total income for a year.

(a) What is her salary (income per year)?

(b) Her electricity bills total £500 in a year. What percentage of her salary is this?

(c) Her mortgage repayments are £450 a month. What percentage of her salary is this?

(d) Sharon aims to give 15% of her income to charity. How much should she give each year?

3 The flow chart (Figure 9:11) illustrates some of the decisions that you will have to reach when you consider buying your first house.

What advice would you give the following couples who all wish to buy a £70 000 house?

(a) Joe and Marissa: annual income £28 000, savings £10 000

(b) Girish and Sharma: annual income £21 000, savings £4000

(c) John and Vicky: annual income £18 000, savings £5000.

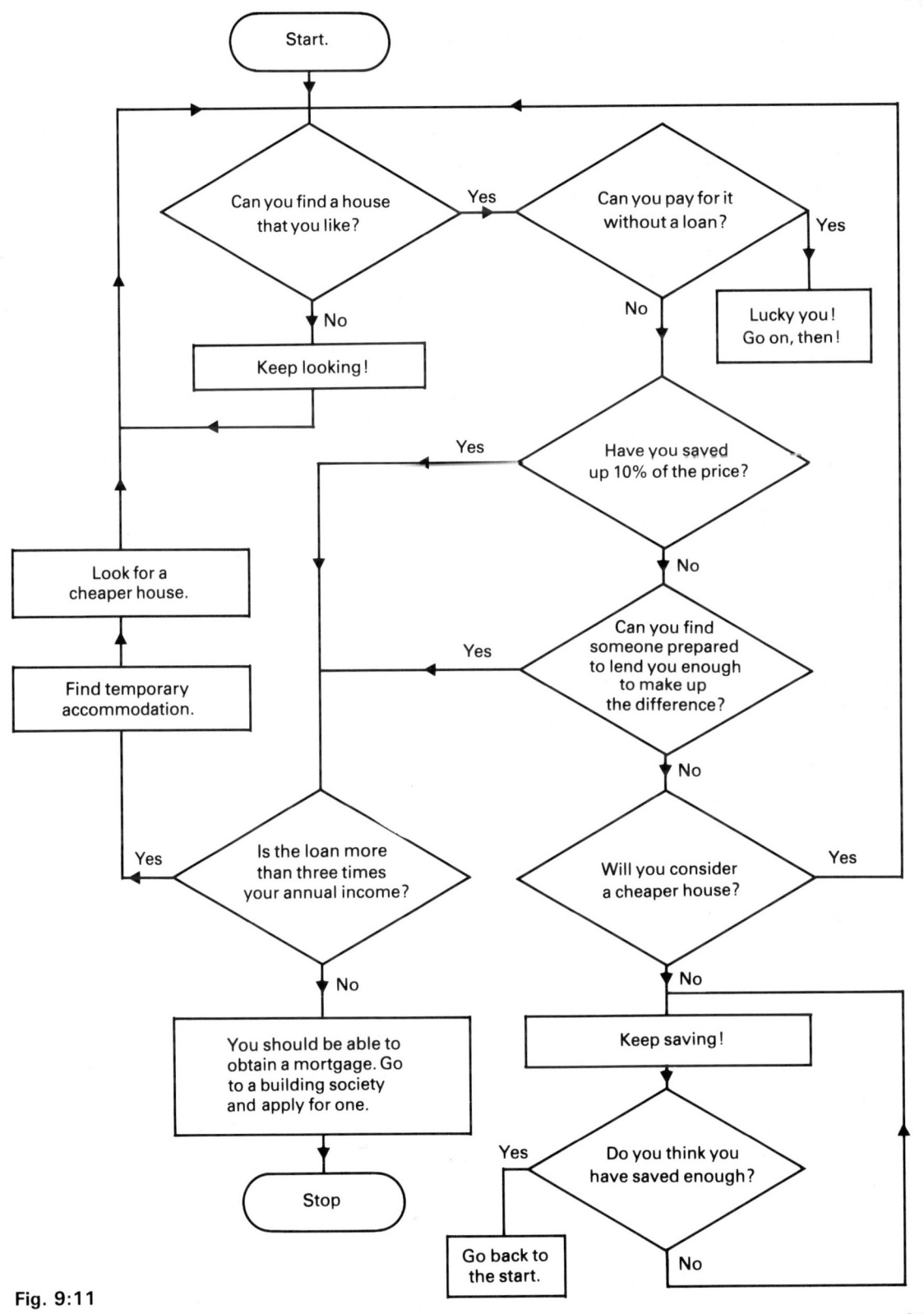

Fig. 9:11

4 Figure 9:12 is a decision tree that identifies parallelograms from facts about their diagonals. Name the parallelograms for each box, chosen from:

parallelogram, rectangle, square, rhombus, isosceles trapezium, trapezium, kite.

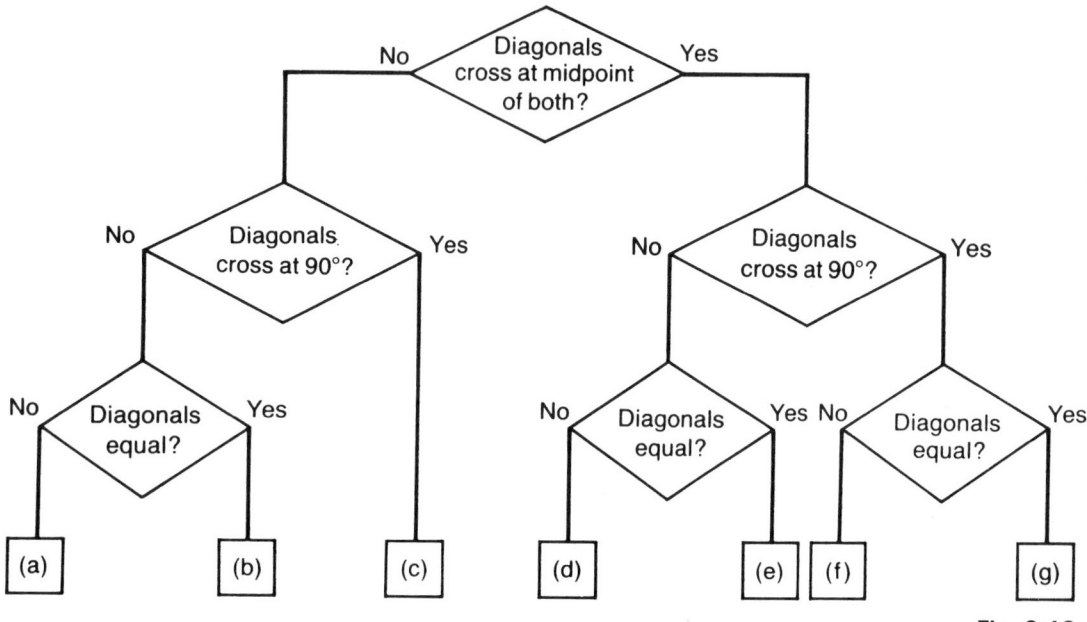

Fig. 9:12

5 Figure 9:13 shows three straight roads, AB, CD and EF. The contour lines indicate the nature of the terrain. Figure 9:14 shows the speed/time graph for a cyclist on one of the roads. Which road do you think he is cycling along: AB, CD or EF?

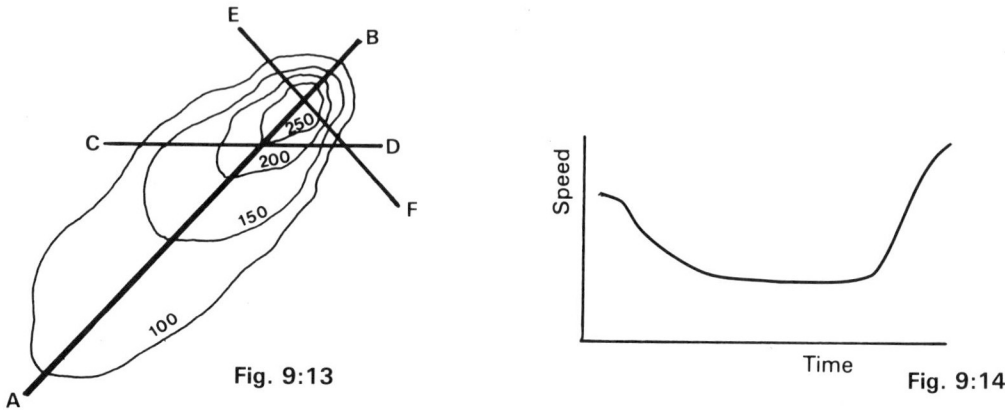

Fig. 9:13

Fig. 9:14

6 A 300-litre cask is to be filled with two liquids, A and B, which come through two pipes at different but constant rates. For 10 minutes one pipe delivers 75 litres of liquid A, then both pipes pour into the tank for a further 10 minutes until the tank is full. The mixture is stirred for 10 minutes, then the cask is emptied at a steady rate in 20 minutes.

Draw a graph to show the changing level of liquid in the cask during the operation. Label the axes clearly.

7 Figure 9:15 shows two containers which are filled from a tap running at a constant rate. Select from the graphs in Figure 9:16 the two which best show the changing depth of water in the containers.

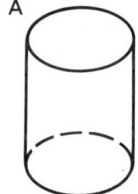 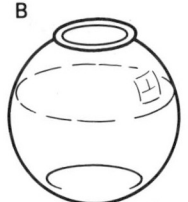

A   B

Fig. 9:15

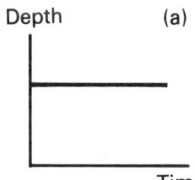 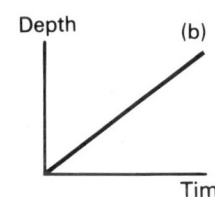 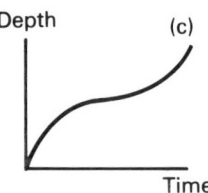 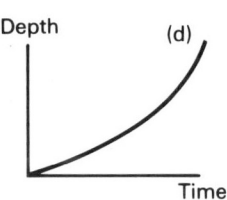

Depth (a)   Depth (b)   Depth (c)   Depth (d)

Time   Time   Time   Time   **Fig. 9:16**

8 Refer back to question 7. Figure 9:17 shows some more containers. Sketch graphs to show how the diameter of the water surface changes for the containers in Figures 9:15 and 9:17 as they are filled at a constant rate.

 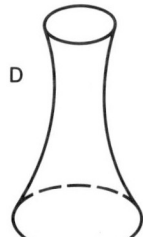

C   D

Fig. 9:17

9 The number of viruses in a culture trebles every minute. Sketch a graph of this to show how the number increases over ten minutes, starting with one virus. Label your axes clearly.

10

Month	July	Aug	Sept	Oct	Nov	Dec
Workers taken on	4	9	6	5	7	4
Output in 1000s	5	15	20	31	40	44

The table shows how the output of a firm changed as more workers were employed, e.g. in July four extra workers were employed and output was five thousand; in August there were thirteen extra workers and output was fifteen thousand.

Plot the information on a suitable graph, draw the line of best fit, and hence estimate the output in January if a further seven workers are taken on.

*11 This table gives the mileage and train times for a journey from London to Bournemouth.

Miles	Station	Time
0	London Waterloo	1105
12	Surbiton	1137
24	Woking	1210
48	Basingstoke	1231
66	Winchester	1243
75	Southampton	1309
93	Brockenhurst	1328
108	Bournemouth	1351

(a) How many minutes for the whole journey?

(b) How many minutes from Winchester to Southampton?

(c) Figure 9:18 shows a common form of distance chart. What distances should be shown in squares A, B and C?

(d) Calculate the average speed of the whole journey in m.p.h. correct to the nearest mile.

(e) The fare from London to Bournemouth is £21.00. How much is this in pence per mile?

(f) Plot distance vertically against time horizontally to show the information given, and hence find which journey between stations is covered at the fastest average speed.

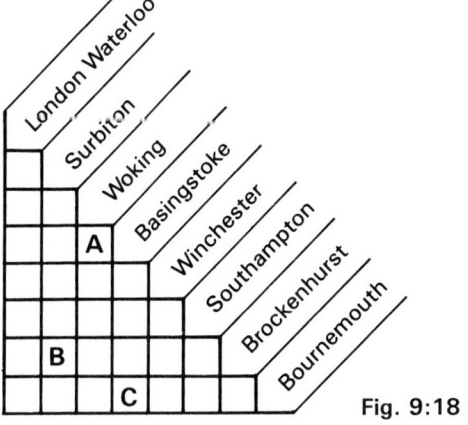

Fig. 9:18

*12 Copy and complete the flowchart in Figure 9:19. You may need to refer to the telephone charges chart in chapter 7 on page 24.

Use these fillers:
Is it before 0900 or after 1300?
Use phone (dear rate).
Is it before 0800 or after 1800?
Can call wait until after 1300?

*13 A computer-company salesman receives a basic wage of £100 a week. For every computer system he sells in a week he receives a commission. The mapping diagram in Figure 9:20 shows how this increases his pay.

(a) Copy and complete the ordered pairs:
(0, 100); (1000, 150); (2000,); (3000,); (4000,); (5000,).

(b) Draw axes, vertical £100 to £600 to represent pay, horizontal £0 to £5000 to represent the sales. Plot your ordered pairs, then join them with a smooth curve.

(c) Estimate how many pounds' worth of equipment he must sell to bring in £400 a week.

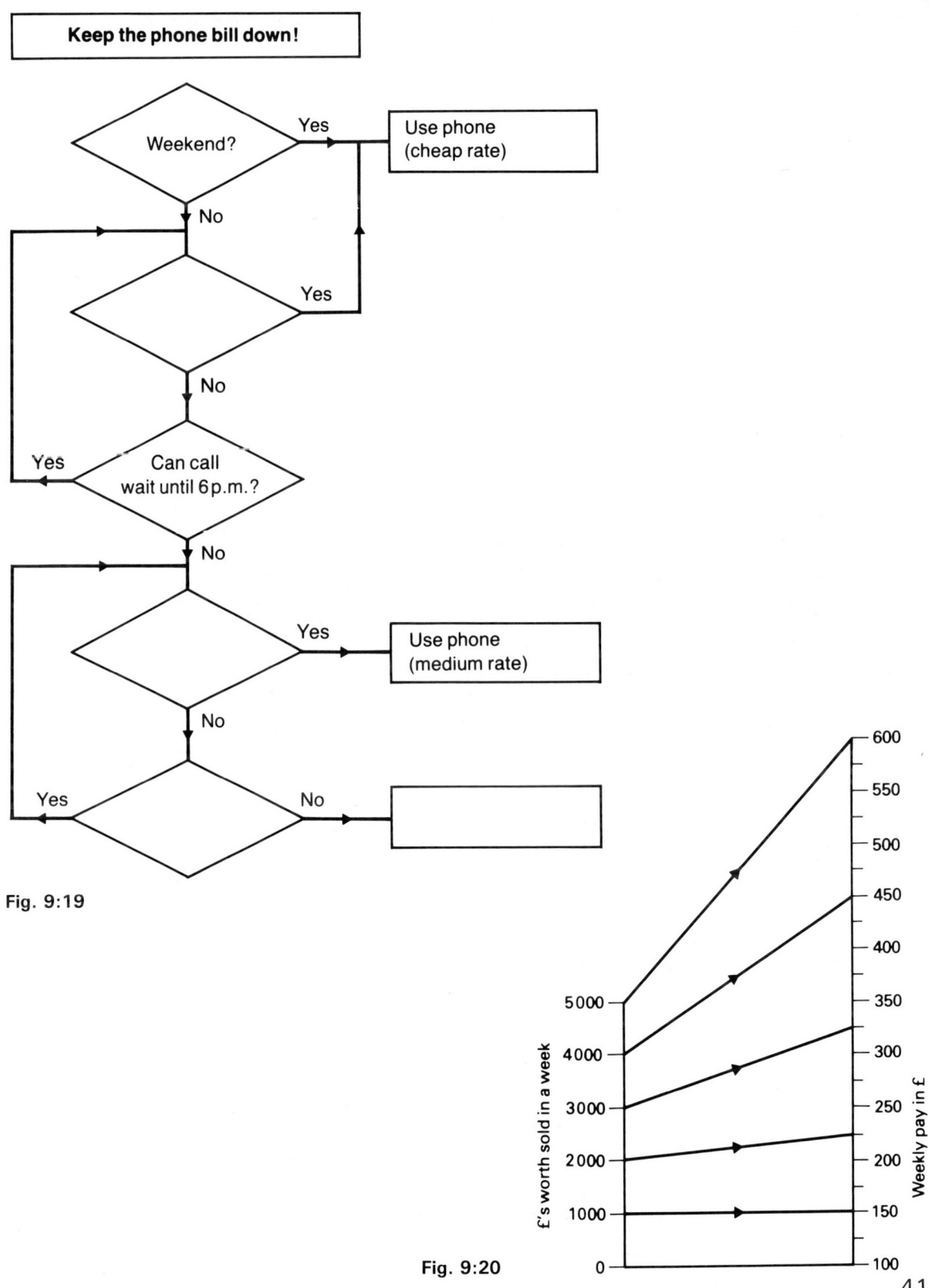

Keep the phone bill down!

Weekend? — Yes → Use phone (cheap rate)

No

Yes

No

Can call wait until 6 p.m.? — Yes

No

Yes → Use phone (medium rate)

No

Yes — No →

Fig. 9:19

Fig. 9:20

£'s worth sold in a week

Weekly pay in £

41

*14 Taking 60 m.p.h. = 88 ft/s, draw a graph to convert between these units. Use a vertical scale of 1 cm to 5 m.p.h. and a horizontal scale of 1 cm to 10 ft/s.

Use your graph to convert:
(a) 45 m.p.h. to ft/s   (b) 50 ft/s to m.p.h.   (c) 15 m.p.h. to ft/s
(d) 78 ft/s to m.p.h.

*15 When a cassette recorder starts to rewind a cassette the counter reads 430. Counter readings (y) after x seconds are shown in the table.

x	0	10	20	30	40	50	60	70	80	90
y	430	408	382	353	317	278	227	174	104	22

(a) Draw axes showing values of x from 0 to 100 and y from 0 to 450. On these axes draw a graph to show the readings in the table. Note that the graph will be a continuous curve.

(b) Estimate from your graph the time taken to rewind until the counter reads:
(i) 300   (ii) 0.   (MEG)

*16 A hot liquid cools at a faster rate than a cooler liquid. Sketch a graph to represent the temperature of a forgotten cup of hot coffee. Label your axes sensibly.

*17 The graphs in Figure 9:21 represent the movement of two athletes. Suggest what sports activity they might be doing.

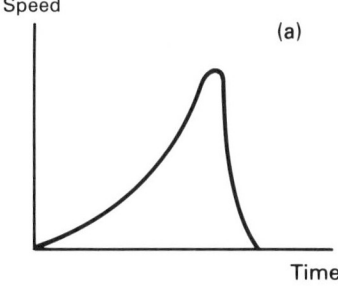

Fig. 9:21

*18 Figure 9:22 shows a graph to find the cost of producing a leaflet.

(a) What is the price per leaflet if 300 are printed?

(b) Will the cost line ever cross either axis? Explain your answer.

*19 Sketch a graph to show the radioactive decay for the element given in chapter 3 question 27 (page 12) for its half-life.

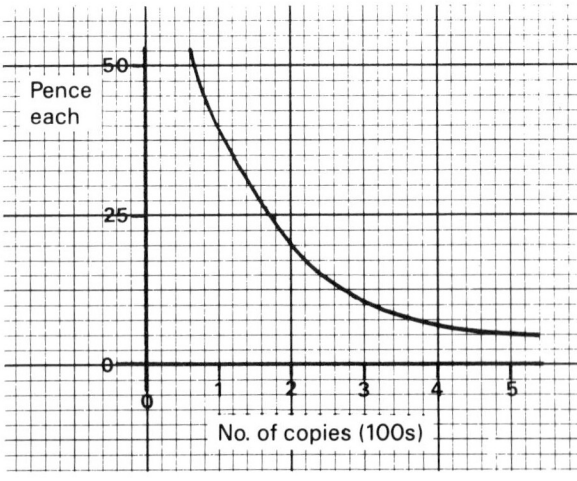

Fig. 9:22

20 Rory joins a health club and undergoes a fitness test. After a strenuous workout her pulse rate is recorded every 30 seconds for four minutes. The results are shown in the table below.

Time (s)	0	30	60	90	120	150	180	210	240
Pulse rate	150	120	100	90	86	84	82	81	80

(a) Plot this information, putting 'Time (s)' horizontal, 30 seconds to 2 cm, and 'Pulse rate' vertical, from 70 to 160, 10 beats to 2 cm. Draw a curve that best fits these points.

(b) From your graph estimate:
   (i) Rory's pulse rate after 45 seconds
   (ii) the number of seconds it takes for Rory's pulse rate to fall to 95.

(c) The gradient of the tangent drawn at a point on the curve gives the rate at which Rory's pulse rate is decreasing each second. Find the rate at which Rory's pulse rate is decreasing after 60 seconds.   (SEG)

H21 The table shows the depth of water in a storage tank over eight minutes.

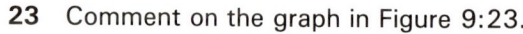

Time (t min)	0	1	2	3	4	5	6	7	8
Depth (d cm)	60	40	30	25	22	27	38	49	57

(a) Using a scale of 2 cm to represent 1 minute on the horizontal t-axis and 2 cm to represent 10 cm on the d-axis, draw a smooth curve to show how the depth of water varies with time.

(b) Say what you think is happening over the eight minutes. Be as precise as you can.

(c) Estimate when the water was 35 cm deep.

(d) By drawing tangents to your curve, estimate the rates at which the depth of water was changing after 2 minutes and after 6 minutes.

H22 Farmer John has a large flock of sheep, quite a few of which are black. Spot the dog drives the sheep at random into a pen, ten at a time, and the farmer removes any black sheep to his barn. The remaining sheep are freed and rejoin the rest. This rather silly process is repeated until the flock is purged of black sheep.

Sketch a graph to show how the number of black sheep in the barn should change with time.

23 Comment on the graph in Figure 9:23.

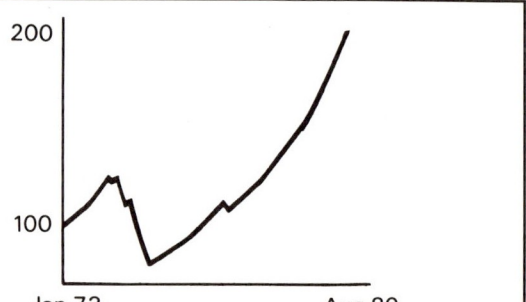

Fig. 9:23   Jan 73   Aug 80

24 When the tide is ebbing, water flows swiftly out of the harbour mouth at Lower Puddlewick. The table shows the rate of flow in cubic kilometres per hour (km³/h) for three hours on a day when high tide is at noon.

Time	12:00	12:30	1:00	1:30	2:00	2:30	3:00
Hours after noon (t)	0	$\frac{1}{2}$	1	$1\frac{1}{2}$	2	$2\frac{1}{2}$	3
Rate of flow (f)	0	24	46	63	75	80	78

Using a scale of 2 cm to represent half an hour on the t-axis (horizontal), and 2 cm to represent 10 km³/h on the f-axis, plot ordered pairs (t, f) and join them with a smooth curve.

(a) It is dangerous for small boats to negotiate the harbour mouth if the flow exceeds 60 km³/h. At what time does it become dangerous on this afternoon?

(b) Estimate the rate at which f is increasing at 1:30 p.m.

(c) Use the trapezoidal method, with intervals of half an hour, to estimate the total volume of water which flows out of the harbour during these three hours.   (SEG)

25 The petrol consumption (y miles per gallon) of a car is related to its speed (x m.p.h.) by the formula $y = 200 - \dfrac{3x}{2} - \dfrac{4200}{x}$.

Copy and complete the following table of values.

x	40	45	50	55	60	70	80
$3x/2$ $4200/x$		67.5 93.3		82.5 76.4			
y		39.2		41.1			

Draw the graph of this relation, taking 1 cm to represent 5 m.p.h. on the x-axis and 1 cm to represent 1 mile per gallon on the y-axis. Label the x-axis from 40 to 80 and the y-axis from 27 to 42.

From your graph estimate:
(a) the petrol consumption at 75 m.p.h.
(b) the most economical speed at which to travel in order to conserve petrol.

26 A cone of maximum volume is required from a circle of 10 cm radius. Using a graphical method find the radius, height and volume of the required cone, correct to 3 significant figures. (Volume of a cone is $\frac{1}{3}\pi r^2 h$, where r is the base radius and h is the vertical height.)

27 The flow chart (Figure 9:24) shows an algorithm (or 'rule') that can be used in a computer program to sort numbers, or words, into order. It is called a **bubble sort**.

Copy and complete the table to show how the example numbers are sorted, then use the flow chart for five numbers of your own, drawing up your own table.

N(1)	7	1				
N(2)	1	7				
N(3)	4	4				
N(4)	8	8				
N(5)	2	2				

Note In our example you start with A = 1, B = 2.
Then N(A) = N(1) = 7
N(B) = N(2) = 1.
The answer to 'Is N(A) < N(B)?' is NO because 7 > 1.
So M = 7
N(1) becomes 1
N(2) becomes 7.

To arrange five numbers into numerical order

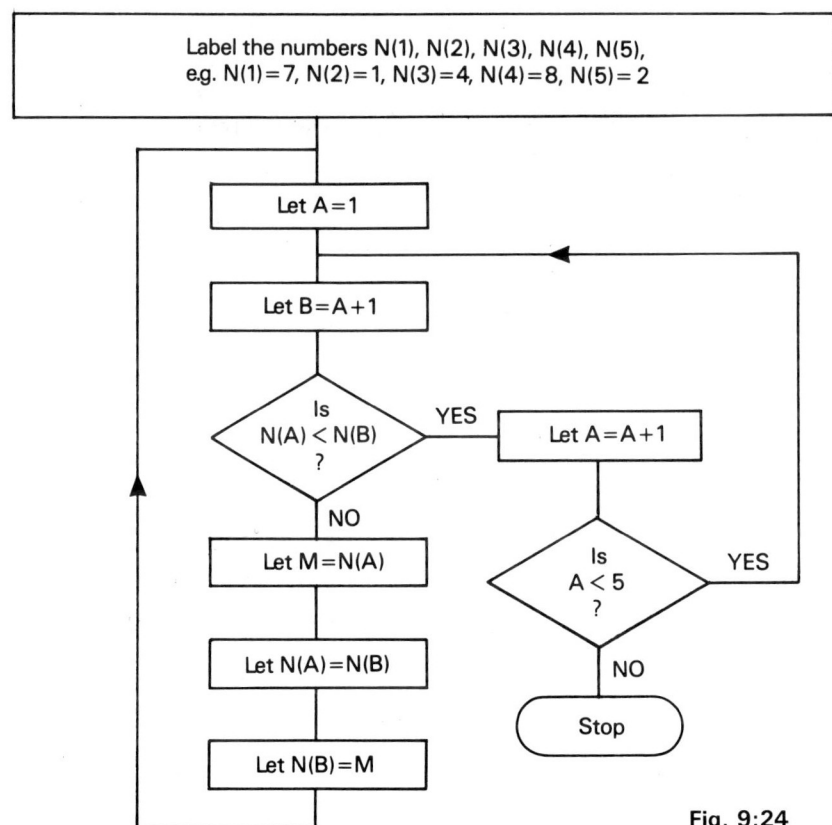

Label the numbers N(1), N(2), N(3), N(4), N(5),
e.g. N(1)=7, N(2)=1, N(3)=4, N(4)=8, N(5)=2

Let A=1

Let B=A+1

Is N(A) < N(B) ? — YES → Let A=A+1

NO

Let M=N(A)

Let N(A)=N(B)

Let N(B)=M

Is A < 5 ? — YES

NO

Stop

Fig. 9:24

10 Formulae and functions

Substitution
Change of subject
Functions

• You need to know . . .

Basic algebra 1 to 3 (page 217)

1 For normal temperature a rough way to change from degrees Celsius to degrees Fahrenheit is 'Double, then add 30'.

About how many degrees Fahrenheit is:
(a) 10 °C (b) 0 °C (c) −5 °C?

2 The surface area of a cube is $6s^2$, where s is the length of one edge. Find the surface area if $s = 1$ m.

3 A car's stopping distance is reckoned to be $v + \frac{v^2}{20}$ feet at v m.p.h. What would be the stopping distance at:
(a) 10 m.p.h. (b) 50 m.p.h.?

4 When $x = 4$, $y = 2$ and $z = -3$, find the value of:

(a) $x + y + z$ (b) $x + y - z$ (c) $x - y + z$ (d) $x - (y - z)$ (e) x^2 (f) $\frac{x^2}{3y}$.

5 Write a simplified expression for the total length of all the edges of each of the solids shown in Figures 10:1 to 10:4.

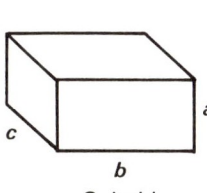

Cuboid

Fig. 10:1

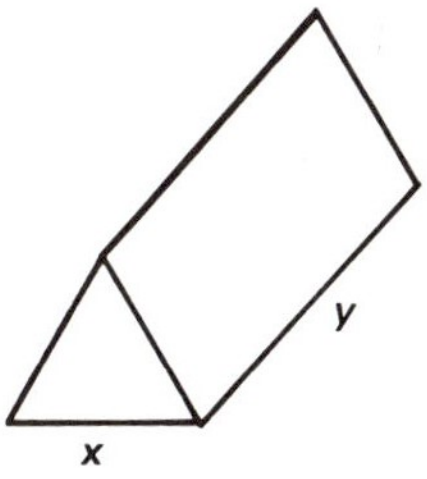

Equilateral triangular prism

Fig. 10:2

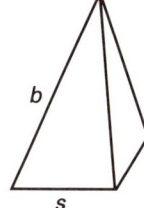

Square-based pyramid

Fig. 10:3

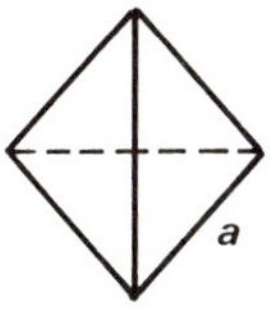

Tetrahedron

Fig. 10:4

6 Jake has £T to spend. He spends £S and has £L left. Write an expression to connect T, S and L starting:
(a) $T =$ (b) $L =$ (c) $S =$

7 Anna buys x apples at y pence each. They cost her z pence. Write an expression to connect x, y and z starting:
(a) $z =$ (b) $x =$ (c) $y =$

8 Transpose the equation to make the term in square brackets the subject:
(a) $V = wbh$ $[h]$ (b) $A = \pi r^2$ $[r]$ (c) $v^2 = u^2 + 2as$ $[s]$
(d) $S = \frac{1}{2}(a + b + c)$ $[c]$ (e) $a^2 + b^2 = 16$ $[a]$ (f) $E = Mc^2$ $[c]$
(g) $S = 2\pi rh + \pi r^2$ $[h]$

*9 A box is three times as long as it is deep and twice as wide as it is deep. Let it be x cm deep.
(a) How long is it?   (b) How wide is it?
(c) What is the sum of all its edges?
(d) What is its surface area?
(e) What is its volume?

*10 The following formula gives a way of multiplying by 21:

'Double the number, then write a zero at the end, then add on the number.'

Example 38×21
$38 \xrightarrow{\ 38 \times 2\ } 76 \xrightarrow{\ \text{write 0}\ } 760 \xrightarrow{\ \text{add 38}\ } 798$

Use the formula to find:
(a) 23×21 (b) 47×21 (c) 102×21.

*11 Make up a formula like the one in question 10 to multiply by 19.

*12 A cookery book gives the formula: 'Number of pounds of jam obtained is found by multiplying the number of pounds of sugar used by five, then dividing by three.'

How many pounds of jam can be made from six pounds of sugar?

*13 A motoring organisation estimates that if a car has x child passengers and y adult passengers then the extra running cost is equivalent to adding $x + 2y$ pence to the price of a litre of petrol.

What is the equivalent extra cost per litre if the passengers are:
(a) 1 child, 1 adult   (b) 3 children, 0 adults   (c) 2 children, 1 adult?

*14 If $E = 2x^2 + x$, what is E when x is 4?

*15 When $x = 3$ and $y = -2$, what is the value of:

(a) $4x + y$ (b) $3x - 2y$ (c) $(4x + y)(3x - 2y)$ (d) $\dfrac{3x - 6y}{3x - 2y}$?

*16 An object is thrown upwards. Its height, h metres, after t seconds is given by the formula $h = 40t - 5t^2$. Find the height of the object after 2 seconds.

*17 A box weighs 1 kg when empty. What is the total weight of 6 boxes each containing p kg of apples?

*18 How long will it take to travel k km at v km h^{-1}?

*19 The length of a man's forearm (f cm) and his height (h cm) are approximately related by the formula:
$$h = \frac{10f + 256}{3}.$$

(a) Part of the skeleton of a man is found and the forearm is 20 cm long. Use the formula to estimate the man's height.

(b) A man's height is 160 cm. Use the formula to estimate the length of his forearm.

(c) Use the formula to find an expression for f in terms of h.

(d) George is 1 year old and he is 70 cm tall. What happens if you use the given formula to estimate the length of his forearm?

(MEG)

*20 For Figure 10:5, write an expression for:
(a) the perimeter of the figure
(b) the area of one square
(c) the area of the whole shape.

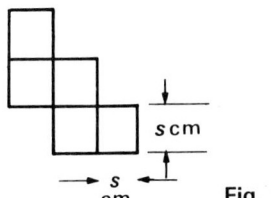

Fig. 10:5

*21 Figure 10:6 shows a cuboid, or 'brick-shape'. The volume of a cuboid is length times width times height. Write a simplified expression for the volume of the cuboid in Figure 10:6.

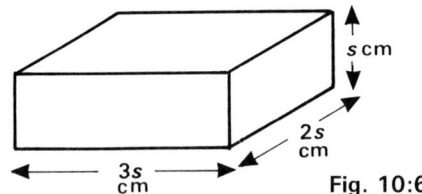

Fig. 10:6

*22 Figure 10:6 represents the 'nominal size' (that is, including an allowance for mortar) of a building brick. The wall it is used for contains 100 bricks. Write an expression for the volume of the wall in terms of s.

*23 A wall is made with the bricks shown in Figure 10:6. It is $300s$ cm long, $2s$ cm wide and $5s$ cm high.

(a) How many bricks wide is the wall?
(b) How many bricks high is the wall?
(c) How many bricks long is the wall?
(d) How many bricks are used altogether?

24 $s = 2.9 \times 10^2$, $t = 8.3 \times 10^3$, $u = 9.5 \times 10^{-1}$

Giving your answer in standard form find:

(a) st　(b) $s + u$　(c) $t - su$　(d) $\dfrac{s}{t}$　(e) $\dfrac{t + 5u}{3s}$　(f) $\dfrac{30u - st}{4us}$.

25 The following rule was always thought to give a prime number.

Think of any positive or negative whole number.
Multiply it by itself.
Add on 41.
Take away the number you first thought of.

(a) By letting x represent the number you first thought of, write this rule as an algebraic expression.

(b) Find the prime number produced when $x = 10$.

(c) Find the two values of x which will give the prime number 83.

(d) By looking at your answer to (a) find a value of x which for this rule does not give a prime number.

(NEAB)

26 A man is paid £12 an hour for normal time and time and a half for overtime. Write a formula to be used in the computer program that works out the total pay due (P pounds) for x hours of normal time and y hours of overtime.

27 Given that $p = 2nx - y$, express:
(a) y in terms of n, x and p (b) n in terms of p, x and y.

28 Transpose the equation to make the term in square brackets the subject:
(a) $p = mv$ $[v]$ (b) $v = s^3$ $[s]$ (c) $a = m^2 - 3f$ $[f]$
(d) $g = \frac{1}{2}(x + y + z)$ $[y]$ (e) $x^2 + y^2 = 4$ $[x]$

29 Given that $5a + 3b = c(4 - b)$, express:
(a) c in terms of a and b (b) b in terms of a and c.

30 Given that $a + b = \dfrac{3a + 7}{b}$, express a in terms of b.

31 Make b the subject of $a = \dfrac{b}{b - 2}$.

32 The mean of n items is q and the mean of m other items is p. Which of the following expressions gives the mean of all the items?

A $\dfrac{p + q}{m + n}$ **B** $\dfrac{mq + np}{(n + m)}$ **C** $\dfrac{m + n}{p + q}$ **D** $\dfrac{nq + mp}{(n + m)}$.

(SEG)

33 $T = 2\pi \sqrt{\dfrac{l}{g}}$, so $g =$

A $\dfrac{2\pi\sqrt{l}}{T}$ **B** $\dfrac{T^2 l}{4\pi^2}$ **C** $\dfrac{l}{(T - 2\pi)^2}$ **D** $\dfrac{4\pi^2 l}{T^2}$.

34 Alex cycles k km in t hours. What is her average speed in m/min?

35 **Example** Given that $f:x \longrightarrow x^2 - 1$ $g:x \longrightarrow 2 - x$

(a) Find $f(2)$.

This means substitute the value $x = 2$ into $x^2 - 1$. Answer: 3.

(b) Find $gf(\frac{1}{2})$.

This means first find $f(\frac{1}{2})$, then use the answer as x in $g(x)$. Sometimes it is written $g \circ f(x)$ or $g * f$.

Answer: $f(\frac{1}{2})$ is $-\frac{3}{4}$, $g(-\frac{3}{4})$ is $2\frac{3}{4}$.

Given that $f:x \longrightarrow 2x$ $g:x \longrightarrow x - 1$ $h:x \longrightarrow s^2$

find:

(a) $f(2)$ (b) $g(-2)$ (c) $h(-3)$ (d) $fg(1)$ (e) $gf(1)$ (f) $g \circ h(-1)$
(g) $h \circ f(-2)$ (h) $f * g * h(1)$.

36 $f(x)$ denotes the sum of the digits of x. For example, $f(75) = 12$.

Given that, in each of the following cases, x is an integer with four digits, find:
(a) x such that $f(x) = 1$
(b) the smallest x such that $f(x) = 10$
(c) the largest x such that $f(x) = 28$.

37 The **inverse** of a function 'undoes' its effect. The inverse of $f(x)$ is written $f^{-1}(x)$. If $f(x) = n$ then $f^{-1}(n) = x$.

Find the inverse of the function:

(a) $f:x \longrightarrow 3 + \dfrac{1}{2x}$ (b) $f:x \longrightarrow \dfrac{x + 2}{x - 4}$ (c) $f:x \longrightarrow 2x^3$.

38 The operation $*$ is defined on the set of real numbers by $p * q = p.q + p$, where $p.q$ means multiply p by q.

For example, $2 * 3 = 2.3 + 2 = 6 + 2 = 8$.

(a) Find the value of $3 * 4$.

(b) When $a = 2$, $b = 3$ and $c = 5$, find the value of:
(i) $a * (b + c)$ (ii) $(a * b) + (a * c)$.

(c) (i) Show that $p * (q * r) = p.q.r + p.q + p$.
(ii) Find a similar expression for $(p * q) * r$.
(iii) Why is it not possible to find the value of $2 * 3 * 5$?

(d) Prove that $a * b \neq b * a$ if $a \neq b$.

(e) Prove that $\{a * (b + c)\} + a = (a * b) + (a * c)$.

(f) Find values of x which satisfy the equation $x * x + 2 * x - 20 = 0$.

(WJEC)

11 Notation and factors

Operations   Indices   Factors

● **You need to know . . .**

Basic algebra 4 to 6 (page 219)
Factors 1 to 5 (page 221)

1 Simplify if possible:
(a) $a + a + a$ (b) $a \times c$ (c) $a + c$ (d) $2a \times 2a$ (e) $2a + 2a$ (f) $2a + 2c$.

2 Multiply out the brackets, then simplify if possible:
(a) $2(a + b)$ (b) $2a(c - 3b)$ (c) $-3(2a - 3b + 2c)$
(d) $6(a - b) + 4(a + b)$ (e) $4(a - 4) - 3(a - 5)$
(f) $3(2t - 5) - 2(2t + 3)$ (g) $a - 2b + (3a - b) - (2a + 3b)$.

3 Factorise by taking out a common factor:
(a) $ax + bx$ (b) $4pq - 8p$ (c) $2a + 4p - 6c$ (d) $2ab - 4bc - 6cd$.

4 Simplify:
(a) $h \times h \times h$ (b) $h \times h^2$ (c) $(h^2)^2$ (d) $h^3 \times h^2$ (e) $h^4 \times h$ (f) $(h^3)^2$.

5 Simplify:
(a) $3a \times b$ (b) $4a^2 \times a$ (c) $2a \times 3a^2$ (d) $4ab \times a$ (e) $3b \times 2ab$.

6 Simplify:
(a) $\dfrac{4a^2b}{6ab^3}$ (b) $\dfrac{3ab^2}{6a^2b}$ (c) $\dfrac{a^3b^2}{a^2b^3}$ (d) $\dfrac{2ab}{2a^2b^2}$ (e) $\dfrac{4km}{km}$.

7 Simplify:
(a) $a^5 \div a^2$ (b) $a^4 \div a$ (c) $a^3 \div a^5$ (d) $2a^2b \div 6a$ (e) $4ab^2 \div 2ab$.

8 Simplify if possible:
(a) $2a^2b + 3a^2b$ (b) $2a^2b + 3ab^2$ (c) $3a^2 - 2a + 2a$ (d) $3a^2 - 2a + 2a^2$
(e) $3a^3 - 2a + 2a^2$ (f) $3ab^2 + 2ab^2 - ab^2$.

9 Simplify:
(a) $2x^3 \times 4x^2$ (b) $\dfrac{2x + 12}{2}$ (c) $\dfrac{4x + 14}{2}$ (d) $7x - y - 3x - 5y$

(e) $2a \times 3b \times 4a$ (f) $\dfrac{30x^5y^2}{16x^2y}$ (g) $\dfrac{10abc}{5ab}$ (h) $\dfrac{8x^2y}{16y^2}$ (i) $\sqrt{25a^4b^2}$.

*10 Simplify:
(a) $2d + d + 2d$ (b) $d \times 3d$ (c) $d \times d \times d$ (d) $c + 4c - 2c$ (e) $4f \times 2f$.

*11 Multiply out the brackets, then simplify if possible:
(a) $5(x - 2y)$ (b) $2c(c + 1)$ (c) $-2(1 - 4r + t^2)$ (d) $2(a + b) + 3(a - 2b)$
(e) $3(q + t) - (q - t)$ (f) $2a - a(1 + a) + 2a(a - 1)$.

*12 Factorise by taking out common factors:
(a) $4b - 2c$ (b) $3rt - 4r$ (c) $2mn - 4m^2$ (d) $ab - a^2b$ (e) $4x^2 - 4xy$.

*13 Simplify:
(a) $4c \times 3$ (b) $2a^2 \times a$ (c) $4c^2 \times 2c$ (d) $5bc \times b$ (e) $a^2b \times ab$.

*14 Simplify:
(a) $x^5 \div x^2$ (b) $5x^2r^3 \div x^2r^2$ (c) $12ab^3 \div 15a^2b$ (d) $a^5 \div a^8$.

*15 Find n if:
(a) $3^4 \times 3^n = 3^7$ (b) $3^6 \div 3^2 = 3^n$.   (SEG)

*16 What expression must be added to $3x + y$ to make it $7x + 4y$?   (ULEAC)

17 Factorise where possible:
(a) $a(c + d) - b(c + d)$ (b) $x(a + b) - y(a - b)$ (c) $3(a + c) - c(a + c)$
(d) $2(a - b) + c(a + b)$ (e) $-d(a - c) + b(a - c)$.

18 Factorise:
(a) $ac + ad - bc - bd$ (b) $mn + mt + pn + pt$ (c) $2b + ab + 2c + ac$
(d) $ax - bx - 2ay + 2by$ (e) $6x + 4y - 9cx - 6cy$ (f) $3x - 3y - x^2 + xy$.

19 Factorise by taking out any common factors, then using the difference of two squares:
(a) $a^2 - 25$ (b) $a^2 - 25b^2$ (c) $4 - 36x^2$ (d) $3c^2 - 12d^2$ (e) $5x^2 - 5y^2$.

20 Expand:
(a) $(x + 2)(x + 1)$ (b) $(x + 2)(x - 1)$ (c) $(x - 2)(x + 1)$ (d) $(x - 2)(x - 1)$
(e) $(2x - 3)(3x - 2)$ (f) $(2w - 1)^2$.

21 Find all possible integral values of the ordered pair (x, y) if $(x - 3)(y - 3) = 6$.

H22 Factorise:
(a) $t^2 - 4t + 3$ (b) $x^2 - x - 6$ (c) $x^2 - 6x + 5$ (d) $2a^2 - a - 10$
(e) $2x^2 + x - 6$ (f) $3a^2 - 22a - 16$ (g) $6a^2 - 5ab + b^2$ (h) $3x^2 + 10x - 8$.

H23 Find the value of k if $x^2 + 4x + k$ is a perfect square.

H24 Find the value of k if $4x^2 - 8x + k$ is a perfect square.

H25 Simplify:
(a) a^{-2} (b) b^{-3} (c) 6^{-2} (d) 2^{-3}.

H26 Simplify:

(a) $100^{\frac{1}{2}}$ (b) $8^{\frac{1}{3}}$ (c) $16^{\frac{1}{4}}$ (d) $64^{\frac{1}{6}}$ (e) $(4d^4)^{\frac{1}{2}}$ (f) $(9d^6)^{\frac{1}{2}}$ (g) $(8c^3)^{\frac{1}{3}}$.

H27 Simplify:

(a) $8^{\frac{2}{3}}$ (b) $16^{\frac{3}{4}}$ (c) $32^{\frac{3}{5}}$ (d) $4^{\frac{3}{2}}$ (e) $81^{\frac{3}{4}}$.

H28 Simplify:

(a) $4^{-\frac{1}{2}}$ (b) $16^{-\frac{1}{4}}$ (c) $8^{-\frac{1}{3}}$ (d) $8^{-\frac{2}{3}}$ (e) $(\frac{1}{4})^{-2}$.

H29 Simplify:

(a) $27^{\frac{1}{3}}$ (b) 8^0 (c) $81^{-\frac{3}{4}}$ (d) $9^{\frac{3}{2}}$ (e) $16^{1\frac{1}{2}}$ (f) $4^{2\frac{1}{2}}$ (g) 4^{-1}
(h) $(\frac{1}{5})^{-1}$ (i) $(\frac{1}{2})^3$ (j) $(\frac{2}{3})^{-2}$.

H30 Assuming the denominator is not zero, factorise then simplify:

(a) $\dfrac{a^2 - 1}{a - 1}$ (h) $\dfrac{2m + 3n}{4m^2 - 9n^2}$ (c) $\dfrac{x^2 + 2x - 8}{x^2 - 2x}$ (d) $\dfrac{a + 4}{a^2 + 3a - 4}$

(e) $\dfrac{3a^2 + 8a + 4}{a^2 - a - 6}$ (f) $\dfrac{a^2 - 9}{a^2 - a - 12}$.

---

31 Given that $n - 4\sqrt{3} = 6\sqrt{3}$, find n.

32 Express as a single fraction in its lowest terms:

(a) $\dfrac{a}{4} + \dfrac{a}{5}$ (b) $\dfrac{3a}{5} + \dfrac{2}{3}$ (c) $\dfrac{a - 1}{2} - \dfrac{a - 2}{3}$

(d) $\dfrac{a - 2b}{4} - \dfrac{a - 3b}{8}$ (e) $\dfrac{2a + 3}{5} - \dfrac{a + 3}{6}$.

33 Simplify:

(a) $\dfrac{3a}{4} \div \dfrac{9a}{16}$ (b) $\dfrac{mnp}{6d} \div \dfrac{dnp}{3m}$ (c) $6a^2b \div \dfrac{3a}{4c}$ (d) $\dfrac{12a^2c}{5d} \div \dfrac{9ac}{10d^2}$.

34 Express as a single fraction in its lowest terms:

(a) $\dfrac{x}{y} + \dfrac{y}{x}$ (b) $\dfrac{3}{x} + \dfrac{4}{2x}$ (c) $\dfrac{3a}{b} - \dfrac{b}{3a}$ (d) $\dfrac{3a}{b} + \dfrac{a}{b^2}$

(e) $\dfrac{4a - 1}{2a} - \dfrac{2a - 1}{3a} - \dfrac{1}{6a}$.

35 **Example** Express as a single fraction in its lowest terms $\dfrac{5}{a-1} - \dfrac{2}{a+3}$.

The LCM of the denominators is $(a-1)(a+3)$.

$$\dfrac{5}{a-1} - \dfrac{2}{a+3} \rightarrow \dfrac{5(a+3)}{(a-1)(a+3)} - \dfrac{2(a-1)}{(a-1)(a+3)}$$

$$\rightarrow \dfrac{5a+15-2a+2}{(a-1)(a+3)} \rightarrow \dfrac{3a+17}{(a-1)(a+3)}$$

Express as a single fraction in its lowest terms:

(a) $\dfrac{a}{a-b} - \dfrac{b}{a+b}$
(b) $\dfrac{1}{a+1} - \dfrac{1}{a+4}$
(c) $\dfrac{5}{a-1} - \dfrac{2}{a+3}$

(d) $\dfrac{5}{3a-1} - \dfrac{3}{2a+3}$
(e) $\dfrac{a}{a-y} + \dfrac{y}{y-a}$ Note: $a-y = -(y-a)$.

36 **Example** Find the product of $(b+2)(b^2-b+2)$.

Each term in the second bracket is multiplied by each term in the first bracket. Multiplying the second bracket by b gives $b^3 - b^2 + 2b$, and multiplying it by 2 gives $2b^2 - 2b + 4$.

Hence $(b+2)(b^2-b+2) \rightarrow b^3 - b^2 + 2b + 2b^2 - 2b + 4 \rightarrow b^3 + b^2 + 4$.

Find the product of:
(a) $(b+1)(b^2+2b-2)$ (b) $(3a-4)(2a^2+2a-6)$
(c) $(2x^2-3x+4)(x^2-3x-4)$.

37 If $(x+a)$ is a factor of f(x) then when $x = -a$, f$(x) = 0$.

Reason Let Q be the result of dividing f(x) by $(x+a)$.
Then $(x+a) \times Q = $ f(x).
If $x = -a$, then $(x+a) = 0$, so $(x+a) \times Q = 0$, so f$(x) = 0$.

Example Find a factor of $x^3 + 5x^2 + 7x + 2$.

f$(1) = 1 + 5 + 7 + 2 \neq 0$
f$(-1) = -1 + 5 - 7 + 2 \neq 0$
f$(2) = 8 + 20 + 14 + 2 \neq 0$
f$(-2) = -8 + 20 - 14 + 2 = 0$, therefore $(x+2)$ is a factor. The quotient may be found by inspection.

$(x+2)(x^2+3x+1) = x^3 + 5x^2 + 7x + 2$.

Factorise:
(a) $x^3 + 2x^2 - x - 2$ (b) $x^3 - 3x + 2$ (c) $x^3 - 6x^2 + 12x - 8$
(d) $2x^3 + 5x^2 - 4x - 12$.

12 Equations and inequalities

Linear
Polynomial by trial and improvement
Inequalities
Simultaneous linear

(Quadratics by factors, formula
**   and iteration)**
(Problems)

● **You need to know . . .**

Equations 1 to 8 (page 225)

1 Find the value of the letter if:

(a) $a - 4 = 3$ (b) $a + 4 = 3$ (c) $4 - x = 7$ (d) $2a - 1 - 8$

(e) $3 - 2a = 7$ (f) $\dfrac{a}{2} + 1 = 9$.

2 Find the value of the letter if:

(a) $3(a + 1) = 12$ (b) $4(2 - a) = 20$ (c) $\dfrac{2}{a + 1} = 1$ (d) $\dfrac{a + 1}{4} + 5 = 0$

(e) $\frac{1}{2}(2a - 3) = 6$ (f) $\dfrac{a}{3} + \dfrac{a}{2} = \dfrac{5}{6}$.

3 Solve:

(a) $4a - 5 = 2a + 1$ (b) $a + 4 = 3a - 5$ (c) $3h + 5 = 14 - 3h$
(d) $3(e - 5) = 2(e + 7)$.

4 Solve the following equations by trial and improvement using a calculator. Accuracy to 2 decimal places is sufficient, but you can be more accurate if you wish.

(a) $p^2 = 12$ (b) $a^3 = 9$ (c) $n^2 - n = 10$ (d) $2b^2 + b = 5$
(e) $x^2 - 2x = 17$.

5 Solve the following equations simultaneously by substitution.

(a) $y = x + 1$ and $3x - y = 3$ (b) $y + 2x = 16$ and $x = y - 4$
(c) $g = 2f - 7$ and $4f - 3g = 15$ (d) $x = 3 - y$ and $2y - 3x + 24 = 0$.

6 Solve the following equations simultaneously by elimination.

(a) $4x - 3y = 15$ and $2x + 3y = 3$ (b) $2x - 3y = 13$ and $4x + 5y = 4$
(c) $3x - 4y = 10$ and $5x + 7y = 3$ (d) $9y = 9 - 4x$ and $3x = 11 + 6y$.

*7 The relative density of a solid (that is, its density compared to water) is given by the formula

$$\text{Relative density} = \frac{\text{mass of the solid}}{\text{mass of water with the same volume as the solid}}.$$

Figure 12:1 shows the two stages in an experiment to find the relative density of a solid. From the masses given calculate the relative density of the solid.

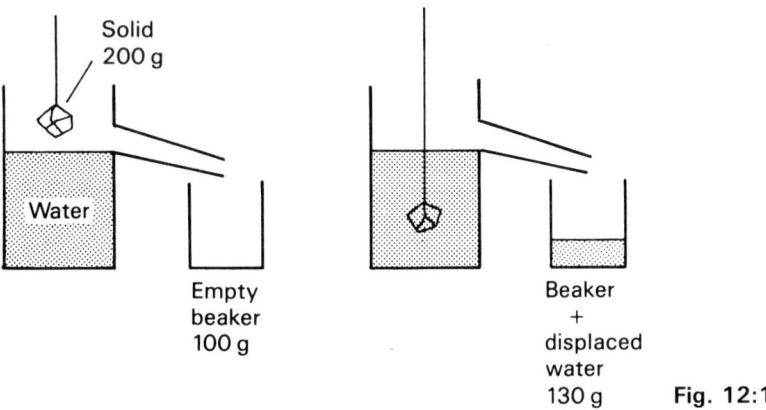

Fig. 12:1

*8 Find the value of the letter if:

(a) $4 + x = 1$ (b) $2x - 7 = 6$ (c) $5(x - 3) = 15$ (d) $\frac{1}{2}(2x + 1) = 9$

(e) $\frac{3}{x} = 6$ (f) $\frac{5}{x} = 15$.

*9 The coefficient of linear expansion of a substance is found from the formula

$$\text{Coefficient of linear expansion} = \frac{\text{expansion}}{\text{original length} \times \text{difference in temperature (}^\circ\text{C)}}.$$

Thirty metres of copper pipe expands to 30.0357 metres when the temperature rises from 8 °C to 78 °C. What is the coefficient of linear expansion of copper?

*10 Lilian asked her uncle how old he was. 'In 13 years I'll be twice as old as I was 7 years ago,' he replied.

(a) Taking his age now to be x years, write down:
(i) his age in 13 years, in terms of x,
(ii) an equation in x.

(b) Solve your equation and find Lilian's uncle's age.   (MEG)

11 Each colour in the table stands for a number. The sum of these numbers for each row is given at the end of that row.

Blue	Green	Green	Red	Yellow	19
Red	Blue	Red	Red	Blue	18
Blue	Red	Red	Blue	Blue	17
Green	Red	Yellow	Blue	Yellow	16

(a) Solve the simultaneous equations:
$$3r + 2b = 18$$
$$2r + 3b = 17.$$

(b) Hence, or otherwise, write down the numbers represented by Red and Blue in the table above.

(c) Find the numbers represented by Green and Yellow.

(d) Investigate the order in which you would place the following so as to obtain a row which when placed below the table of colours above will produce totals for the five columns which are equal to the totals for the five rows.

| Yellow | Yellow | Yellow | Green | Red |

(NEAB)

12 Solve:

(a) $2(x + 1) = x + 4$ (b) $3a = 2(4 + a)$ (c) $2a - 5 = 5a - 17$

(d) $\frac{1}{2}(a - 4) = 3$ (e) $\frac{3}{a} = 4$ (f) $5 - 3(2 - a) = 1$ (g) $2 + 2(4y + 11) = 21$

(h) $p(p - 1) - p(p + 4) + 15 = 0$ (i) $3(p - 2) - 2(3 - p) = 3p - 2.$

H13 Solve:
(a) $t^2 + 3t = 0$ (b) $2x^2 = 7x$ (c) $2x^2 + 3x = 0$ (d) $4a^2 = 16a$
(e) $3b^2 - 9b = 0$ (f) $3x^2 = 12x.$

H14 Solve by factorisation:
(a) $x^2 - 3x + 2 = 0$ (b) $x^2 + 2x = 3$ (c) $x^2 + 6x + 9 = 0$
(d) $x^2 - 7x - 60 = 0$ (e) $2x^2 + 6 = 7x$ (f) $8t^2 - 6 = 13t.$

H15 Solve:
(a) $(x - 2)(x - 3) = 20$ (b) $(x + 2)(x + 3) = 12$ (c) $(x + 4)(x - 1) = 36$
(d) $x(x - 5) = 6$ (e) $x(x + 4) = 32$ (f) $2(x + 1)(x - 3) = 24.$

H16 Solve by the formula, or by completing the square:
(a) $x^2 - 4x + 2 = 0$ (b) $x^2 + 2x = 5$ (c) $x^2 - 3 = 6x$
(d) $2x^2 - 7x - 4 = 0$ (e) $3d^2 + 2d = 4.$

H17 Find the range of values of x for which $9 - (x - 3) \leqslant 2x.$

H18 The e.m.f. (E) and internal resistance (r) of a cell may be found by measuring the current flowing (I) for different external resistances (R).

The formula connecting these variables is $I = \dfrac{E}{R + r}$.

I is in amperes, E in volts, and R and r in ohms.

Find the e.m.f. and the internal resistance of a cell which produces 0.25 amperes when the external resistance is 5 ohms and 0.15 amperes when the external resistance is 9 ohms.

H19 Given that $w = 3y$, $x = 4y$, and $x = z$, express z in terms of w.

H20 **Example** Solve $\dfrac{c}{2} - \dfrac{c}{3} = \dfrac{7}{5}$.

First remove the fractions by multiplying each term by the common denominator, 30. Then $15c - 10c = 42 \rightarrow c = 8\frac{2}{5}$.

Solve:

(a) $\dfrac{a}{3} - \dfrac{a}{4} = \dfrac{1}{6}$ (b) $y + \dfrac{3y}{2} = 1$ (c) $\dfrac{3a}{2} - \dfrac{2a - 1}{4} = \dfrac{1}{8}$ (d) $\dfrac{2y - 4}{3} + 2 = \dfrac{3y + 3}{2}$

(e) $\dfrac{3(2y - 1)}{4} - \dfrac{2(y + 2)}{3} = 1$.

H21 Solve:

(a) $\dfrac{4}{a - 2} = 8$ (b) $\dfrac{9}{y} = \dfrac{4}{3}$ (c) $\dfrac{2a + 7}{a - 4} = 5$ (d) $\dfrac{4}{a} = \dfrac{5}{a + 2}$ (e) $\dfrac{a + 4}{a - 3} = \dfrac{a + 2}{a - 4}$.

H22 Given that $p = \dfrac{2y - 1}{y - 2}$ and $q = \dfrac{y + 2}{y - 1}$, show that $p - q = \dfrac{y^2 - 3y + 5}{y^2 - 3y + 2}$, then calculate the values of y, if any, which satisfy the equation:
(a) $q = 0$ (b) $p = 1$ (c) $p - q = 1$ (d) $p + q = 7\frac{1}{2}$.

H23 Find the solution set of integers if:
(a) $2 < x + 5 \leqslant 12$ (b) $-6 < 2x + 4 \leqslant 6$ (c) $3 < -3x < 9 - 4x$.

H24 Find the equation whose roots are:
(a) 2 and -3 (b) 0 and 4   (c) $2\frac{1}{2}$ and -2.

H25 Given that $\dfrac{y}{x} - \dfrac{y - 2}{x - 3} = \dfrac{1}{x + 5}$, find the two values of x for which $y = 3$.

H26 You are given that $6t = 11 - t^2$.

(a) Show that $t = 1$ is too small to be a solution and that $t = 2$ is too big.

(b) Use the iterative formula $t_{n+1} = \dfrac{11 - t_n^2}{6}$ starting with $t = 1.5$ to find t correct to 3 significant figures. As part of your answer, list the values of t you obtain.

H27 To solve the equation $x = x^2 - 7$ the following step-by-step method is used:

$x_{n+1} = \sqrt{(x_n + 7)}$

Choosing $x_1 = 4$, then $x_2 = \sqrt{(4 + 7)} \approx 3.316\,624\,8$

(a) Calculate x_3, x_4, x_5, x_6.

(b) Solve $x = x^2 - 7$ correct to 3 decimal places.

(SEG)

H28 (a) Given that $f(x) = 2x^2 - 4x - 1$, calculate the value of: (i) $f(2)$ (ii) $f(3)$.

(b) Explain briefly what the results obtained in (a) show about one of the roots of the equation $2x^2 - 4x - 1 = 0$.

(c) Show that the quadratic equation $2x^2 - 4x - 1$ can be written in the form
$$x = 2 + \frac{1}{2x}.$$

(d) Use the iterative formula $x_{n+1} = 2 + \dfrac{1}{2x_n}$ to find:

(i) the value of x_2 if $x_1 = 2.5$
(ii) the value of x_3, giving your answer to three places of decimals.

(e) The solutions of the quadratic equation $ax^2 + bx + c = 0$ are $x = \dfrac{-b \pm \sqrt{b^2 - 4ac}}{2a}$.

Use this formula to obtain the negative solution of the equation $2x^2 - 4x - 1 = 0$, giving your answer correct to three decimal places.

(NEAB)

---

29 An allotment which is in the shape of a square of side b metres has a neighbouring allotment which is in the shape of a rectangle b metres long and $(b - 3)$ metres wide. The rectangular allotment is smaller than the square allotment by $24\,\text{m}^2$. Find the dimensions of each allotment.

30 In an examination, the lowest and highest marks were 36 and 61 respectively. In order to change any mark y into a mark N on a new scale the formula $N = 4(y - 36)$ was used. Calculate the lowest and the highest mark on the new scale and the mark which would remain unchanged.

31 In a factory a chargehand earns £40 per week more than a packer, and three packers and two chargehands together earn £430. Find the weekly wage of a packer.

32 You are given that $y = 2x^2 - 3x$.
(a) Calculate, correct to 2 decimal places, the two values of x for which $y = 1$.
(b) Show that there is one value of x for which $y = -\frac{9}{8}$ and find this value.
(c) Find the set of values of x for which $y < 0$.

33 The formula $h = 12t - t^2$ gives the height h metres of a ball above the ground, t seconds after it is thrown into the air.
(a) Find h when $t = 1.2$.
(b) Find when the ball subsequently hits the ground.
(c) Find at what times the height of the ball is 4 metres.

34 A rectangle has sides of length $(2x + 1)$ and $(x + 4)$ cm. Write down simplified expressions in x for the perimeter and for the area.

Given that the area of this rectangle is $63 \, \text{cm}^2$, form an equation in x and solve it to find the positive value of x. Hence find the perimeter of the rectangle.

35 Achmed bought some calculators for £200. Two were stolen, but, by selling each of the remaining ones for £1 more than he paid for it, he made an overall profit of £40.

(a) If he sold each calculator for £x, how many did he sell?

(b) How many did he buy?

(c) Show that x must satisfy the equation $x^2 + 19x - 120 = 0$, and hence find the price at which he sold each calculator.

36 A solid rectangular block has a square base of side x cm and a height of y cm. Given that the total length of the edges of the block is 28 cm and the total surface area of the block is $32 \, \text{cm}^2$, write down two equations in x and y, and by elimination prove that $3x^2 - 14x + 16 = 0$. Hence find the values of x and y and the two possible values for the volume of the block.

37 (a) Show that $x = \dfrac{1}{x^2} + 2$ is equivalent to $x^3 - 2x^2 - 1 = 0$.

(b) Starting with $x_0 = 2$, use the iterative formula

$$x_{n+1} = \frac{1}{x_n^2} + 2$$

to calculate $x_1, x_2, \ldots$ etc. correct to 4 decimal places, continuing the process until two successive values agree. Hence find an approximate root of the equation $x^3 - 2x^2 - 1 = 0$.

(c) Construct a flow chart or computer program to carry out the above iterative process, which will stop when two successive values of x_n agree to 4 decimal places, and which will print out that value.

38 If f(x) is a continuous function, then a change in value of f(x) from positive to negative, or vice versa, means that f(x) must become zero during the change.

Using this fact an equation may be solved by trial and error. A programmable calculator or a computer will be useful.

Example $f(x) = x^3 - 5x + 1$. Find the solutions of f(x) = 0.

x	−3	−2	−1	0	1	2	3
f(x)	−11	3	5	1	−3	−1	13

From the table, solutions will be found
   (i) between −3 and −2
   (ii) between 0 and 1
   (iii) between 2 and 3.

(i)   f(−2.5) is −ve ∴ solution is between −2.5 and −2
      f(−2.3) is +ve ∴ solution is between −2.3 and −2.5
      f(−2.4) is −ve ∴ solution is between −2.3 and −2.4
      f(−2.35) is −ve ∴ solution is between −2.35 and −2.3
      f(−2.32) is +ve ∴ solution is between −2.32 and −2.35
      f(−2.33) = 0.000 663, which means it is very close to a solution.
      The process can be continued to any degree of accuracy.

Find the other two solutions to the given function.

39 Find the three solutions of f(x) = 0 where $f(x) = 2x^3 - x^2 - 13x - 4$, given that they all lie between −3 and 3.

40 There was an old man, much revered,
Who said, 'What's the length of my beard?
If I double it, then
Add its square minus ten
I get minus two feet; very weird!'

13 Function graphs

● **You need to know . . .**

Algebraic graphs 1 to 8 (page 234)

1   Look at Figure 13:1.

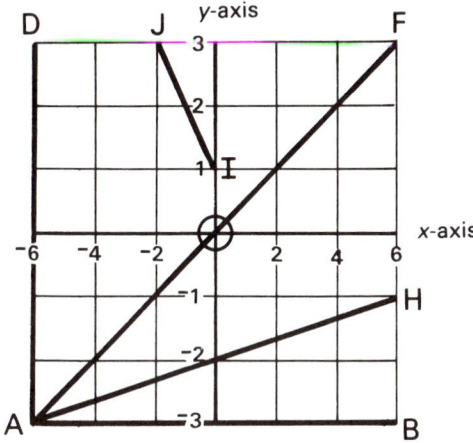

   (a) Write the co-ordinates of the ends of:
      (i) line AB   (ii) line IJ.

   (b) State the equation of:
      (i) line AB   (ii) line AD   (iii) line AF.

   (c) State the gradient (slope) of:
      (i) line AF   (ii) line AH   (iii) line IJ.

   (d) Which line has the description:
      (i) $x \rightarrow 1 - x$ (ii) $x - 6y = 12$?

   Fig. 13:1

2   Draw four sets of axes, x from -3 to 3 and y from -5 to 9. On your axes draw the graphs:
   (a) $y = 2x + 1$ (b) $y = x^2$ (c) $x \rightarrow \frac{1}{2}x^2$ (d) $x \rightarrow -\frac{1}{2}x^2$.

3   On your answers to question 2 shade the regions:
   (a) $y \leqslant 2x + 1$, $y \geqslant 0$ and $x \leqslant 2$ (b) $-2 \leqslant x \leqslant 2$, $y \leqslant x^2$ and $y \geqslant 0$
   (c) $2 \geqslant y \geqslant \frac{1}{2}x^2$ (d) $0 \leqslant x \leqslant 2$ and $0 \geqslant y \geqslant -\frac{1}{2}x^2$.

4 (a) Copy the tables for $y = 1/x$, then use a calculator to help you to complete them, giving y correct to 2 d.p.

x	-5	-4.5	-4	-3.5	-3	-2.5	-2	-1.5	-1	-0.5
y										

x	5	4.5	4	3.5	3	2.5	2	1.5	1	0.5
y										

(b) A problem arises when $x = 0$. Why?

(c) $\dfrac{1}{0}$ is taken to have the value infinity

(∞), although such a value does not exist except as a mathematical idea. Figure 13:2 shows a sketch of the graph of $y = 1/x$. Use your table values to construct an accurate graph.

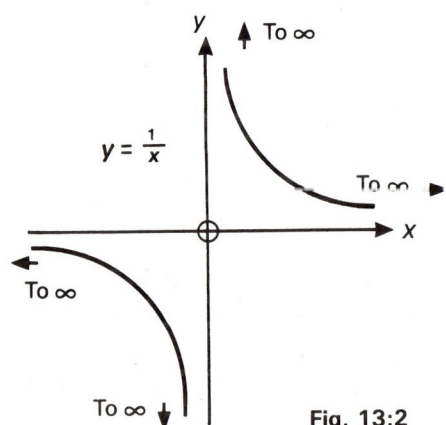

Fig. 13:2

5 By drawing pairs of graphs on axes from -6 to 6, solve the following pairs of equations simultaneously.
(a) $y = x + 3$ and $y = -2x + 6$
(b) $y = x$ and $y = -\frac{1}{3}x + 4$
(c) $y = x + 2$ and $y + 2x + 4 = 0$.

6 Draw axes, x from -2 to 4, y from -5 to 5.

(a) State the gradient and the crossing point on the y axis for:
(i) $y = 3x - 1$ (ii) $y = x + 2$.

(b) By drawing their graphs, solve $y = 3x - 1$ and $y = x + 2$ simultaneously.

(c) Shade on your grid the region where $y \geqslant 0$, $y \leqslant x + 2$ and $y \geqslant 3x - 1$.

7   The region R shown in Figure 13:3 is bounded by three sloping lines and the two axes.

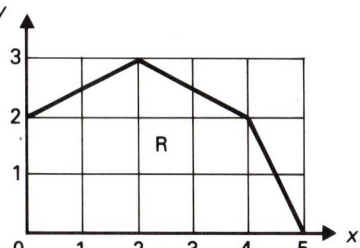

Fig. 13:3

(a) State the five inequalities which define region R.

(b) For the points (x, y) in region R find:
   (i) the greatest possible value of $x + y$, and the co-ordinates of the point at which this greatest possible value occurs.
   (ii) the least possible value of $x - 2y$, and the co-ordinates of one point at which this least possible value occurs.
   (MEG)

8   By drawing suitable sketches, find the slope (m) of the lines passing through:
   (a) $(3, -5)$ and $(5, -1)$ (b) $(1, 1)$ and $(5, 3)$ (c) $(-6, -1)$ and $(-4, -4)$.

9   The equations of all straight-line graphs can be written in the form $y = mx + c$. Using the value of m, and one pair of co-ordinates to find c, find the equation of each line in question 8.

10 Sketch axes and the general shape of the following. Draw all the graphs in each part on the same grid. Plotting is not required. Label each line clearly.

   (a) The linear graphs $y = x$, $y = -x$, $y = x - 2$, $y = 2x$ and $y = \frac{1}{2}x - 2$.

   (b) The quadratic graphs $y = x^2$, $y = 2x^2$, $y = \frac{1}{2}x^2$ and $y = -x^2$.

   (c) The quadratic graphs $y = x^2$, $y = -x^2$ and $y = x^2 - 2$.

   (d) The reciprocal graphs $y = \dfrac{1}{x}$ and $y = \dfrac{4}{x}$.

*11 If $x - y = 5$, state the value of x when y is:
   (a) 0   (b) 3   (c) 5   (d) -1 (e) -2 (f) -4.

*12 Draw one pair of axes, both from -6 to 6. Draw on your grid the straight lines from:
   (a) $(0, 3)$ to $(3, 6)$ (b) $(3, -5)$ to $(5, -1)$ (c) $(1, 1)$ to $(5, 3)$
   (d) $(0, -3)$ to $(1, 0)$ (e) $(0, -6)$ to $(6, -5)$ (f) $(0, 0)$ to $(-3, 3)$
   (g) $(-4, 5)$ to $(0, 4)$ (h) $(-6, -1)$ to $(-4, -4)$ (i) $(-3, -3)$ to $(0, -5)$.

*13 By plotting on axes from -6 to 6 each, or otherwise, find (i) the gradient and (ii) the equation of the line through:
   (a) $(0, 3)$ and $(3, 6)$ (b) $(0, -3)$ and $(1, 0)$
   (c) $(0, -6)$ and $(6, -5)$ (d) $(0, 0)$ and $(-3, 3)$
   (e) $(-4, 5)$ and $(0, 4)$ (f) $(-3, -3)$ and $(0, 5)$.

*14 Draw one pair of axes from -6 to 6 each. Draw on your grid the straight lines:
   (a) $x = -1$ (b) $y = x$ (c) $y = x - 4$ (d) $y = x + 4$.

*15 Repeat question 14 for:
(a) $y = 6$ (b) $x + y = 3$ (c) $x + y = 6$ (d) $y - x = 4$ (e) $y = \frac{1}{2}x - 6$
(f) $y + \frac{1}{2}x = -6$.

*16 Repeat question 14 for:
(a) $y = 2x$ (b) $y = 2x - 3$ (c) $x + y = 2$ (d) $x + y = -2$.

*17 (a) On axes from -6 to 6 each, shade the regions A and B defined as follows.
A: $-4 \leqslant x \leqslant 5$ and $1 \leqslant y \leqslant 3$
B: $y \leqslant x$ and $-4 \leqslant x \leqslant 5$

(b) State the co-ordinates of the region which is formed where A and B cross.

*18 On axes x from -3 to 3, f(x) from -20 to 20, draw graphs of $x \rightarrow 2x^2$ and $x \rightarrow -2x^2$.

*19 (a) The area of a circle is given by the formula $A = \pi R^2$. By considering values of A for values of R from 0 to 5 at half-unit intervals, draw a graph of A against R. Take π as 3.14.

(b) Read from your graph, as accurately as possible:
(i) the area of a circle of radius 3.2 cm
(ii) the radius of a circle of area 20 cm^2.

*20 On axes from -6 to 6 each, shade the region where:
$3 > x \geqslant 0$ and $\frac{2}{3}x + 3 > y > -\frac{1}{3}x + 2$.

H21 In an experiment Pierre was given a fixed length of time to memorise a list of items. Pierre was tested and the number of errors noted.

This experiment was repeated for other lists with different lengths of time. The results were as follows.

Length of time in seconds, x	5	6	7	8	9	10	11	12	13	14
Number of errors, y	9	7	8	5	6	6	4	5	3	2

(a) Plot these points, using 1 cm to 2 units on each axis, and draw the line of best fit.

(b) Give an interpretation of the point where the line meets the x-axis.

(c) Explain why there is no sensible interpretation of where the line meets the y-axis.

(d) Calculate the gradient of the line, correct to one decimal place.

(e) Find the equation of the line.

(f) The experiment was repeated by Kerith. He made the same number of errors as Pierre when the time was 5 seconds, but for longer times his memory was a little worse than Pierre's. On your diagram draw a possible line of best fit for Kerith.

H22 Figure 13:4 shows the parabola
$y = x^2 + x - 1$.

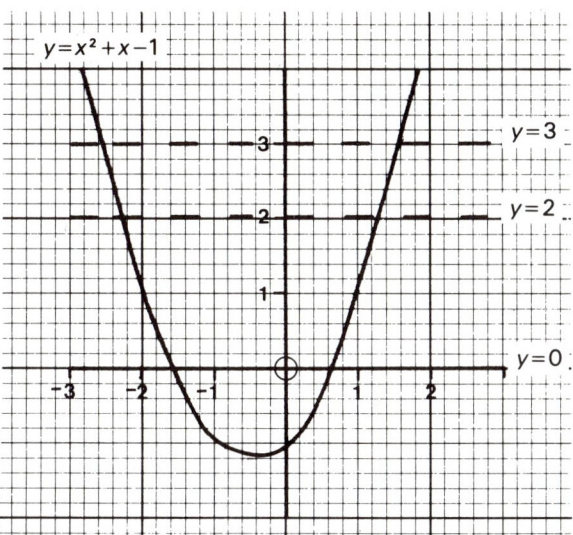

Fig. 13:4

The line $y = 0$ crosses this parabola at about $x = 0.6$ and $x = -1.6$. These are the solutions of $x^2 + x - 1 = 0$.

(a) Read the values of x where $y = 3$ crosses the parabola. Show that these are the solutions of $x^2 + x - 4 = 0$.

(b) Read the values of x where $y = 2$ crosses the parabola. Show that these are the solutions of $x^2 + x - 3 = 0$.

H23 (a) Copy and complete the table for $y = 2x^2 + 4x - 5$.

x	-4	-3	-2	-1	0	1	2
$2x^2$	32						
$4x$	-16						
-5	-5						
y	11						

(b) Draw the graph of $y = 2x^2 + 4x - 5$ using scales: x from -4 to 2, 2 cm to 1 unit; y from -7 to 11, 1 cm to 1 unit.

(c) What is the minimum value of $2x^2 + 4x - 5$? At what value of x does this minimum value occur?

(d) Read the values of x where the following lines cross your parabola, and state the equations solved by these values.
  (i) $y = -5$ (ii) $y = 1$ (iii) $y = 11$ (iv) $y = 5$ (v) $y = -1$
  (vi) $y = -4$ (vii) $y = 3$.

(e) By drawing a suitable line on your graph find the values of x that make $2x^2 + 4x - 5 = 0$.

H24 If you had drawn the parabola $y = x^2 - 3x - 6$, how would you solve $x^2 - 3x - 6 = 0$?

H25 State the equation solved where the following lines cross $y = 2x^2 - 3x - 1$. Do not draw graphs.
(a) $y = 3$ (b) $y = -1$ (c) $y = x$ (d) $y = x + 2$ (e) $y = -x + 3$
(f) $y = -2x + 1$

H26 **Example**   To solve the equation $2x^2 - 3x - 1 = 0$ using the parabola $y = 2x^2 - 3x - 1$, simply read the solutions where $y = 0$.

To solve $2x^2 - 3x - 5 = 0$ using the same parabola $(2x^2 - 3x - 1 = y)$, the given equation must be changed until one side of it is the same as the equation of the parabola. This is done by adding 4:

$$2x^2 - 3x - 5 = 0 \xrightarrow{\text{add 4}} 2x^2 - 3x - 1 = 4$$

Now it is clear that the solutions will be found on the line $y = 4$.

Example   Having drawn the parabola $y = x^2 - 2x + 1$, solve the equation $x^2 + 4x + 3 = 0$.

The equation must be changed so that the $+4x$ becomes $-2x$, and the $+3$ becomes $+1$:

$$x^2 + 4x + 3 = 0 \xrightarrow{\text{subtract 6x}} x^2 - 2x + 3 = -6x \xrightarrow{\text{subtract 2}}$$
$$x^2 - 2x + 1 = -6x - 2.$$

The solution will be found where the line $y = -6x - 2$ crosses the parabola. This is shown in Figure 13:5.

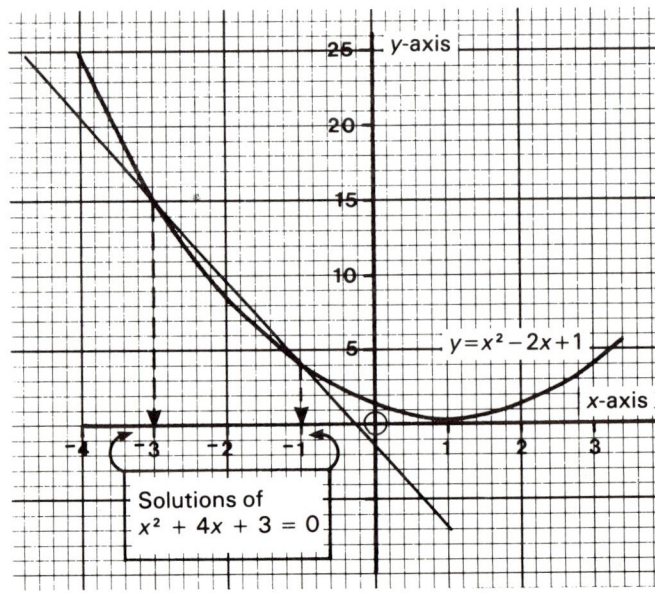

Fig. 13:5

Imagine you have drawn the parabola $y = x^2 - 2x + 1$. What line should be drawn to solve:
(a) $x^2 + 2x - 3 = 0$ (b) $x^2 + 3x - 1 = 0$ (c) $x^2 - 3x + 1 = 0$
(d) $x^2 + x + 2 = 0$ (e) $x^2 - x - 4 = 0$?

H27 Draw another copy of the graph of $y = 2x^2 + 4x - 5$ (see question 23). Find the slope of the tangent to your parabola at:
(a) $x = 1$ (b) $x = -4$ (c) $x = -2$.

H28 Solve graphically the equation $x^3 - x^2 - 6x = 0$, taking values of x from -3 to 4.

H29 On one pair of axes draw the graphs of $y = x^3$ and $y = -x^2 + 6x$ for values of x from -3 to 2. Read off the values of x at the intersections of the two graphs. For what equation are these values the solution?

H30 The consumption of electricity in a house is measured by the number of rotations of the disc in the electricity meter. One unit of electricity is equivalent to 150 rotations of the disc.

(a) When an immersion heater is switched on, the disc rotates at a constant rate of 12 revolutions per minute. Calculate the number of units of electricity used in one hour.

(b) Figure 13:6 shows the speed of rotation of the disc, w, in revolutions per minute (r.p.m.) over a three hour period, where t is the time in hours after 7 p.m.

The curved part of the graph is given by the equation $w = \dfrac{15}{t} - t$.

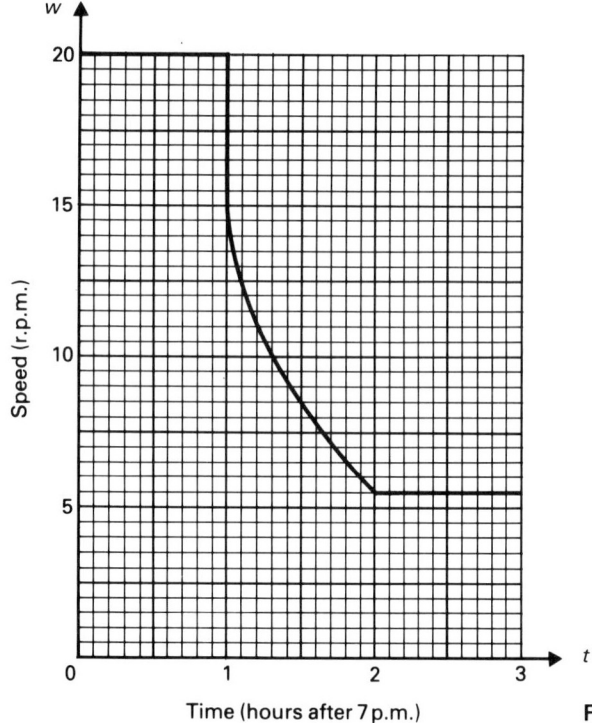

Fig. 13:6

(i) From the graph estimate the value of t when the disc is rotating at 8 r.p.m.

(ii) Use this estimate and the iterative formula
$t_{n+1} = \frac{1}{8}(15 - t_n^2)$
to calculate this time to the nearest minute.

(c) (i) Estimate as accurately as possible the area of the region between the graph and the t-axis from $t = 0$ to $t = 3$.

(ii) Electricity is charged at 6p per unit. Use your answer to part (c)(i) to estimate the cost of the electricity used between 7 p.m. and 10 p.m.

(MEG)

You will find a graphics calculator or a computer graph-drawing program useful in the following questions.

31 $f(x) = x - x^2$. Sketch:
(a) $y = f(x)$ (b) $y = f(x) - 3$ (c) $y = f(x - 3)$ (d) $y = 3 \times f(x)$.

32 A graph is drawn of some function $y = f(x)$. Explain the similarities and differences of the following graphs derived from $f(x)$:
(a) $y = f(x) + 5$ (b) $y = f(x) - 5$ (c) $y = f(x - 5)$ (d) $y = 5 \times f(x)$.

33 Sketch roughly the graphs:
(a) $y = x^2$ (b) $y = \frac{1}{2}x^2$ (c) $y = -x^2 + 4$ (d) $y = \dfrac{1}{x^2}$ (e) $y = \dfrac{1}{x^3}$

(f) $y = \dfrac{1}{x^4}$.

34 Sketch roughly the graphs of:
(a) $y = (x - 2)(x - 3)$ (b) $y = x^3 - 2$
(c) $y = 10^x$ (an 'exponential' graph).

35 Draw the graph of:
(a) $x = y^2 - 3$ for values of y from -3 to 3
(b) $y = 10/x^2$ for values of x from -10 to 10
(c) $y^2 = 9 - x^2$ for values of x from -3 to 3
(d) $y = x^3 - x^2 - x - 1$ for values of x from -2 to 3.

36 Example   Find the range of values of x for which $2x^2 - 5x - 3 \geqslant 0$.

$2x^2 - 5x - 3 \equiv (2x + 1)(x - 3)$
$\therefore (2x + 1)(x - 3) \geqslant 0$

Sketch $f(x) = (2x + 1)(x - 3)$, using the fact that $f(x) = 0$ when $x = -\frac{1}{2}$ and when $x = 3$, and that $f(x)$ is a 'cup way up' parabola. See Figure 13:7.

From the diagram it is clear that $f(x) \geqslant 0$ when $x \leqslant -\frac{1}{2}$ and when $x \geqslant 3$, that is $-\frac{1}{2} \geqslant x \geqslant 3$.

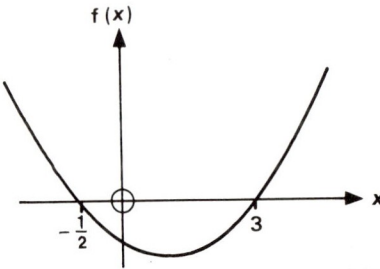

Fig. 13:7

Find the range of values of x for which:
(a) $3x^2 + 5x - 2 \geqslant 0$ (b) $2x^2 - 7x - 15 < 0$ (c) $(2 + x)(4 - x) \geqslant 0$.

14 Sequences

nth terms
Summing sequences

● **You need to know . . .**

Sequences 1 to 3 (page 239)

1 Write the first six terms of the sequence whose nth term is:
   (a) $3n$ (b) $n + 1$ (c) n^2 (d) n^3 (e) $n^2 + n$ (f) $\frac{1}{2}(n^2 + n)$

   (g) $3n^2 - 2n$ (h) $n^3 + n^2 + n$ (i) $2n^3 - 3n^2 + 4n$ (j) $\dfrac{n}{n+1}$.

2 **Example   Searching for a common difference**

		5		28		87		200		385		660
1st difference			23		59		113		185		275	
2nd difference				36		54		72		90		
3rd difference					18		18		18			

   (a) For each of your sequences in question 1, write rows of differences until you find a common one.

   (b) Can you see any connection between the number of rows of differences and the nth term?

*3 A sequence starts 2, 3, 5, . . . Suggest, with reasons, at least two possible next terms.

*4 Figure 14:1 shows some patterns formed by adding matches to make repeating polygons. Investigate the rules for how many matches are used to make n polygons.

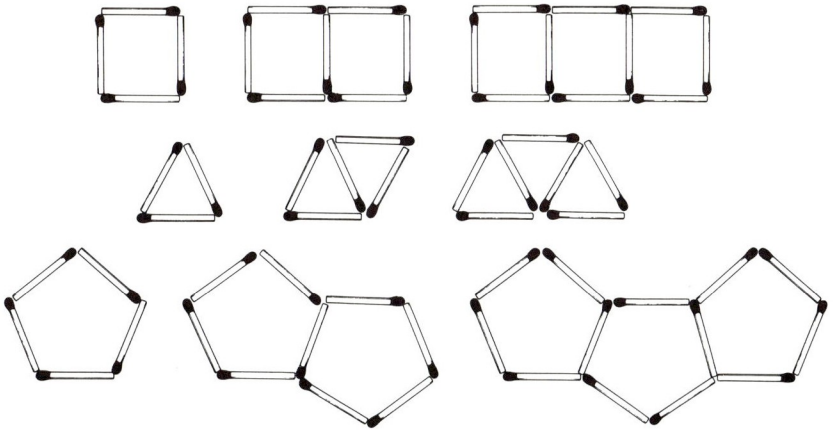

Fig. 14:1

*5 The odd numbers 1, 3, 5, 7, 9, . . . are written in a triangular block as follows:

1st row   1
2nd row   3   5
3rd row   7   9   11
4th row   13   15   17   19
5th row   21   ..   ..

(a) Write down the five numbers which will appear in the 5th row.

(b) For the 7th row, write down:
(i) the first number   (ii) the last number.

(c) Copy and complete the following table.

Row	1	2	3	4	5	6
Total	1					
Mean (average)	1					

(d) For the nth row of the triangular block, write down (in terms of n):
(i) the total of the numbers   (ii) the mean of the numbers.   (MEG)

H6 Consider the sequence $\frac{1}{2}$, $\frac{1}{4}$, $\frac{1}{8}$, $\frac{1}{16}$, . . .
(a) What is the nth term?
(b) The sum of the terms of the sequence converges. What does it tend towards?

H7 (a) Show that the equation $x^2 - x - 1 = 0$ can be written as both $x = \sqrt{(x + 1)}$ and
$x = \dfrac{1}{x - 1}$.

(b) Use the iteration formulae $x_{n+1} = \sqrt{(x_n + 1)}$ and $x_{n+1} = \dfrac{1}{x_n - 1}$ to find the two
solutions of $x^2 - x - 1 = 0$, correct to three decimal places.

H8 (a) Investigate the iteration formula $x_{n+1} = \dfrac{1}{1 - x_n}$ to see if it converges.

(b) Show that this iteration develops from $x^2 - x + 1 = 0$ and explain by considering the
quadratic equation formula or by completing the square why it has no solutions.

H9 (a) Show that the equation $4 - x^2 = x^3$ can be rearranged to:
(i) $x = \dfrac{4}{x^2} - 1$ (ii) $x = \dfrac{2}{\sqrt{(1 + x)}}$

(b) Using your solutions in (a) as x, determine whether the sequences generated by the
iterative formulae below are convergent.
(i) $x_{n+1} = \dfrac{4}{(x_n)^2} - 1$ (ii) $x_{n+1} = \dfrac{2}{\sqrt{(1 + x_n)}}$

(c) Write down, correct to one decimal place, the positive solution of the equation
$4 - x^2 = x^3$.
                                                                          (ULEAC)

H10 A purple ball is placed on a table and is represented by T_1 as shown in Figure 14:2. It is then surrounded by white balls to form a triangular shape T_2. Shape T_3 is formed by surrounding T_2 by purple balls, and so on.

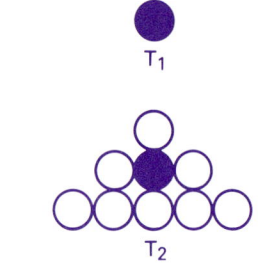

T_1

T_2

(a) What kind of triangles are formed by the centres of the outside balls?

(b) How many rows of balls will there be in:
(i) shape T_7 (ii) shape T_n?

(c) How many balls will there be in:
(i) shape T_7 (ii) shape T_n?

(d) What colour balls will be added:
(i) to T_{17} to make T_{18} (ii) to T_{2n-1} to make T_n?

(e) How many balls will be added to T_{n-1} to make T_n?

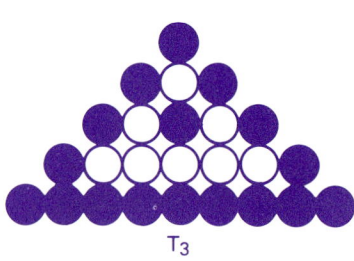

T_3

Fig. 14:2

(NEAB)

H11 When seeking the nth term of a sequence, the common difference method can be a big help.

First we find some terms of the sequences with nth terms of the form:
(a) $an + b$ (b) $an^2 + bn + c$ (c) $an^3 + bn^2 + cn + d$.

Here is (c):

1st term	2nd term	3rd term	4th term	5th term
$a + b + c + d$	$8a + 4b + 2c + d$	$27a + 9b + 3c + d$	$64a + 16b + 4c + d$	$125a + 25b + 5c + d$
1st diff. $7a + 3b + c$		$19a + 5b + c$	$37a + 7b + c$	$61a + 9b + c$
2nd diff. $12a + 2b$		$18a + 2b$	$24a + 2b$	
3rd diff. $6a$		$6a$		

You can see why a sequence whose nth term is, say, $3n^3 + 2n^2 + 7n - 2$ will have a common difference at the 3rd line.

In the example in question 2:
$$6a = 18$$
$$12a + 2b = 36$$
$$7a + 3b + c = 23$$
$$a + b + c + d = 5$$

so $a = 3$, $b = 0$, $c = 2$, $d = 0$. The nth term is $3n^3 + 0n^2 + 2n + 0 \rightarrow 3n^3 + 2n$.

Write out the first five terms, and the differences until they become common, for the sequences in parts (a) and (b).

H12 Use the method of common difference to find the nth terms of the following sequences:
(a) 2, 5, 8, 11, 14, 17   (b) 5, 11, 21, 35, 53, 75   (c) 4, 10, 18, 28, 40, 54
(d) 4, 15, 32, 55, 84, 119   (e) 1, 16, 57, 136, 265, 456   (f) 1, 4, 10, 20, 35, 56

H13 Other common sequences are triangular numbers, arithmetic progressions and geometric progressions. Their nth terms and sum to n terms are given on page 240, but before you look, try these:

(a) Find the nth term of triangular numbers, i.e. 1, 3, 6, 10, 15, . . . (Hint: 1×2, 2×3, 3×4, . . .)

(b) Find the sum of the first n terms of triangular numbers, i.e. 1, 3, 6, 10, 15, . . . (Hint: 1, 4, 10, 20, 35, . . . ; find differences . . .)

(c) Find the nth term of an arithmetic sequence with first term a, increasing by adding d each time.

(Hint: a	$a + d$	$a + 2d$	$a + 3d$	
$a + 3d$	$a + 2d$	$a + d$	a	ADD

(d) Find the sum of the first n terms of the arithmetic sequence. (Hint: a, $2a + 1d$, $3a + 3d$, $4a + 6d$, . . . Now where have you seen 1, 3, 6, . . . before?)

(e) Find the nth term of the geometric progression with first term a, increasing by multiplying by r each time. (Hint: a, ar, ar^2, ar^3, . . .)

(f) Find the sum of the first n terms of a geometric progression.
[Hint: a, $a(r + 1)$, $a(r^2 + r + 1)$, $a(r^3 + r^2 + r + 1)$. Expand: $(r - 1)(r^2 + r + 1)$ and $(r - 1)(r^3 + r^2 + r + 1)$.]

H14 State the nth term and the sum of the first n terms for the sequence:
(a) 1, 3, 6, 10, 15, . . .   (b) 5, 7, 9, 11, 13, 15, . . .   (c) 4, 12, 36, 108, . . .

---

15 The table shows the start of the infinite set of numbers that can be made by adding a multiple of 3 to a multiple of 4.

	4	8	12	16	20
3	7	11	15	19	23
6	10	14			
9	13				
12	16				

(a) Copy the table and fill in the missing totals.

(b) Which numbers cannot be made by adding a multiple of 3 to a multiple of 4?

(c) Investigate for other number pairs, say multiples of 3 and 5, then 3 and 6, followed by 4 and 5, 4 and 6, 4 and 7, 4 and 8. Write about your findings. Here are some ideas:

- What happens when both numbers are odd, or both are even?

- What pairs do have an 'all numbers from n are possible' property? What is special about them? What rule gives the largest 'not possible' number? Can you prove your rule?

16 Explain why the three factors of the square of any prime number (p) are 1, p and p^2. Hence find the rule for the sum of the factors of p^n.

17 Explore $\dfrac{a}{b} \rightarrow \dfrac{a+b}{a+2b}$.

18 $\sigma(n)$, for a positive whole number n, is defined as the sum of the factors of n. For example, $\sigma(6) = 1 + 2 + 3 + 6 = 12$.

(a) When t is a power of 2, find an expression for $\sigma(t)$ in terms of t.

(b) p is a prime number.
    (i) Show that $\sigma(p) = p + 1$.
    (ii) Find $\sigma(p^2)$.
    (iii) Find $\sigma(p^n)$.

(ULEAC)

19 A tower is formed from cubical bricks. The tower is formed from a column of single bricks buttressed on all four sides by stepped one-brick-wide triangles. Figure 14:3 shows a plan and elevation of one such tower. How many bricks are needed when the central column is n bricks high?

Investigate other ways of building towers.

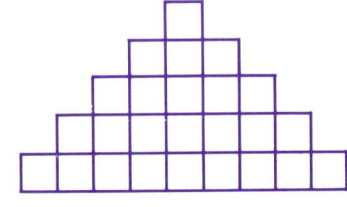

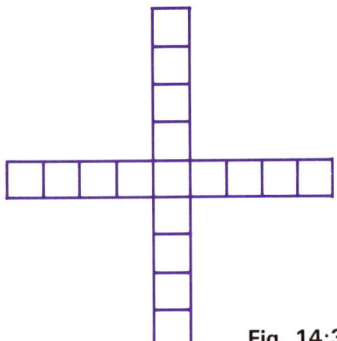

Fig. 14:3

20 A rectangle is divided into squares. Through how many of these squares does a diagonal of the rectangle pass? Generalise your result for different size rectangles.

21 In a square made up of n^2 small squares can be seen:
(a) how many squares
(b) how many rectangles (including squares)?

22 In a rectangle made up of x squares by y squares can be seen:
(a) how many rectangles (including squares)
(b) how many squares?

Direct and inverse proportion

● You need to know . . .

Variation 1 to 4 (page 240)

The volume of a cylinder is given by $V = \pi r^2 h$.

Changing the subject gives $h = \dfrac{V}{\pi r^2}$.

$\dfrac{1}{\pi}$ is a constant; call this k. Then $h = k\dfrac{V}{r^2}$ or $h \propto \dfrac{V}{r^2}$.

Think about this!

We say that h varies directly as V and inversely as the square of r.

If r remains constant but V doubles, then h doubles.

If V remains constant but r doubles, then h decreases by $\frac{1}{4}$.

---

1 Write in k form and in $\propto$ form:

   (a) The circumference of a circle (C) of diameter d is given by $C = \pi d$.

   (b) The surface area of a sphere (A) of radius r is given by $A = 4\pi r^2$.

   (c) $y = \dfrac{2}{x}$ (d) $y = \dfrac{3x}{z^2}$

   (e) For an outing of n people, the cost per person (£C) is given by $C = £\dfrac{120}{n}$.

2 The volume (V) of a prism varies directly as its cross-section area (A) and its height (H). Write an expression connecting V with A and H.

3 The height (H) of a cone varies directly as its volume (V) and inversely as the square of its radius (R). Write an expression connecting H with V and R.

4 The acceleration (A) of a body is proportional to the force (F) and inversely proportional to the mass (M). Write an expression connecting A with F and M.

5 $1\,m^2$ of carpet costs £k. If s square metres cost £y, write an equation and an expression connecting y and s.

6 Painters are paid £k per hour. If Jethro earns £E in h hours, write an equation and an expression connecting E with h.

*7 Given that $A \propto b$ and that $A = 9$ when $b = 27$:
  (a) write an equation involving the constant k
  (b) substitute for A and b to find the value of k
  (c) find A when $b = 15$
  (d) find b when $A = 4$.

*8 Given that $P \propto \dfrac{1}{q}$ and that $P = 6$ when $q = 5$:
  (a) write an equation involving the constant k
  (b) substitute for P and q to find the value of k
  (c) find P when $q = 25$
  (d) find q when $P = 8$.

*9 Given that $x \propto \dfrac{1}{y^2}$ and that $x = 4$ when $y = 9$:
  (a) write an equation involving the constant k
  (b) substitute for x and y to find the value of k
  (c) find x when $y = 6$
  (d) find y correct to 3 decimal places when $x = 12$.

10 State whether each proportion statement given at the start of questions 7, 8 and 9 is an example of direct proportion, inverse proportion, or inverse square.

11 When a stone is dropped from rest the time to reach the ground, t seconds, is proportional to the square root of the distance, d metres, through which it falls.

  (a) A stone takes 2 seconds to reach the ground when it is dropped from a height of 20 metres. Obtain an equation which expresses t in terms of d.

  (b) Glyn estimates the depth of an empty well by measuring the time it takes a stone to fall from the top until it is heard striking the bottom. If the time taken is 1.2 seconds, find Glyn's estimate of the depth of the well.

  (WJEC)

12 The variables x, y and z are connected by the equation $y = \dfrac{kx^2}{z}$ where k is a constant.

  (a) State in words how y varies with x and z.
  (b) Given that $y = 8.1$ when $x = 9$ and $z = 3$, find the value of k.
  (c) Find x when $y = 6$ and $z = 5$.

13 The number (n) of solid rubber balls which can be made from a fixed volume of rubber varies inversely as the cube of the diameter (d) of a ball. A certain volume of rubber makes 250 balls of diameter 2 cm. How many 3 cm balls could be made?

14 The cost of printing a leaflet varies inversely with the square of the number ordered. The cost is 60 pence each when 500 are ordered.

  (a) What will 1000 cost?

  (b) To bring the cost down to under 10 pence each, how many should be ordered, rounded up to the next hundred?

15 A conker is suspended on a string. When another conker hits it at u metres per second, it swings upwards until it is h centimetres above its height at rest.

Various readings of u and h give this table of results:

u (m/s)	1.5	1.9	2.5	3.4
h (cm)	12	18	32	60

Would you say that h varies as u, as u^3, as u^2, or as $\sqrt{u}$? Explain the reasons for your choice.

16 The volume of a cylinder, radius r and height h, is $\pi r^2 h$. What change will be made in the volume if:
(a) the radius is halved   (b) the height is halved
(c) both the radius and the height are halved
(d) the radius is doubled and the height is halved
(e) the radius is halved and the height is doubled?

17 The volume (V) of a cone is directly proportional to the square of its base radius (R) and its height (H).

(a) Write an expression connecting V with R and H.

(b) If H is constant, then $V \propto R^2$. If R becomes four times as long, how many times will V increase?

(c) How will V change if H is doubled and R is trebled?

(d) How will V change if both H and R are trebled?

18 A variable s is inversely proportional to a second variable t. If t is increased by 25% find the percentage change in s.

(SEG)

19 The heat (H) in calories developed in an electrically heated wire varies directly as the square of the voltage (V), directly as the time (T) and inversely as the resistance (R ohms).

(a) Write an expression connecting H with V, T and R.

(b) Using constant k, write an equation developed from your answer to part (a).

(c) Given that 4 calories are produced by 20 volts in 1 second for a wire of resistance 25 ohms, find the value of k.

(d) A longer length of the same wire has a resistance of 50 ohms. How many calories will be produced in 10 seconds by 5 volts?

20 The square of the orbital period (T years) of a planet varies as the cube of its mean distance (D km) from the sun.

For Earth, T is 1 year and D is 1.5×10^8 km. Calculate the orbital period of Jupiter, 7.82×10^8 km from the sun.

16 Lines and shapes

Basic terms   **Polygons**
Lines and angles **Symmetry**

● You need to know . . .

Basic geometry 1 to 3 (page 243)
Plane symmetry 1 to 3 (page 244)

1   Name the shapes used for the road signs in Figure 16:1.

(a)    (b)    (c)

Fig. 16:1

2   Calculate the sum of the interior angles of each shape used in Figure 16:1.

3   ABCD is an isosceles trapezium. AB is parallel to DC with DC longer than AB. Sketch the trapezium.
   (a)   The acute angles are 50°. Calculate the size of the obtuse angles.
   (b)   From vertex A draw AX perpendicular to DC, meeting DC in X. Calculate angle DAX.

4 (a)   Figure 16:2 has a vertical line of symmetry. Calculate angle a, giving reasons.
   (b)   Figure 16:3 has a horizontal line of symmetry. Calculate angle b, giving reasons.
   (c)   Figure 16:4 has a vertical line of symmetry. Calculate angle c, giving reasons.

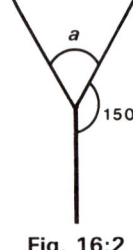      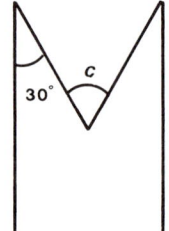

Fig. 16:2   **Fig. 16:3**   **Fig. 16:4**

5   The bisectors of the angles of any triangle are concurrent. Draw a sketch to illustrate this.

6   State the order of rotational symmetry for:
   (a) an equilateral triangle   (b) a rhombus   (c) a regular pentagon.

7 Sketch Figure 16:5.

(a) Has your sketch line symmetry? If so, draw the lines of symmetry.

(b) Has your sketch point symmetry? If so, mark the point about which it is symmetrical.

SOS
Fig. 16:5

8 Name all possible special quadrilaterals if they have the following properties:
(a) two lines of symmetry and rotational symmetry of order 2
(b) one line of symmetry and no rotational symmetry (that is, only order 1)
(c) no lines of symmetry and order 2 rotational symmetry
(d) four lines of symmetry and rotational symmetry of order 4.

9 In Figure 16:6, find the sizes of angles x, y and z. Give reasons for your working.
(WJEC)

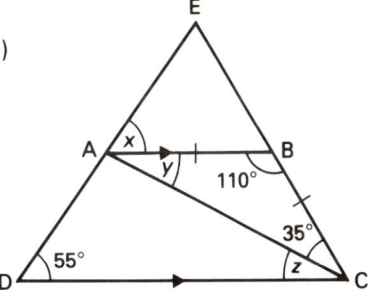

Fig. 16:6

10 In Figure 16:7, all lines are equal.
(a) Write the special names of the shapes in the diagram.
(b) Calculate, with reasons, the angles p, q, r and s.

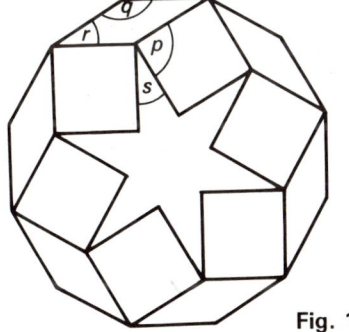

Fig. 16:7

16

11 Figure 16:8 shows part of a tessellation made up from three tile shapes.
(a) Name the three shapes used.
(b) Calculate the size of the angle α.

12 Describe the symmetry of Figure 16:9.

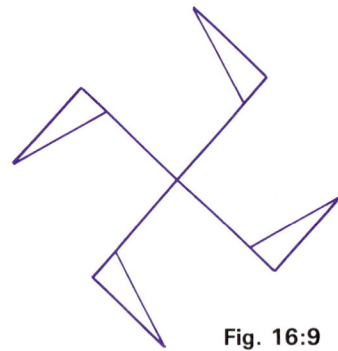

Fig. 16:9

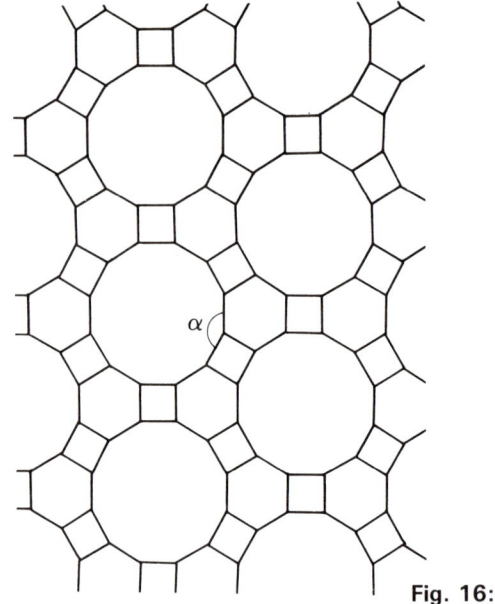

Fig. 16:8

13 (a) Sketch Figure 16:10. Add two straight lines to your diagram to give it one vertical line of symmetry. Name the quadrilateral you have drawn.
(b) If angle HXY was 65°, what would be the size of angle HYX?

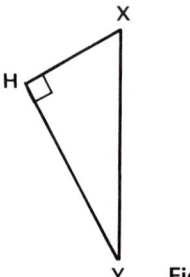

Fig. 16:10

14 Triangle ABC is isosceles. XA is the bisector of angle A with X lying on BC. Calculate the size of angle AXC if:
(a) angle B is 100° (b) angle B is 20°.

15 What is strange about a 3 cm-, 6 cm-, 9 cm-sided triangle?

16 The angles of a pentagon are in the ratio 2:3:4:5:6. Calculate the largest angle.

17 In Figure 16:11, FHBD is a parallelogram. Calculate angle HBD, giving full reasons.

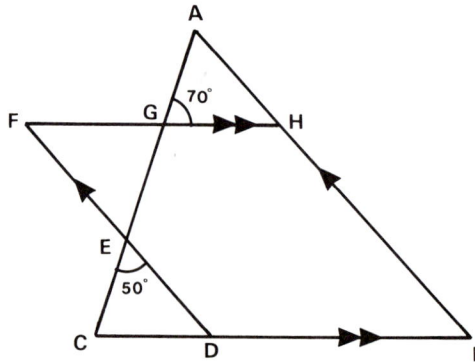

Fig. 16:11

18 Calculate angle θ in Figure 16:12, giving full reasons.

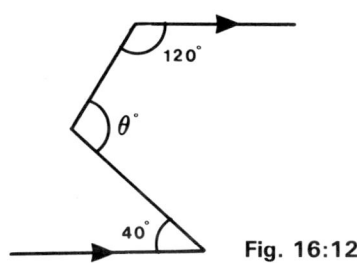

Fig. 16:12

19 Refer to Figure 16:13.

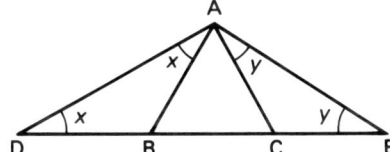

Fig. 16:13

(a) Calculate in terms of x the size of angle ABC.
(b) Calculate in terms of x and y the size of angle BAC.
(c) Explain clearly why DE is equal in length to the perimeter of the triangle ABC.
(d) Use the results obtained in parts (a) to (c) to construct a triangle ABC with perimeter 12 cm and with angle ABC = 60°, angle BCA = 70° and angle CAB = 50°. You may use a protractor.

20 A regular polygon X has twice as many sides as polygon Y. The interior angle sum of Y is 540°. What is the interior angle sum of X? (SEG)

21 In triangle XYZ, angle X is 100° and angle Z is 35°. P is the foot of the perpendicular from X to YZ. Show that triangle XPY is isosceles.

22 When the number of sides of an n-sided polygon is increased by 2, each interior angle is increased by 2°. How many sides has the polygon?
(WJEC)

23 Investigate the greatest possible number of 90° interior angles in various polygons.

24 Investigate tessellations made with pentagons. Describe any features which ensure a given pentagon will tessellate.

Grid references
Bearings
Networks

Loci constructions
LOGO

● **You need to know . . .**

Movement 1 to 5 (page 244)

1 State as three-figure bearings:
(a) NW   (b) NNE   (c) S48°E   (d) N40°W.

2 (a) In Figure 17:1, a yacht is sinking in the square with grid reference 2241. What is the grid reference of the squares in which are the cruiser and the two lifeboat stations?

(b) The cross marking the yacht's position is at about six-figure grid reference 228412. What are the six-figure grid references of the other three crosses?

(c) Copy and complete the following tables.

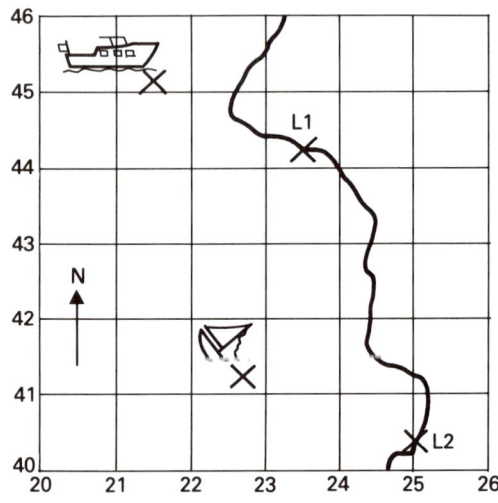

Fig. 17:1

Three-figure bearings

Bearing of	From			
	Cruiser	Yacht	L1	L2
Cruiser	XXX			
Yacht		XXX		
L1			XXX	
L2				XXX

Cardinal bearings

Bearing of	From			
	Cruiser	Yacht	L1	L2
Cruiser	XXX			
Yacht		XXX		
L1			XXX	
L2				XXX

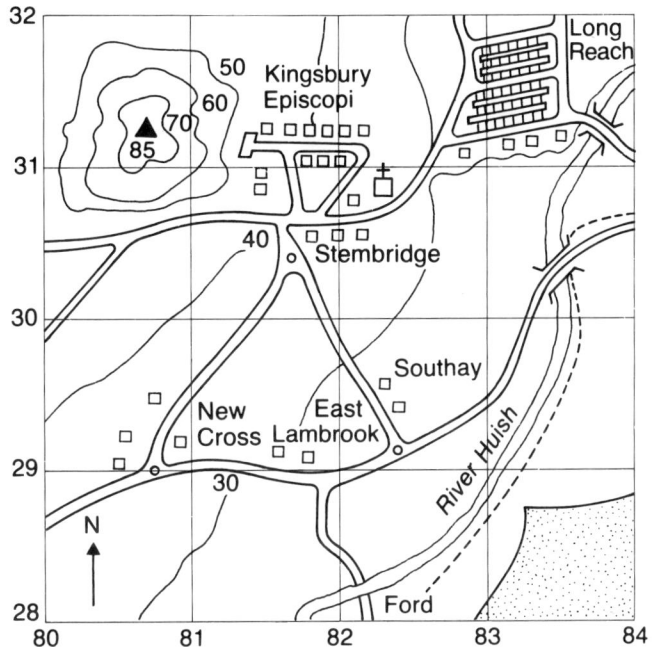

Fig. 17:2

3 Figure 17:2 is a section of a 1 : 50 000 map.

(a) How many square kilometres are represented by the map?

(b) Give the grid reference of the square in which Southay lies.

(c) Give the six-figure reference of the centre of the cross showing Kingsbury church tower.

(d) A balloonist is over Kingsbury Episcopi church. The wind is blowing from the north-east. Give the approximate grid reference of the point at which the balloon is likely to come off the map.

(e) Rachel is at grid reference 820290. What is the bearing from her of the top of Burrow Hill (marked with a trig point triangle)?

4 (a) Sketch a diagram to show a point A on a bearing of 060° from a point B. State and illustrate the bearing of B from A.

(b) Sketch a diagram to show a point C on a bearing of 250° from a point D. State and illustrate the bearing of D from C.

(c) In parts (a) and (b) your answers are called back-bearings. Back-bearings may easily be calculated without drawing a diagram. Find the rule.

5 Peter is walking the triangular road route from New Cross through Southay, into Stembridge, and back to New Cross.

(a) State the three-figure bearings he should take at the start and at the two junctions.

(b) Rewrite your answers to part (a) as cardinal bearings (like N40°W).

(c) About how many kilometres does Peter walk?

6 Figure 17:3 represents the positions J, K, L and M of the homes of four friends, Jatinder, Krishna, Lynda and Maria, respectively, and the roads between them.

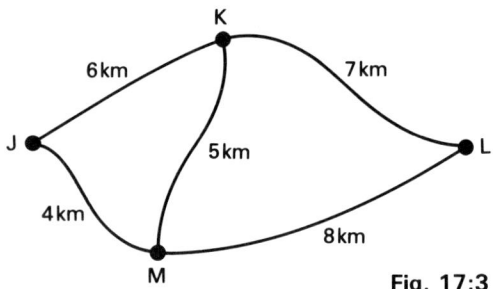

Fig. 17:3

(a) What is the shortest distance by road from J to L?

(b) Jatinda has some disco tickets which she wants to deliver to her three friends as quickly as possible. She cycles from her home to the homes of her three friends. She does this so that the total distance she cycles from her home to the home of the third friend is as small as possible. Which route does she take?

(MEG)

7 (a) Figure 17:4 shows five company main offices. The master computer at Bristol North is to have a direct landline to the other four offices, each of which must be in contact with one other office besides Bristol North. The map is to scale, and the landlines are to be straight. Draw a network diagram to show the shortest landlines that are needed.

(b) On the system in part (a), some centres would be cut off from others if the Bristol North computer 'crashed'. Suggest the best way to overcome this problem.

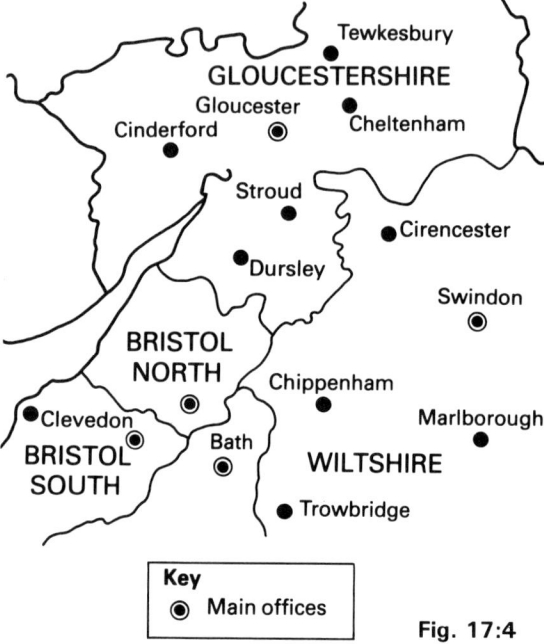

Fig. 17:4

8 The network in Figure 17:5 shows the distances in miles between five towns, P, Q, R, S and T.

(a) How far is it from town P to town T passing through S?

(b) What is the shortest distance from town P to town T?

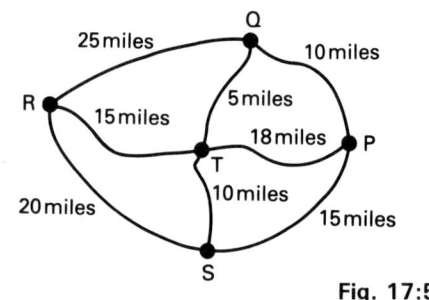

Fig. 17:5

(c)

	A	B	C	D
A	—	5	10	3
B	5	—	14	4
C	10	14	—	20
D	3	4	20	—

The table gives the times taken, in minutes, to travel between four villages A, B, C and D. Draw a network to show the times of the journeys between the four villages.

(WJEC)

9 Spend some time using the LOGO computer language. You should at least be able to draw regular polygons efficiently. Perhaps you could reproduce Figures 16:7, 16:8 and other tessellations?

10 In order to answer the following loci questions you should be able to construct an angle of 60°, bisect an angle (including a 180° angle) and draw the perpendicular bisector of a line. Practise as necessary!

11 Two marker buoys, X and Y, are 100 metres apart. A boat is 150 metres from both buoys. Construct a scale diagram to show the course of the boat if it is to pass midway between the buoys.

12 (a) Construct a triangle, sides 9 cm, 7.5 cm and 6 cm.

    (b) By bisecting two angles of your triangle, find the centre of the incircle of the triangle and draw this circle.

13 Figure 17:6 represents part of a rugby pitch. XY is the goal. A player touches down at P; he may then try to 'convert' by kicking the ball between the goalposts from anywhere on the line PA. Copy the diagram, the same size, then use instructions (1) to (4) to find the best position from which to arrange the conversion.

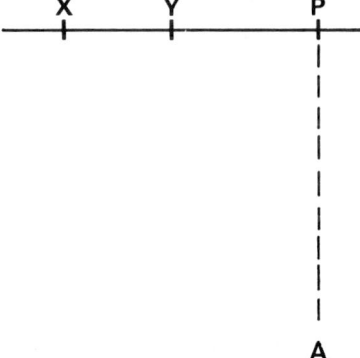

Fig. 17:6

    (1) Draw the perpendicular bisector of XY, crossing XY at M.

    (2) Draw an arc, centre Y, to cross the perpendicular bisector at O, where YO = PM.

    (3) Draw a circle, centre O, passing through X and Y, and touching PA at C.

    (4) Point C is the best position from which to attempt the conversion, as angle YCX is then a maximum. Measure angle YCX. Mark a point D and a point E, each 1 cm from C on PA, and check that both angle YDX and angle YEX are smaller than angle YCX.

14 Figure 17:7 shows a square KLMN with sides of 20 mm. A 100 mm thread is attached to K and pulled to P so that PKN is a straight line. The thread is kept tight and wound round the square in a clockwise direction.

(a) Copy the diagram and draw the locus of P as the thread is wound.

(b) Calculate the length of this locus (circumference of a circle ≃ 3.14 × diameter).

(ULEAC)

Fig. 17:7

15 Figure 17:8 shows a piece of string wound tightly around a wooden equilateral triangle of side 2 cm. The end of the string is at one vertex of the triangle.

Draw the locus of the end of the string as 6 cm of it is unwound, keeping the string tight at all times.

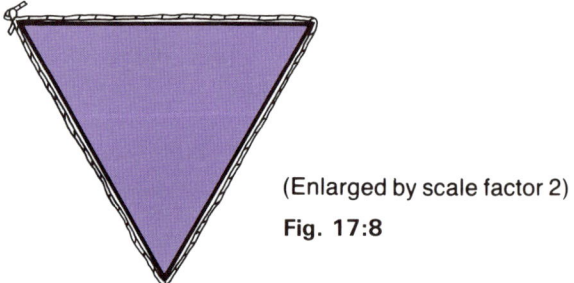

(Enlarged by scale factor 2)

Fig. 17:8

16 (a) Copy Figure 17:9 using a scale of 1 cm to 100 m.

(b) Due east of port O are two buoys. At 0600 a ship at A is 850 metres from both. It sets a course to take it between the two buoys, coming no closer than 550 m to the inshore buoy. Mark the position of the buoys and the planned course of the ship.

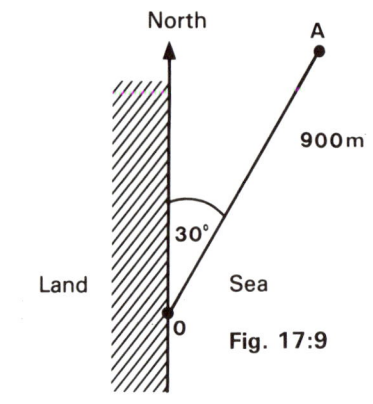

Fig. 17:9

(c) State the bearing the ship takes from A and how close it comes to the offshore buoy.

(d) The ship travels at 10 knots (10 nautical miles per hour). About how far from O is it 4 minutes after leaving A?

1 nautical mile is about 1850 metres.

17 Draw a line PQ, 8 cm long. Construct and label clearly:
(a) the locus of points R equidistant from P and Q
(b) a point S, equidistant from P and Q and such that angle SPQ = 45°
(c) the locus of points T which are 4 cm from S.

(Leave all your construction lines and arcs clearly visible.)

18 Using a scale of 1 cm to 10 m, draw a plan of a rectangular field ABCD, with AB = 100 m, and BC = 40 m. On your plan draw accurately the path of a man who walks over the field in the following way:

(a) He starts at A and walks, keeping always 40 m from D, until he reaches a point which is 40 m from A.

(b) He then walks towards CB, keeping always the same distance from C as he is from B, until he reaches a point where he is equally far from the boundary BC as he is from the boundary CD.

(c) He then walks to C, keeping as far from BC as he is from CD.   (MEG)

19 A plane takes off from Aberdeen. It develops a fault while over the North Sea and the pilot ejects. A search is made over an area from 350 km to 500 km from Aberdeen, and between the bearings 020° and 060° from Aberdeen.

(a) Using a scale of 2 cm to 100 km make an accurate scale drawing of the area to be searched. Indicate clearly the position of Aberdeen and the direction of north from Aberdeen.

The pilot actually comes down 400 km from Aberdeen on a bearing of 040° and his life-raft drifts east at a rate of 5 km/h.

(b) Show the position of the pilot when he lands on the sea.

(c) (i) Draw in a line to show how the raft drifts across the search area.
    (ii) Use this line to calculate how long it will be after the crash before the raft drifts out of the search area. Answer to the nearest hour.   (NEAB)

1 knot
= 1 nautical mile
per hour

20 Four ships, A, B, C and D, receive a distress signal from a ship at point S. Ship A is 40 nautical miles due east of point S, and B is 9 nautical miles due south of S.

(a) Using 1 cm to represent 5 nautical miles, draw a scale diagram showing the positions of S, A and B.

(b) The ship C is 21 nautical miles due west of S, and the ship D is 45 nautical miles from S on a bearing of 300°. Mark the positions of C and D on your diagram.

(c) Measure the angle DCS.

(d) What is the bearing of ship D from ship C?

(e) Ships A and C answer a distress signal from ship S. Ship A travels at a speed of 12 knots and ship C travels at 6 knots.
    (i) Which ship arrives first at S?
    (ii) How many minutes earlier does it arrive than the second ship?
                                                                      (SEG)

21 Figure 17:10 shows a view looking down on a door, hinged at B to the door-post BC. XP and YP are the arms of a spring-loaded door closer, fixed to the door at X and to the door-post at Y.

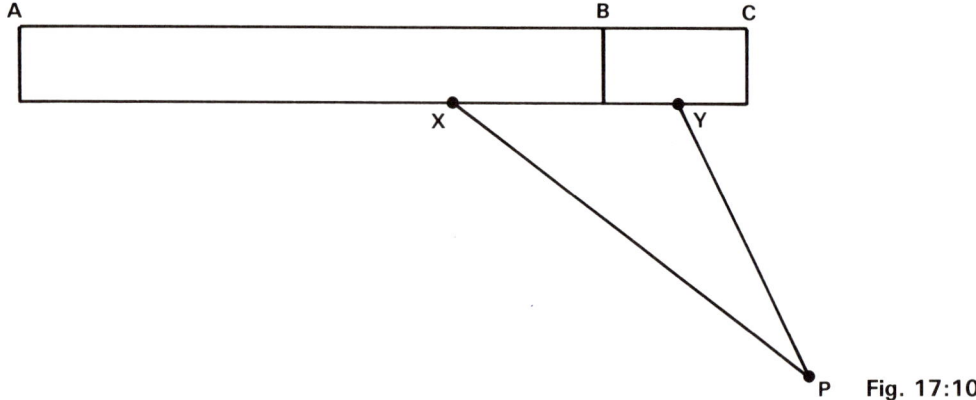

Fig. 17:10

Copy the diagram to the size drawn, leaving 8 cm above ABC, then draw about six different positions of bar PX as the door opens.

Find the angle ABC when the door is as wide open as the closer will allow.

22 Three searchers have radios with a maximum transmitter range of 5 km. They can all just hear each other. A fourth searcher can hear two of them, but not the third. Draw a diagram to show the region within which the fourth searcher could be.

23 Figure 17:11 shows distances in miles between towns.

(a) Bridget is an inspector of roads. Can she inspect every road once without going over the same road twice? If she can, where should she start and finish?

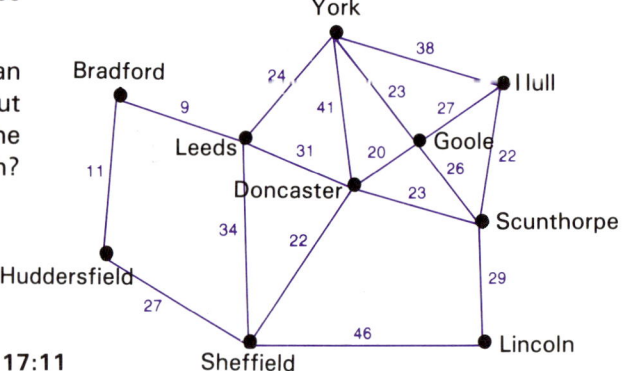

Fig. 17:11

(b) If Bridget has to start and finish at York, what is her shortest route? Which roads will she have to cover twice?

(c) Norma needs to visit each town once. She starts out from York but does not mind where she finishes. What is her shortest route?

(d) Bob lives in York and also wants to visit each town once, but he wants to finally return to York. What is his shortest route?

24 In Figure 17:12, OX = 3 cm, XY = 1 cm and angle AOX = 60°.

(a) Construct an accurate full-size drawing of this diagram.

(b) A circle is to be drawn to touch the given line and to touch the given circle at the point A. Explain how this can be done (without trial and improvement) and carry out the construction on your drawing.

(MEG)

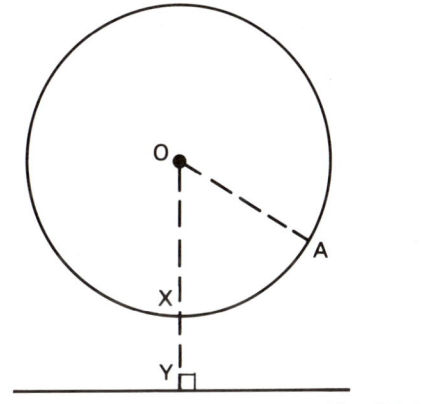

Fig. 17:12

25 At the end of an exam 90 desks have to be returned from the gym and the hall to rooms 1, 2 and 3. Figure 17:13 shows the distances and desks involved.

Five pupils have been caught and one pupil can carry one desk. Which desks should be sent where to wear the pupils out as little as possible?

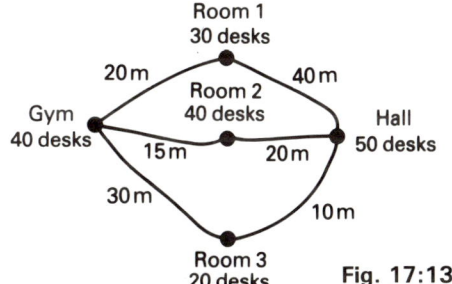

Fig. 17:13

26 Figure 17:14 shows the road mileage between four towns. The Post Office plans a new sorting centre to serve the towns, built somewhere on the marked roads. The codes like 3D show how many deliveries from the sorting office will be made to each town post office each day.

Where should the sorting centre be built to keep the total daily mileage as small as possible, assuming each delivery van only delivers to one town per trip?

Fig. 17:14

27 What difference would there be to your answer to question 26 if:

(a) the roads joining the towns made a quadrilateral PQRS, with PS (i) 6 miles or (ii) 4 miles (the sorting centre must still be on one of the roads)

(b) the roads are as in Figure 17:14, but one van could take deliveries to several different towns during one trip

(c) both changes in parts (a) and (b) are made

(d) the roads are as in Figure 17:14, but the sorting centre does not have to be along them (you must allow for the distance along any extra roads you draw in)

(e) both changes in parts (a) and (d) are made?

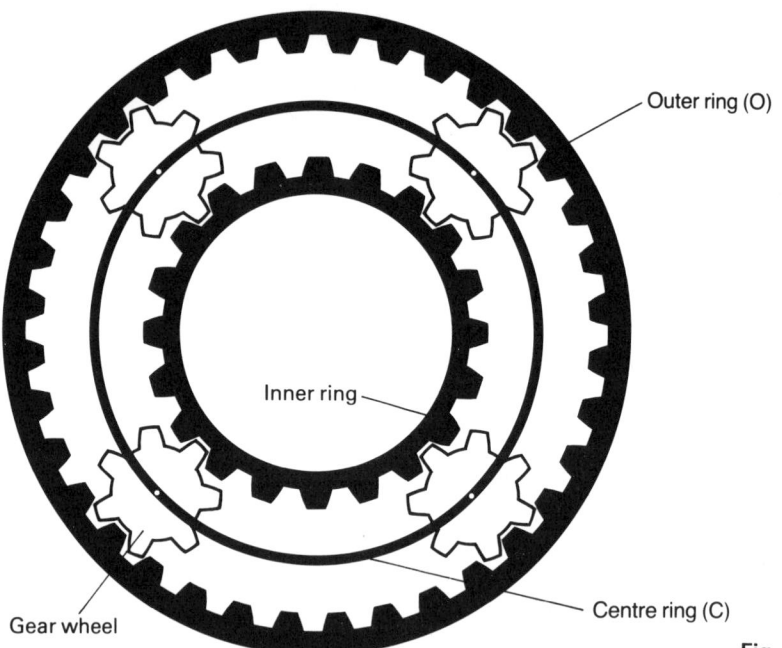

Outer ring (O)

Inner ring

Centre ring (C)

Gear wheel

Fig. 17:15

28 Figure 17:15 shows the type of gearing used in a cycle gear box.

The inner ring is part of the axle and does not turn.

In 'High' the centre ring is driven and the outer ring turns the road wheel.

In 'Normal' the outer ring is both driven and drives the road wheel.

In 'Low' the outer ring is driven and the centre ring drives the road wheel.

Describe the direction of rotation and speed of turn of each of the three rings and the gear wheels in all three gears. How is the gear ratio calculated for High and Low gears? (Ignore the chain drive sprockets.)

(Please note: If you take your cycle hub to pieces to investigate this, keep the pieces you take out in order so that you can, hopefully, put it back together again!)

29 Investigate the locus of a point on a rolling circle, taking the point either on the circumference or inside the circle. The circle can roll on a straight line, or round another circle, or round a triangle, or whatever you like!

Investigate the locus of a point on the outer circumference of a flanged wheel, like that on a locomotive.

Consider the locus of a point on the circumference of a circle rolling on a straight line (called a 'cycloid'). What is the ratio of the area of the circle to the area between the cycloid and the line? (Use the trapezoidal rule, measuring to obtain the heights.)

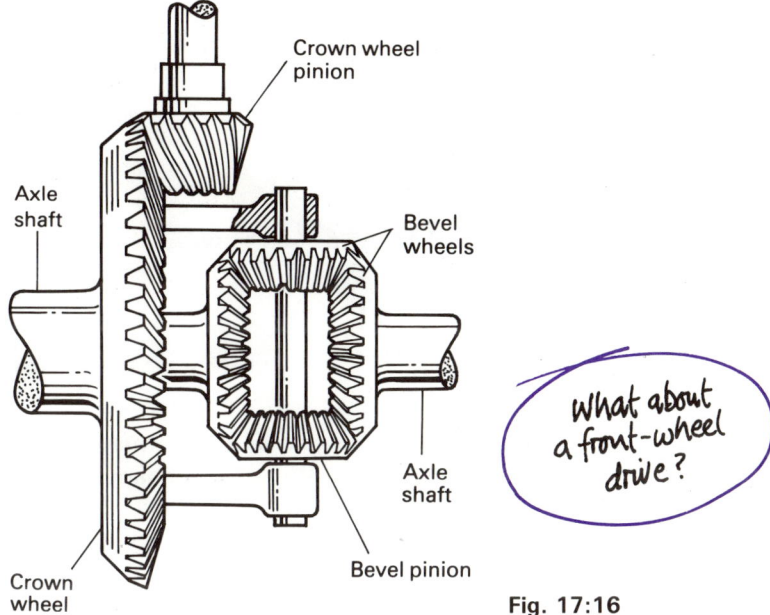

Fig. 17:16

30 Figure 17:16 shows the mechanism used to drive the rear wheels of a car. It is called a differential.

(a) Why cannot the wheels be fixed to a single axle?

Consider the direction of rotation and comparative speed of each gear when:

(b) both road wheels are turning at the same rate

(c) the offside wheel turns faster than the nearside (When will this happen?)

(d) the offside wheel is stationary

(e) the offside wheel loses its grip in soft mud.

31 Figure 17:17 shows James Watt's link-motion, by which he attempted to keep a point P moving in a straight line as rods AC and BD pivot about A and B.

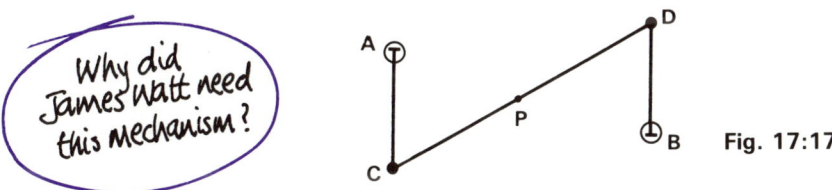

Fig. 17:17

(a) Let AC = BD = 5 cm, CD = 1 cm, and AB = 9 cm. Find the locus of P, the midpoint of CD.

(b) Let AB = CD = 10 cm, and AC = BD = 7 cm. Construct the locus of P.

32 Design and make some rod-linkages.

18 Similarity and congruence

Recognising congruent and
    similar shapes
Sides ratio

(Area and volume ratio)
(Congruent triangles)

● **You need to know . . .**

Similarity 1 to 5 (page 247)

1 Congruent shapes are exactly the same in size and shape, although one shape may be a reflection of another ('turned over'). Which two shapes in Figure 18:1 are congruent?

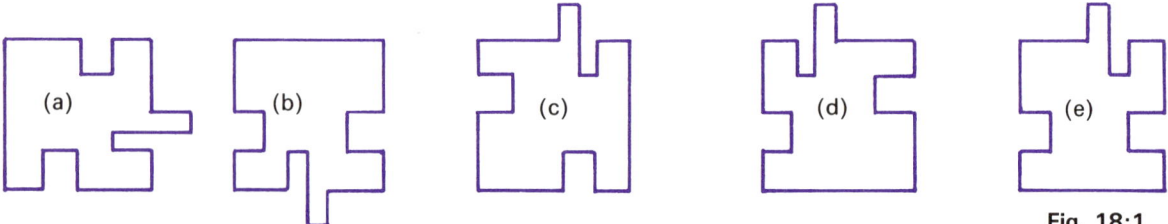

(a)   (b)   (c)   (d)   (e)

Fig. 18:1

2 Which of the following must be mathematically similar?
   (a) A negative of a photo and an enlarged print made using it
   (b) A large painting and a reproduction of the painting in a book
   (c) Two brothers   (d) Two spheres   (e) Two mugs
   (f) Equilateral triangles   (g) Isosceles trapeziums   (h) Cones

3 Figure 18:2 represents a scientific experiment in which a lens casts an image of a match on a screen.

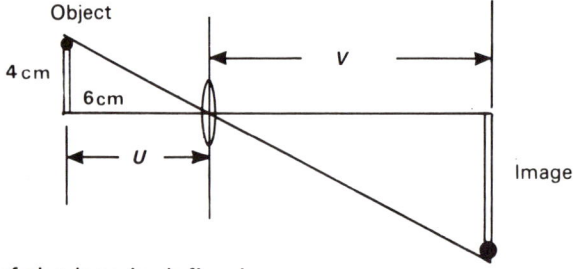

Object
4 cm
6 cm
v
U
Image

Fig. 18:2

The magnification of the lens is defined as

$$\frac{\text{height of image}}{\text{height of object}}$$

The lens shown has a magnification of $\times 2\frac{1}{2}$.

Calculate:
(a) the height of the image   (b) the distance v.

4 Figure 18:3 shows two similar triangles.
   (a) Their sides are in the ratio 12:8. Simplify this ratio.
   (b) Calculate the value of (i) *x*   (ii) *y*.

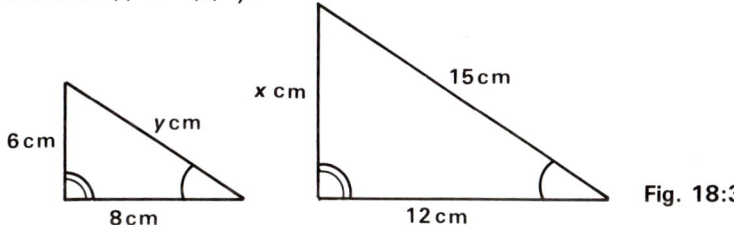

Fig. 18:3

5 A model of a car is to a scale of 1:50. The model has five doors and a total length of 8 cm. How long is the real car, and how many doors has it?

*6 (a) Write the mathematical definitions of 'similar' and 'congruent'.

   (b) Which shapes in Figure 18:4 are similar to each other and which are congruent?

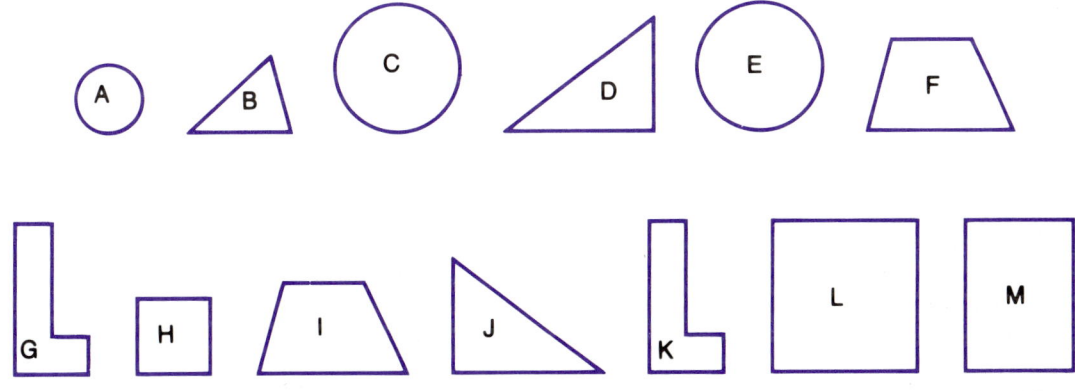

Fig. 18:4

7 Three of the quadrilaterals in Figure 18:5 are similar.

   (a) Which quadrilateral cannot be similar to the others? Why?

   (b) Calculate the length of KN.

   (c) Calculate the length of OP.

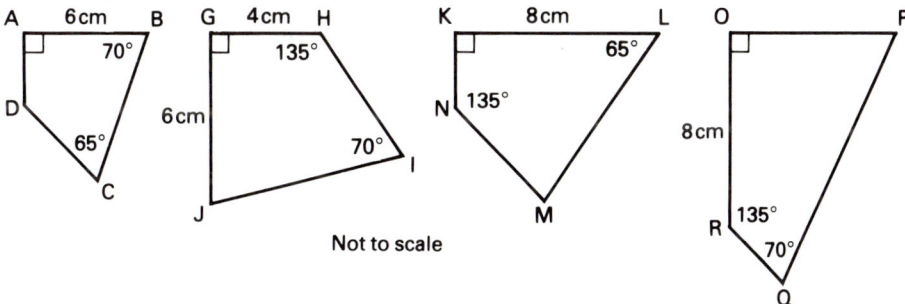

Not to scale

Fig. 18:5

8 (a) The squares in Figure 18:6 have 0.5 cm sides. Using 1 cm squared paper, draw a similar shape to that shown, with sides twice as long.

(b) State the ratio of the area of the shape in Figure 18:6 to the area of the shape that you have drawn.

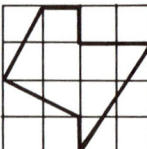

Fig. 18:6

Fig. 18:7

H9 Refer to Figure 18:7.
(a) By measurement check that the rectangles are similar.
(b) State the ratio of their lengths.
(c) State the ratio of the areas of the two rectangles.

H10 In Figure 18:8, angle A is 90°, and AC is an altitude.

(a) Name an angle equal to the angle marked θ.

(b) Name an angle equal to the angle marked α.

(c) Write equal angles over each other to complete the statements:

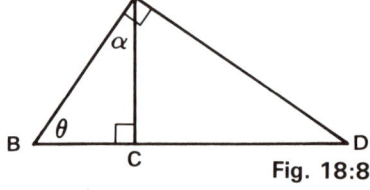

Fig. 18:8

(i) △s $\begin{array}{ccc} A & B & C \\ \_ & A & \_ \end{array}$ are similar

(ii) △s $\begin{array}{ccc} A & B & D \\ \_ & B & \_ \end{array}$ are similar

(iii) △s $\begin{array}{ccc} A & C & D \\ \_ & \_ & D \end{array}$ are similar.

(d) Using your answers to part (c), copy and complete:

(i) $\dfrac{AB}{} = \dfrac{}{DC} = \dfrac{BC}{}$ (ii) $\dfrac{}{CB} = \dfrac{AD}{} = \dfrac{}{BA}$ (iii) $\dfrac{AC}{} = \dfrac{}{BD} = \dfrac{}{AD}$.

H11 In Figure 18:9, AX : XY = AP : PQ.
Calculate XY if AY is 12 cm, AP is 5 cm and PQ is 4 cm.

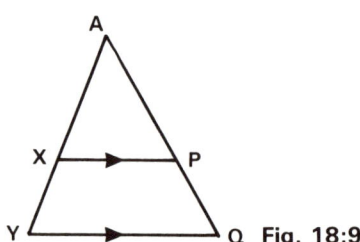

Fig. 18:9

H12 In Figure 18:10, RQ is parallel to BC and QP is parallel to AB. AQ = 2 cm, QC = 6 cm, RQ = 4 cm and RB = 9 cm.

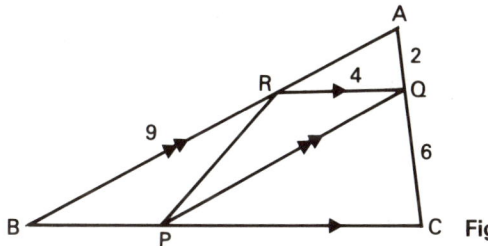

Fig. 18:10

(a) Name a triangle congruent to triangle PQR.

(b) Name two triangles, each of which is similar to triangle ARQ.

(c) Calculate (i) PC   (ii) AR   (iii) $\dfrac{\text{area } \triangle ARQ}{\text{area } \triangle ABC}$.

(d) Given that area $\triangle ARQ = k$ cm², express area $\triangle PRQ$ in terms of k.   (MEG)

H13 For Figures 18:11 to 18:15, state if each pair of triangles is congruent, giving your reason.

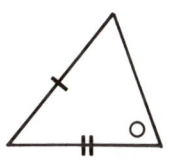

Fig. 18:11

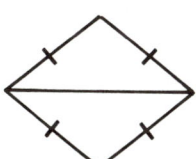

Fig. 18:12

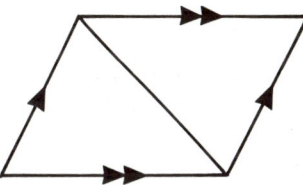

Fig. 18:13

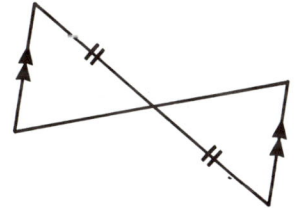

Fig. 18:14

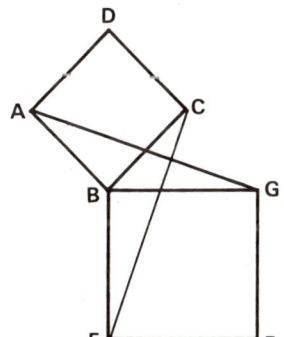

△ s ABG and BCE?

ABCD, BGFE are squares.

Fig. 18:15

H14 A child has a set of wooden bricks. Each brick is a cube of side 5 cm. The bricks are to be put together to make larger cubes.

(a) How many bricks are needed to make a cube of side:
    (i) 10 cm   (ii) 15 cm   (iii) 20 cm?

(b) What is the surface area of each cube in part (a)?

(c) What is the volume of each cube in part (a)?

(d) Find the ratio of surface area to volume for each cube in the form 1:n.

H15 Two cans are mathematically similar. One holds 0.125 litres, the other holds a full litre. The smaller can is 3 cm high. How tall is the larger can?

H16 Two cuboids are mathematically similar. One is 12 cm long by 10 cm wide by 8 cm high. The other is 9 cm long. State as simply as possible the ratio of:
(a) their heights (b) their surface areas (c) their volumes.

---

17 A cone of height 18 cm and base radius 9 cm is cut by a plane parallel to the base circle and 9 cm from it. Find the ratio of the volumes of the two resulting pieces.

18 Figure 18:16 shows a triangle, right-angled at B, with BD perpendicular to AC. BD is of length 2 cm and CD is of length x cm.

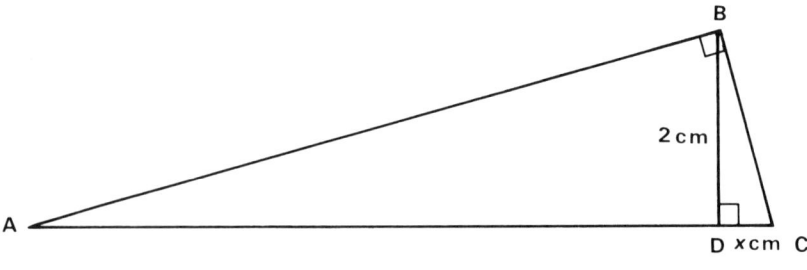

Fig. 18:16

Using similar triangles, show that the length of AD is $\dfrac{4}{x}$ cm and hence that the area of triangle ABC is $x + \dfrac{4}{x}$ cm^2.

Find, correct to 3 significant figures, the value of x when the area of triangle ABC is 7 cm^2
(Scottish)

19 An important application of the effect of enlargement is the ratio of surface area to volume for various sizes of animals and plants. For animals, heat loss depends on surface area (of the skin); heat production depends on muscle volume. An idea of the effect can be found in comparing the surface area and volume of cubes:

1 cm cube: area 6 cm^2; volume 1 cm^3; ratio 6:1.

10 cm cube: area 600 cm^2; volume 100 cm^3; ratio 3:5.

Clearly a large animal is in far less danger of hypothermia (excessive loss of body heat) than a similar-shaped small one.

Compare the surface area to weight ratio of a man of 25 weighing 68 kg with a skin area of 1.8 m^2, and a baby of one year weighing 10 kg with a skin area of 0.47 m^2. Discuss the implications of your answer.

19 Solid geometry

Nets
Solid symmetry

Representation
3-D co-ordinates

● You need to know . . .

Solid geometry 1 to 4 (page 250)

1 Select from the following list the correct mathematical name for each solid shown in Figure 19:1.

cone, cube, cuboid, cylinder, pyramid,

sphere, square, circle, tetrahedron

(a) OXO   (b) Polycell
(c) Nestlé   (d) golf ball

(Robin Lox)

Fig. 19:1

2 Figure 19:2 shows the net of a cube of edge 10 cm. The net is folded to make the cube.

(a) How many edges has the cube?

(b) Which line will come together with the line AK?

(c) Which points will come together with the point B?

(d) State the shortest distance from P to E on the cube.

(e) State the shortest distance from A to M along the edges of the cube. (MEG)

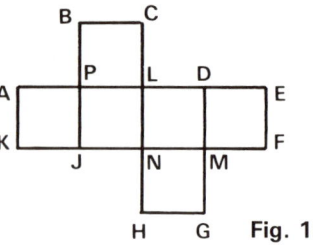

Fig. 19:2

3 In Figure 19:3, one net will not make a cube. Which one? (ULEAC)

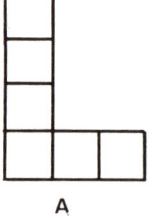  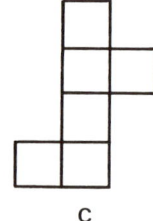 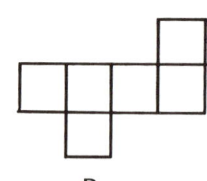 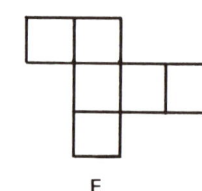

A   B   C   D   E

Fig. 19:3

4 (a) Draw full size a net for a cuboid which is to be 4 cm long, 3 cm wide and 2 cm deep. You must be able to cut the net from a rectangle of card 10 cm long and 8 cm wide.

(b) State the dimensions of the minimum area rectangle of card from which two such nets could be cut.

5 Figure 19:4 shows the net of a solid with one face missing.

(a) By measuring Figure 19:4, state the size and the shape of the missing face.

(b) Name the solid.

(c) Sketch the solid in three dimensions.

(d) If the solid is laid on a table with face AJED downwards, what is the height of point I above the table?

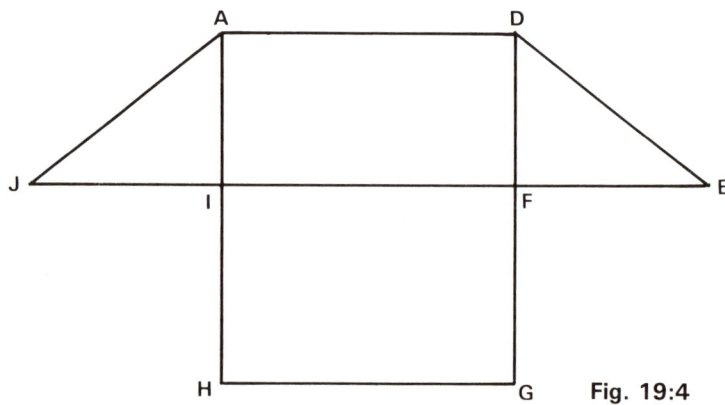

Fig. 19:4

6 For each of Figures 19:5 to 19:9, state:
   (a) the number of axes of symmetry
   (b) the number of planes of symmetry
   (c) the symmetry number (that is, the number of different ways the solid can be placed in a mould of itself).

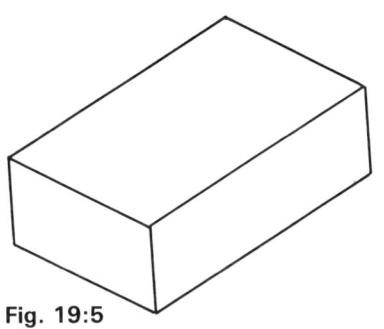

Fig. 19:5

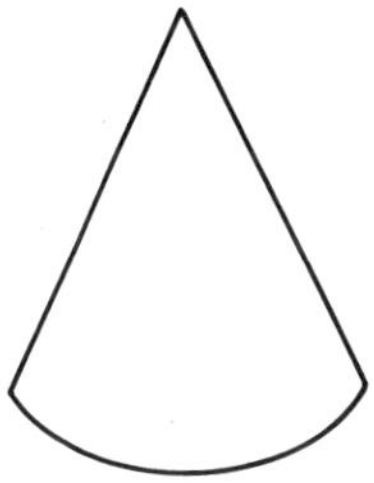

Fig. 19:6

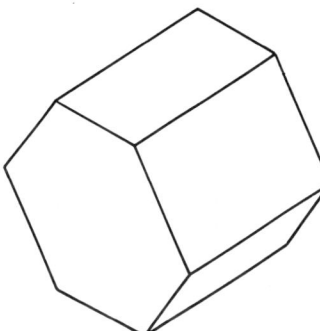

Fig. 19:7

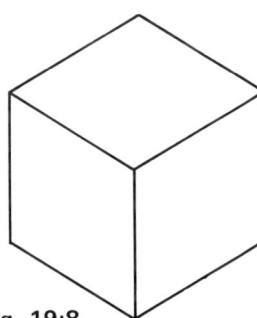

Fig. 19:8

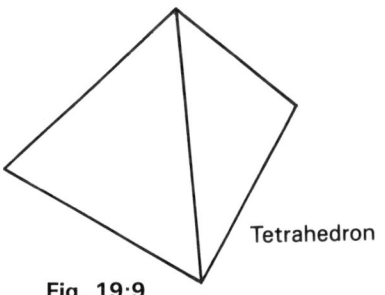

Tetrahedron

Fig. 19:9

7 Figure 19:10 shows a cylindrical water tank. There are six marks on the tank; these are equally spaced from the bottom.

(a) When the water level is at mark 4, fifteen litres are poured in, and the water level rises to mark 5.
   (i) How much water was in the tank originally?
   (ii) How much water is there in the tank now?
   (iii) How much water will be in the tank when it is at mark 6?

(b) Water is now drawn off so that the level falls from mark 5 to exactly half-way between mark 1 and mark 2. How much water has been drawn off?

(MEG)

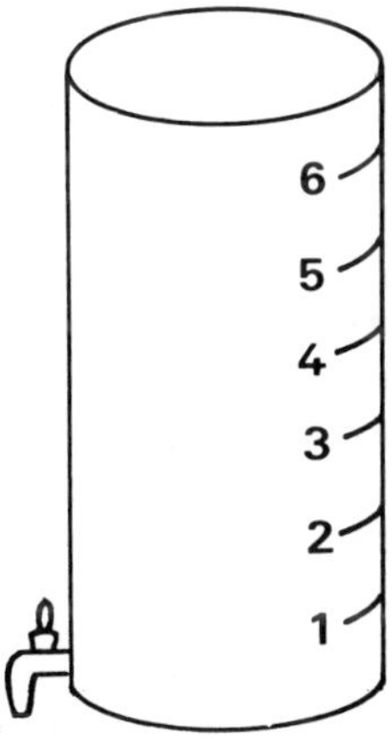

Fig. 19:10

8 Figures 19:11 to 19:15 show the five regular Platonic solids. They are, in increasing order of number of faces: tetrahedron, cube, octahedron, dodecahedron, icosahedron.

(a) Write the name of the solid shown in each figure.

(b) State the number of faces (F), the number of vertices (V), and the number of edges (E) for each solid. Show that in each case $F + V = E + 2$.

(c) Sketch two different (non-congruent) nets for a tetrahedron.

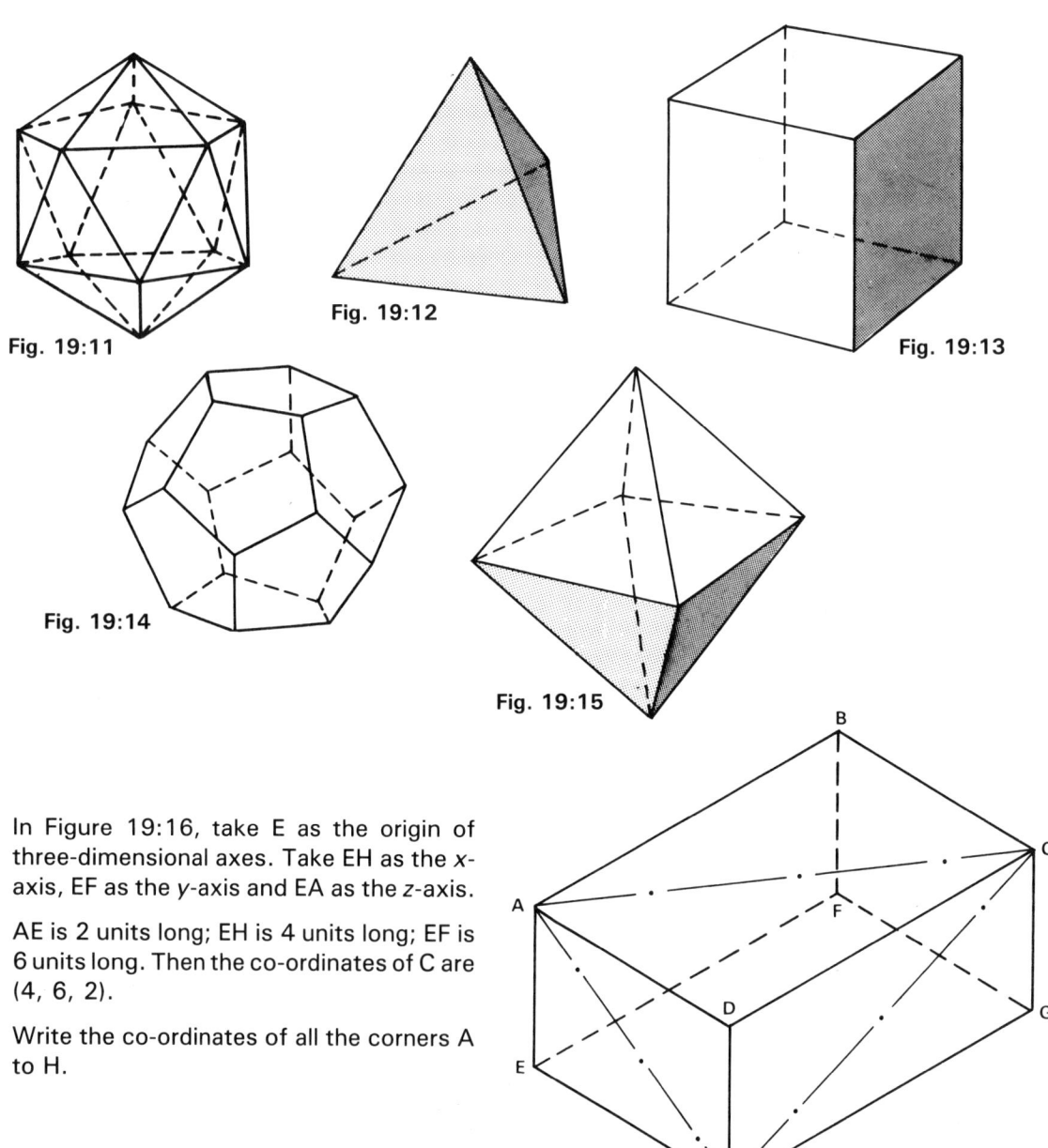

Fig. 19:11

Fig. 19:12

Fig. 19:13

Fig. 19:14

Fig. 19:15

9 In Figure 19:16, take E as the origin of three-dimensional axes. Take EH as the x-axis, EF as the y-axis and EA as the z-axis.

AE is 2 units long; EH is 4 units long; EF is 6 units long. Then the co-ordinates of C are (4, 6, 2).

Write the co-ordinates of all the corners A to H.

Fig. 19:16

10 Figures 19:17 to 19:24 show the plans and elevations of some solids.
(a) Name the solids. (b) Sketch a 3-D view of each solid.

Elevation

Plan

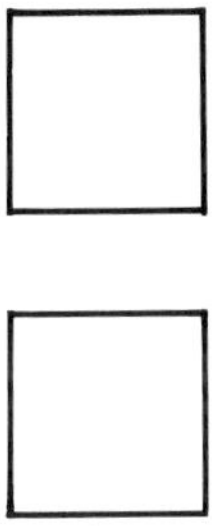

Fig. 19:17

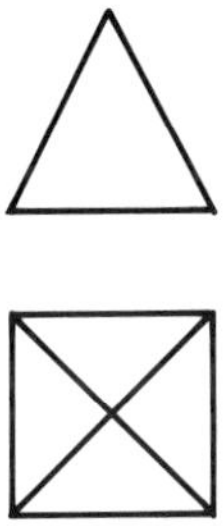

Fig. 19:18

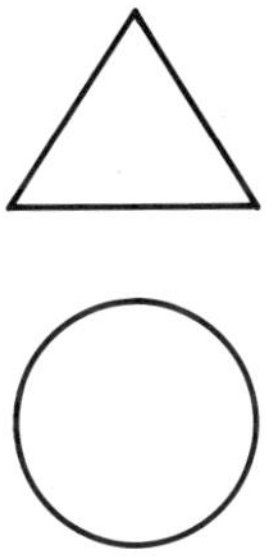

Fig. 19:19

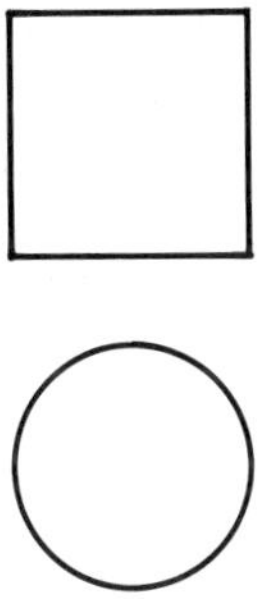

Fig. 19:20

Elevation

Plan

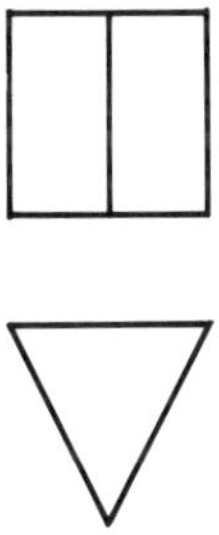

Fig. 19:21

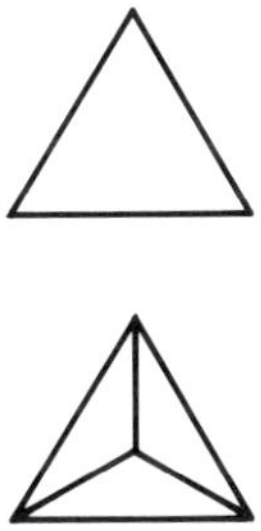

Fig. 19:22

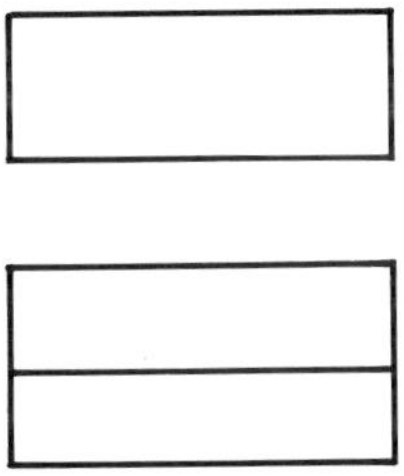

Fig. 19:23

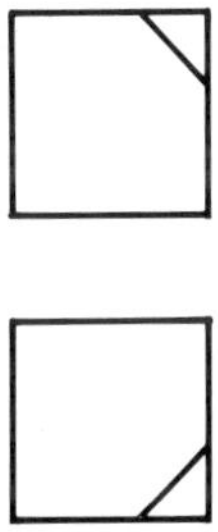

Fig. 19:24

11 The distance between the point P (3, 4, 0) and the point Q (3, 4, 12) is 12 units.

(a) Write down the co-ordinates of three other points which are 12 units from Q.

(b) Describe fully the figure formed by the set of points which are 12 units from Q.

(c) Calculate the length of the line from the origin O (0, 0, 0) to the point (3, 4, 12).

(d) Write down the co-ordinates of the point R on the positive z-axis such that RQ = OQ.

(e) Calculate the size of angle OQR.

(MEG)

12 Figure 19:16 shows a cuboid ABCDEFGH. Which of the following angles are right angles in the actual cuboid?
(a) $\angle$ABC (b) $\angle$AEH (c) $\angle$ACG (d) $\angle$AHC (e) AHG (f) $\angle$BAH
(g) $\angle$ACG (h) $\angle$ACH.

13 An ant crawls from E to C on the cuboid in Figure 19:16. It takes the shortest possible route. By drawing find the length of this route. (The cuboid is 6 cm by 4 cm by 2 cm.)

14 When two cubes are joined together on a table you can only see eight of their twelve faces. Investigate!

20 Pythagoras' theorem

Right-angled triangle   Distance between co-ordinates

● You need to know . . .

Mensuration 1 (page 253)

1 Calculate the unknown side in each triangle in Figure 20:1, giving each answer correct to 2 d.p.

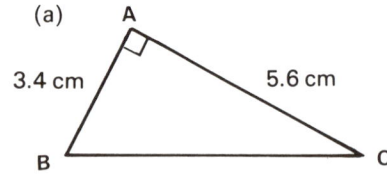

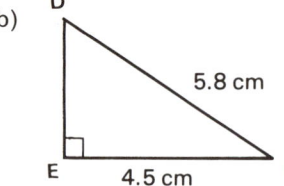

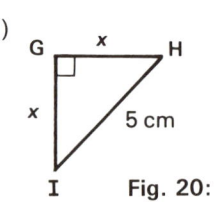

Fig. 20:1

2 (a) Copy Figure 20:2 exactly.
Join the points to make kite ABDC.
Measure AB and BD.
(b) Check the accuracy of your construction by calculating AB and BD.

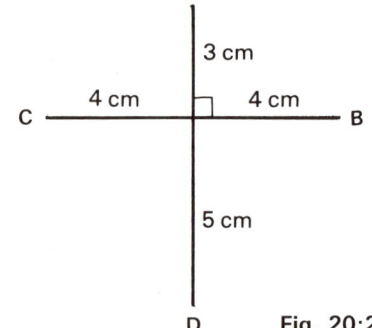

Fig. 20:2

3 At a school, pupils have to walk on the path round the perimeter of a rectangular lawn, 16 metres long and 12 metres wide. Teachers are allowed to walk across the lawn! How much farther than a teacher does a pupil have to walk to get from the gate at one corner to the main door at the opposite corner?

4 Some of the triangles with the following sides have three acute angles, some have an obtuse angle, some have a right angle, and some do not exist. Use Pythagoras' theorem to calculate which is true for each.
(a) 5 cm, 6 cm, 7 cm   (b) 3 cm, 5 cm, 7 cm   (c) 5 cm, 9 cm, 10 cm
(d) 2 cm, 3 cm, 6 cm   (e) 8 cm, 15 cm, 17 cm   (f) 2 cm, 9 cm, 11 cm
(g) 6 cm, 6 cm, 6 cm   (h) 4 cm, 4 cm, 5 cm

5 A rhombus has diagonals of length 18 cm and 13 cm. Calculate the length of its sides.

6 A circle, centre O and radius 13 cm, has two parallel chords of lengths 10 cm and 24 cm. Calculate the two possible perpendicular distances between the chords.

7 Calculate the shortest distance between the following points on a graph. Leave your answers with a $\sqrt{}$ sign where they are not rational.
(a) (5, 9) and (8, 5)   (b) (7, 1) and (8, 0)   (c) (−1, 0) and (4, 0)
(d) (−3, −2) and (2, −1)   (e) (2, −3) and (−5, 2)   (f) (X, Y) and (x, y)

8 Figure 20:3 shows a cuboid measuring 6 cm by 3 cm by 2 cm. Calculate the length of:
(a) AF   (b) DF.

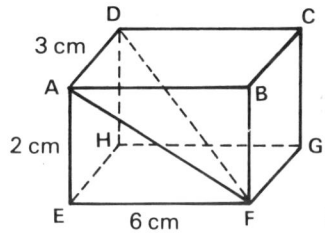

Fig. 20:3

9 Figure 20:4 represents a frame made from 15 cm, 12 cm, and 8 cm lengths. It is to be strengthened with a strut from corner A to corner B. Calculate the length of the strut.

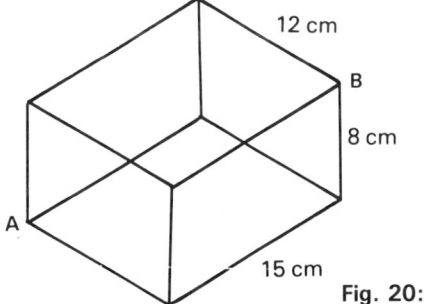

Fig. 20:4

10 Show that the altitude of an equilateral triangle of side 2 units is $\sqrt{3}$ units.

11 In Figure 20:5, O is the centre of the circle and AB is a tangent. AO is to be 30 cm when AB is twice the radius of the circle. Find this radius.

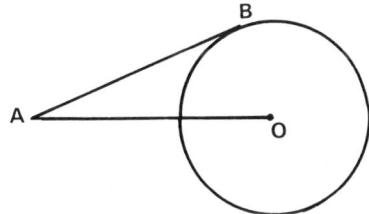

Fig. 20:5

12 A right-angled triangle has to have sides of lengths x cm, $x + 2$ cm, and $x + 4$ cm. Calculate the value of x.

13 In Figure 20:6, AOB and COD are two straight lines intersecting at right angles at the point O. The area of the triangle AOD is 15 cm², that of the triangle COB is 75 cm², and the length of the line AOB is 20 cm. Taking the length of AO as x cm, show that the length of OD is $30/x$ cm and that the length of OC is $150/(20 - x)$ cm. Given also that the length of COD is 16 cm, obtain an equation in x and solve it to find two possible values of x.   (MEG)

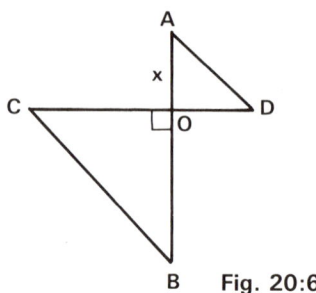

Fig. 20:6

14 (a) Write any fraction.

(b) Invert the fraction, then double the result.

(c) Add 2.

(d) Multiply by the LCM of the two numbers used to make the original fraction. Call the answer x.

(e) Add 2 to the fraction you started with, converting your answer to a top-heavy fraction.

(f) Multiply by the LCM of the two numbers used to make the original fraction. Call the answer y.

Let x and y be the shorter sides of a right-angled triangle. Calculate the hypotenuse. It should be integral. Can you prove why this must be so?

15 If n is one side of a right-angled triangle and $n + 1$ is the hypotenuse, then $\sqrt{2n + 1}$ is the third side. Show that this is so.

If n is an integer then $2n + 1$ must be an odd number. If $\sqrt{2n + 1}$ is rational then it too must be odd. Letting $\sqrt{2n + 1}$ be any odd number, there must then be a right-angled triangle with integral sides having that odd number as the length of one side.

Find the integral right-angled triangles with one side:
(a) 3   (b) 5   (c) 7   (d) 9   (e) 11   (f) 13.

16 Figure 20:7 illustrates a proof of Pythagoras' theorem written by the USA's President Garfield.

Express the area of the trapezium in two ways:

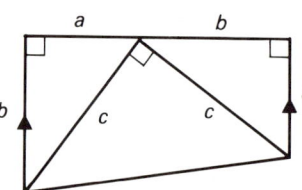

(a) by using the usual trapezium formula

(b) as the sum of the areas of the three triangles.

Fig. 20:7

The expressions in parts (a) and (b) must be equal. They simplify to show that $a^2 + b^2 = c^2$. Neat, ain't it!

21 Mensuration

Calculating length, area and volume
Distinguishing formulae

● **You need to know . . .**

Mensuration 2 to 5 (page 253)

Find out from your teacher which formulae you need to know and learn them!

1 (a) Figure 21:1 shows a running track. Taking π as 3.14 calculate the length of one lap.

   (b) A runner completes one lap in 42 seconds. What is his average speed in m/s correct to 2 d.p.?

   (c) What is the area of the running track, if it has 5 one-metre wide lanes and Figure 21:1 shows the inside edge?

   (d) How much stagger is needed between the outside and inside tracks if runners are assumed to run on the inside of their lane and to have covered the same distance after one lap?

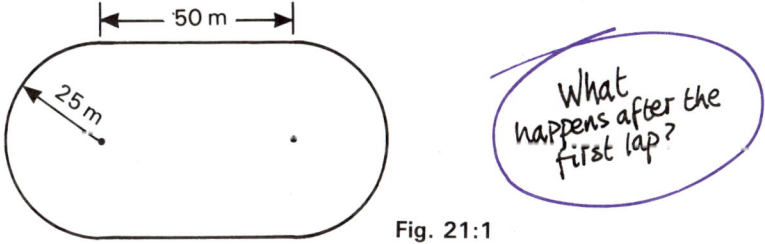

Fig. 21:1

2 A circular tree-trunk has a circumference of 1.8 metres. What is its diameter in cm? (Take π as 3.14, and give your answer correct to 1 d.p.)

3 Figure 21:2 shows a cycle milometer. The eight-toothed wheel is turned one tooth each time the cycle wheel rotates. The cycle wheel is 26 inches in diameter. How many turns must the toothed wheel make to record 1 mile? (1 mile = 1760 yards; 1 yard = 36 inches. Take π as 3.142.)

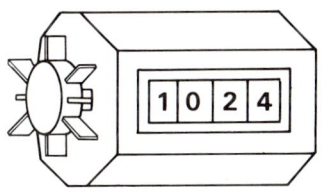

Fig. 21:2

4 Pots of jam, each of 500 g gross weight, are to be packed in boxes of 24. Figure 21:3 shows a plan of a full box. Figure 21:4 shows an elevation of one jar.

   (a) Calculate the weight of a full box in kg, allowing 500 g for the packing material.

   (b) The box top is formed of four flaps, ACDB, GEFH, AIJG and BIJH, as shown in Figure 21:3. The base of the box is formed in the same way. Calculate the area of card used in making the box, allowing 100 cm² for the vertical join.

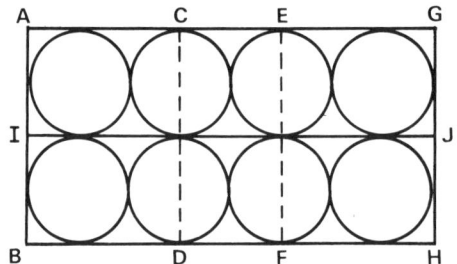

Fig. 21:3

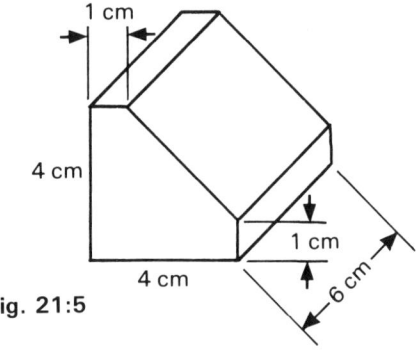

6.5 cm

12 cm

Fig. 21:4

   (c) Calculate the volume of one cylindrical pot.

   (d) An 8 cm high label surrounds the pot, with an overlap of 1 cm. How many labels can be printed on an A1 sheet (594 mm by 841 mm)? Take π as 3.142.

5 Figure 21:5 shows the design for a metal block. The density of the metal is 8.6 g/cm³. Calculate the weight of the block.

1 cm

4 cm

4 cm

1 cm

6 cm

Fig. 21:5

6 Explain how you decide whether a given formula involving lengths is to find a length, an area, a volume, or none of these.

7 In the following formulae x, y and z are lengths. State whether F is a formula for length, area, volume, or is impossible.

   (a) $F = 2(x + y)$ (b) $F = 2xy$ (c) $F = xyz$ (d) $F = 3z^2y$ (e) $F = \frac{1}{2}x(y + z)$

   (f) $F = x(y^2 + z)$ (g) $F = x^2y^2$ (h) $F = \dfrac{xy}{z}$ (i) $F = 4x^2y - 3yz^2$

*8 A wall 3.6 m long and 2.4 m high is to be tiled with 10 cm square tiles. How many tiles are needed, and what is their total cost at 15p each?

An alternative is to use 15 cm square tiles at 24p each. Would this be cheaper?

*9 A rectangular flower-bed, 15 m by 9 m, is to be surrounded by a 1.5 m-wide path, made from paving slabs 30 cm square costing £1.25 each. Find the cost of paving the path.

10 Figure 21:6 shows how a well rope is wound round a cylindrical axle of 7 cm radius. How many times must the handle be turned to lower the bucket 22 metres? (Take $\pi = 3\frac{1}{7}$.)

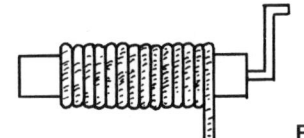
Fig. 21:6

*11 A shed floor, 2.5 m long and 2 m wide, is to be concreted to a depth of 10 cm. One cubic metre of concrete mix requires one cubic metre of ballast and 3 bags of cement. Ballast costs £6.50 per cubic metre and cement is £6.50 a bag.

(a) Write down the cost of 1 m³ of concrete mix.

(b) Find the area of the shed floor in m².

(c) Find, in cubic metres, the volume of concrete mix required for the floor.

(d) Find the cost of concrete used for the floor.

(e) If ballast may be bought by the $\frac{1}{2}$ cubic metre, but cement has to be bought by the bag, find the total cost of material for the floor.　　　　　(ULEAC)

12 Figure 21:7 shows the plan and elevation of a cone to be made from a sector of card as shown in Figure 21:8. A 5° sector is allowed as a joining tab. Take π as 3.14. Calculate:
(a) the slant height of the cone
(b) the volume of the cone
(c) the circumference of the base of the cone
(d) the radius of the sector
(e) the length of arc ABC
(f) the angle θ
(g) the total area of card used in making the cone.

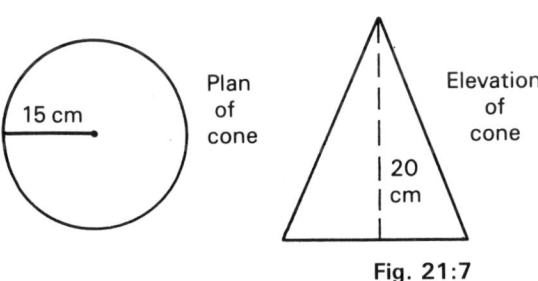

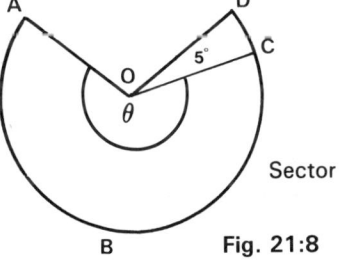

Fig. 21:7

Fig. 21:8

13 A metal ball 7 cm in diameter is placed in a cylindrical can of 9 cm diameter. Water is poured in until it is 10 cm deep.
(a) What volume of water has been poured into the can?
(b) If the ball is removed, how deep will the water be?

14 An expression for the volume and an expression for the surface area of the capsule shown in Figure 21:9 are contained in the following list. Select the two formulae, giving reasons for your choice.

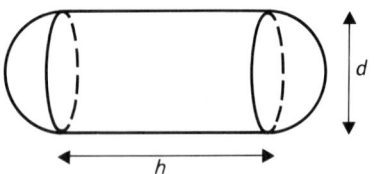

$\pi d + 2h$ $\pi d^2 + \frac{1}{4}\pi d^2 h$ $\pi d^2 + \pi dh$
$\frac{1}{6}\pi d^3 + \frac{1}{4}\pi d^2 h$ $\frac{1}{6}\pi d^3 + \pi dh$

(MEG)

Fig. 21:9

15 (a) Calculate the volume of the capsule in Figure 21:9 when $h = d = 5$ mm.

(b) What is the percentage increase in the volume of the capsule when h is doubled, d remaining the same?

(c) Is your answer to (b) the same for any size of h and d? Give reasons for your answer.

16 An evaporation unit to distil pure water from sea water is made from Perspex curved in an arc of a circle. Figure 21:10 shows a cross-section of the unit, which is 20 metres long.

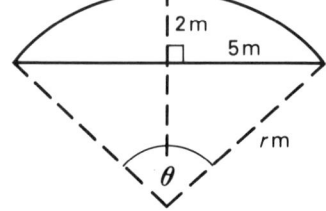

Not to scale

Fig. 21:10

(a) Calculate r, the radius of the arc.

(b) Calculate angle θ at the centre of the sector.

(c) Calculate the surface area of the Perspex.

(d) Calculate the volume of the evaporation unit.

(e) Would the idea work? Where would the distilled water be collected? Why not make a model and try it out! You do not have to use salt water.

17 Figure 21:11 represents a solid tetrahedron ABCD, where BC = BD = 8 cm, and AB = 7 cm.

Given that angle ABC = angle CBD = angle ABD = 90°, calculate:

(a) the length of the perpendicular from B to CD
(b) the length of the perpendicular from A to CD
(c) the volume of the tetrahedron
(d) the area of the triangle ACD.
(e) Use your answers to (c) and (d) to find the length of the perpendicular from B to face ACD, correct to 3 significant figures.

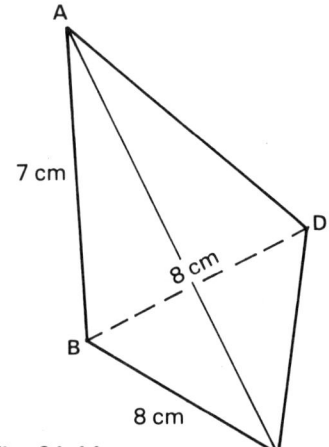

Fig. 21:11

18 A base for a table lamp is to be made by cutting tetrahedrons from each of the corners of a 15 cm-sided cube of wood, leaving faces that are regular octahedrons. Figure 21:12 shows one corner.

Calculate:
(a) the length of one side of the octahedron
(b) the volume of one tetrahedron
(c) the percentage of the original cube that remains.

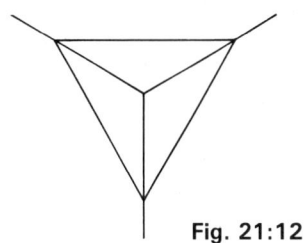

Fig. 21:12

19 An emergency reservoir in a forest is in the shape of a frustum of a cone, as shown in Figure 21:13. The reservoir has a radius of 12 m at the top and a radius of 9 m at the bottom. It is 4 m deep.

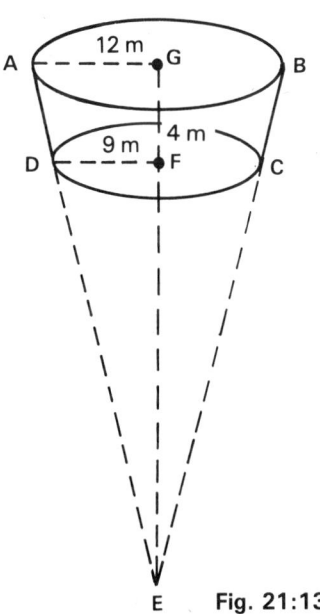

(a) Use similar triangles to find the length FE correct to 3 d.p.

(b) Calculate:
   (i) the volume of the cone DCE
   (ii) the volume of the cone ABE, and hence the volume of the frustum ABCD.

(c) Calculate the capacity of the reservoir in litres.

(d) Calculate the lengths:
   (i) AE   (ii) AD.

(e) The sides and the base of the reservoir are to be coated with a paint containing a herbicide. One litre of the paint will cover 6 m^2 to the nearest m^2. Calculate:
   (I) the surface area of the walls
   (ii) the area of the base
   (iii) the number of litres of paint needed to coat the reservoir.

Fig. 21:13

Volume of cone = $\frac{1}{3}\pi r^2 h$;
curved surface area of cone = $\pi r l$, where l is the slant height.

20 A 20 mm-diameter water pipe delivers 5 litres in 30 seconds. Calculate the speed of the water in the pipe, in cm/min, assuming that the water completely fills the pipe.

Right-angled triangle
(Sine and cosine rules)

(Area formulae)
(Generating functions)
(Graphs)

● **You need to know . . .**

Trigonometry 1 to 6 (page 256)

1 Calculate side x or angle θ in the triangles in Figure 22:1.

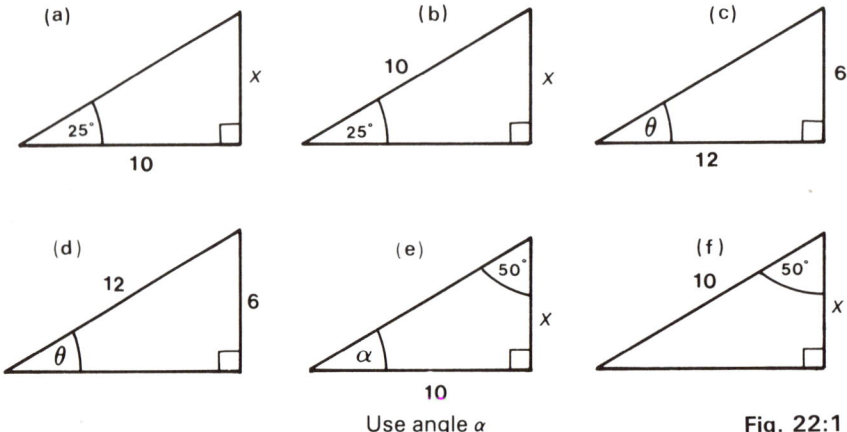

Fig. 22:1

2 Three different triangles are all right-angled, have one angle of 67°, and one side of length 4 cm. Sketch the three triangles, then calculate the lengths of the other two sides for each.

3 Two different triangles are both right-angled and have one 5 cm side and one 3 cm side. Sketch both triangles, then calculate the other angles in each.

4 The top of a tree has an angle of elevation of 25° from an observer 25 metres from its foot. How high is the tree?

5 Looking down from her prison window in a tower, Rapunzel sees a prince 100 metres from the foot of the tower on level ground, at an angle of depression of 15°. About how long does her hair have to be if the prince is to climb up it to rescue her?

6 At noon one day a post 6 metres high has a shadow 3 metres long. What is the altitude of the sun at this moment?

7 A hill has a gradient of 20% (1 in 5). What is its inclination to the horizontal in degrees?

8 An isosceles triangle has two sides of 8 cm and one of 6 cm. Calculate its vertical angle.

9 A ship is 3 km from a 150 m-high cliff. What is its angle of depression from the top of the cliff?

10 A rocket travels 5 km at an inclination of 80° to the horizontal, then a further 10 km at an angle of 32.6° with the vertical. How far down-range is it then?

11 A ship sails at 8 knots (8 nautical miles per hour) for 2 hours on a bearing of 135° from a buoy A. It then turns to a bearing of 200° and continues at the same speed for 30 minutes. How far south and how far east of A is it then?

*12 Find side x or angle θ in the triangles in Figure 22:2.

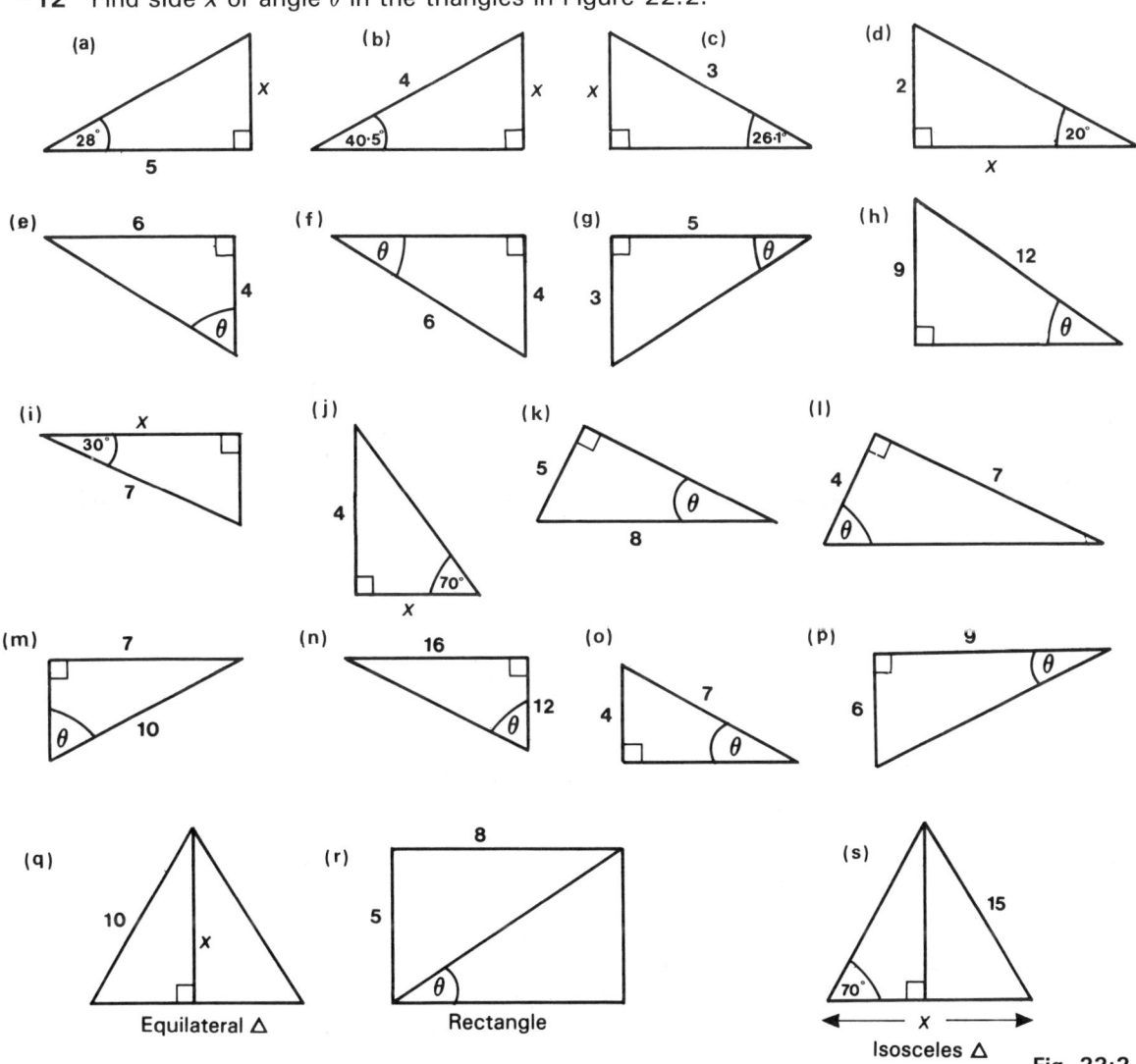

Fig. 22:2

*13 ABCD is a rectangle with AB > BC. X is a point on AB such that DX = XB = 10 cm and angle CXB = 37°. Y is a point on XC such that angle BYX is 90°. Calculate:
(a) XY   (b) BC   (c) ∠AXD.

14 Calculate the semi-vertical angle of a cone of height 6 cm and base radius 6 cm.

15 Calculate the length *x* for each triangle in Figure 22:3.

(a)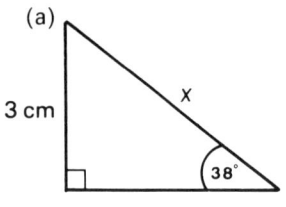

3 cm, *x*, 38°

(b)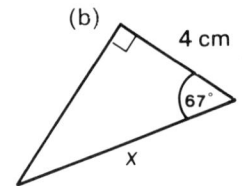

4 cm, 67°, *x*

(c)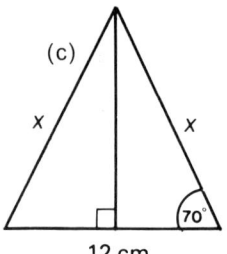

x, *x*, 12 cm, 70°

Fig. 22:3

H16 (a) Copy the table, then use your calculator to help you complete it, giving values correct to 2 significant figures.

θ	0°	10°	20°	30°	40°	50°	60°	70°	80°	90°	100°	etc. to 360°
sin θ		0.17									0.98	
cos θ		0.98									−0.17	
tan θ		0.18								∞	−5.7	

(b) Setting degrees along the horizontal axis, draw the graphs of $y = \sin \theta$, $y = \cos \theta$, and $y = \tan \theta$. For the first two graphs, take *y* from −1 to 1 with a scale of 2 mm to 0.1. For the third, take *y* from −10 to 10 with a scale of 1 mm to 0.1. Figure 22:4 shows you how to plot the tangent graph as the curve approaches ∞ (infinity) at 90° (and at 270°).

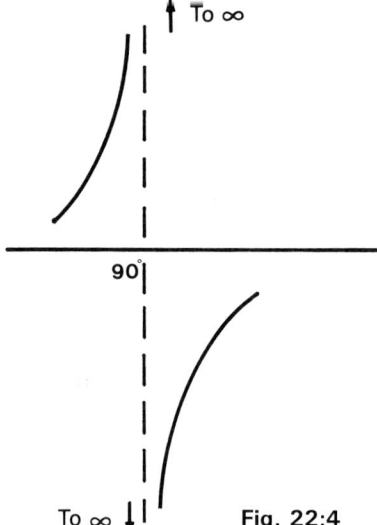

To ∞

90°

To ∞

Fig. 22:4

H17 Draw graphs of:
(a) $y = \sin \theta + \cos \theta$ for θ from $0°$ to $360°$
(b) $y = 3 \sin \theta - 1$ for θ from $0°$ to $360°$.

H18 Figure 22:5 shows a CD radio mast erected at O and supported by four equal guy-wires fixed at E, 4.5 metres below the top of the mast, and at A, B, C and D on the ground such that ABCD is a rectangle of length 25 metres and width 20 metres. M is the midpoint of BC and the angle between the planes EBC and ABCD is $40°$.

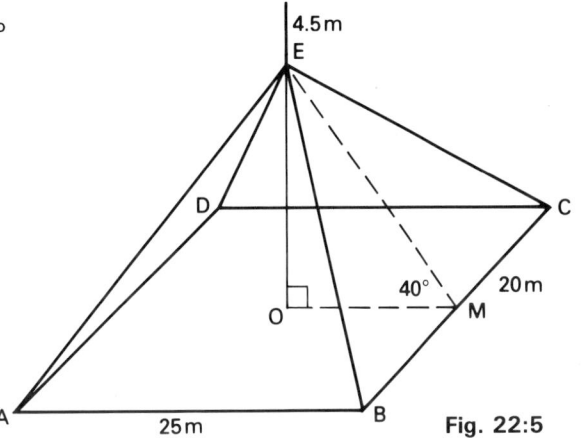

Fig. 22:5

(a) Calculate the height of the mast.

(b) Calculate the length of a guy-wire.

(c) Calculate the angle between the planes AEB and ABCD.

H19 Figure 22:6 represents three points A, B, C on horizontal ground. The point C is due south of A. The point B is on a bearing of $050°$ from C and 1350 m due east of A. A light aircraft, flying south at a constant height and speed, passes over A at the point L_1 where its angle of elevation from B is $22°$.

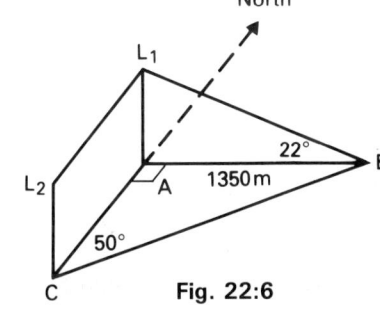

Fig. 22:6

(a) Calculate correct to three significant figures the height of the aircraft.

(b) Twenty seconds after passing over A the aircraft is at L_2, vertically above C. Calculate correct to three significant figures the ground speed of the aircraft in km/h.

(c) Calculate correct to the nearest degree the angle of elevation of the aircraft from B when the aircraft is at L_2.   (NEAB)

H20 (a) Draw the graphs of $y = 10 \sin x°$ and $20y - x = 60$ for x from 0 to 180.

(b) State the equation solved where the graphs cross and give its two solutions in the range 0 to 180.

(c) Explain with a sketch why there must also be a negative solution to the equation in part (b).

H21 Use a calculator to find the sine, cosine and tangent of the following angles. Draw up a two-way table to show each set of results.
(a) $30°, 150°, 210°, 330°$ (b) $45°, 135°, 225°, 315°$

H22 Explain the use of Figure 22:7, referring to your tables in your answer to question 21 to help you.

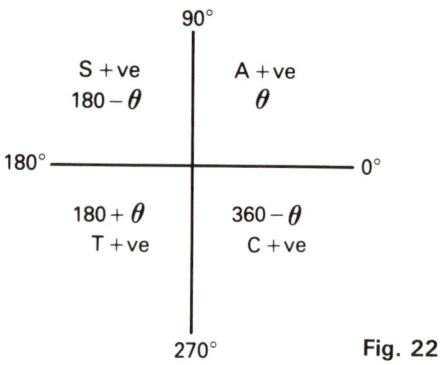

Fig. 22:7

H23 State all values of θ in the range of 0° to 360° when:
(a) $\sin \theta = 0$ (b) $\sin \theta = 1$ (c) $\sin \theta = -1$ (d) $\cos \theta = 0$ (e) $\cos \theta = 1$
(f) $\cos \theta = -1$ (g) $\tan \theta = 0.5$ (h) $\tan \theta = -0.5$ (i) $\tan \theta = -3$
(j) $4 \sin \theta = 3$ (k) $5 \cos \theta = -2$ (l) $2 \tan \theta = 3$.

H24 After 360° the trig. functions repeat the cycle. So if $\sin \theta = 0.5$, then $\theta = 30°, 150°, 390°, 510°, 750°, 870°, \ldots$

For the values in question 23, state the values of θ between 360° and 720°. Check your answers with your calculator; (a) has three answers, (b), (c) and (f) have one, the rest have two.

H25 Use first the cosine rule then the sine rule to solve fully triangle ABC where:
(a) AB = 7 cm, AC = 3 cm, $\angle A = 25°$
(b) AB = 6 cm, AC = 4 cm, $\angle A = 50°$
(c) AB = 7 cm, BC = 8 cm, AC = 6 cm
(d) AB = 5 cm, BC = 6 cm, AC = 7 cm
(e) $\angle B = 71.3°$, AB = 2.7 cm, BC = 3.1 cm
(f) AB = 2.75 m, AC = 4.6 m, BC = 3.9 m
(g) AB:BC:CA = 2:4:5, perimeter = 25.3 cm.

Use the cosine rule to find the bigger of two unknown angles.

H26 A ship's navigator observes two landmarks 5 km apart with one due east of the other. They bear 045° and 330° from his ship. Calculate how far he is from each landmark and from the straight line joining them.

H27 A surveyor needs to find the height of a watch-tower on the other side of a mined border strip, as shown in Figure 22:8. The angles of elevation from two points 50 metres apart, level and in line with the base of the tower, are 7.2° and 10.1°. Calculate the height of the tower, to the nearest tenth of a metre.

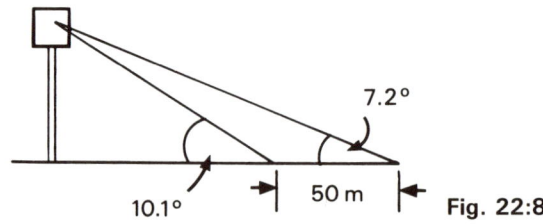

7.2°

10.1°   50 m   Fig. 22:8

H28 Triangle ABC has BC = 6 cm, AC = 5 cm, and angle B is 50°. Find the possible values for angle BAC.

29 Triangle ABC has AB = 8 cm, BC = 7 cm and AC = x cm. Angle BAC = 60° and the angles ABC and BCA are acute.

(a) Show that x satisfies the equation $x^2 - 8x + 15 = 0$.

(b) The equation is satisfied by two values of x. State the value of x which is a solution for the given triangle.

(c) Sketch a labelled diagram for which the other value is a solution.   (WJEC)

30 Calculate the area of the triangle in Figure 22:9.

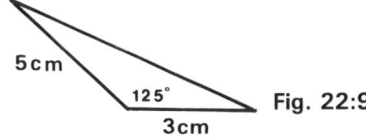

Fig. 22:9

31 Calculate the area of a triangle with sides 5 cm, 7 cm and 8 cm.

32 Two ships set out at 0630 from ports A and B, 50 nautical miles apart, A being due west of B. The ship leaving A is on a course of 037°, travelling at 12 knots. The other ship is travelling at 15 knots on a course of 330°.

Calculate the distance between the ships at
(a) 0730   (b) 0830   (c) 0930   (d) 1030.

33 Figure 22:10 represents a kite K flown from a point A on the ground. It has a tail KT of length 10 m. The length of the string AK is 40 m. The angle of elevation of K from A is 61° and that of T from A is 49°.

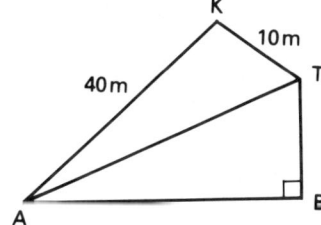

Fig. 22:10

(a) Show that the angle AKT = 111.73°.

(b) Calculate the height BT of T above the horizontal ground at the same level as A.

(c) The string AK is now extended so that its length is x m. The angle between the tail and the string, and the angle between the string and the ground, remain unchanged. Obtain an expression for the height $B_1 T_1$ in this case.

(d) Find the length of the string when the end of the tail is 50 m above the ground.
                                                                            (WJEC)

34 ABCD is a tetrahedron. AB = AC = AD = 8 cm and angles BAC, BAD and CAD are all 90°.
(a) Sketch the solid standing on face ABC.
(b) Calculate the lengths DC and BC.
(c) Calculate angle BDC.
(d) Calculate the angle between faces BDC and ABC.

35 Figure 22:11 shows a cone with circular base of radius 5 cm. The slant height of the cone is 20 cm. AB is a diameter of the base and V is the vertex of the cone. P is a point on VB such that VP:PB = 1:3. Find the shortest distance from A to P along the surface of the cone.

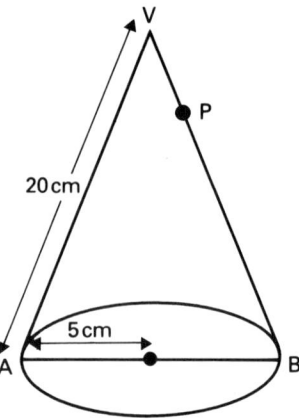

Fig. 22:11

36 (a) At a certain location in the British Isles the number of hours of daylight per day, t months after the spring equinox (21 March), is given approximately by the formula $y = 12 + 4 \sin (30t)$. What are the greatest and least amounts of daylight per day during the year?

   (b) At a certain location in Greenland the formula becomes $y = 12 + 24 \sin (30t)$, but only for certain times of the year. Why is the formula not valid when $t = 3$?

(SEG)

37 The depth of water in a harbour is given by the formula $d = 6 + 4 \cos (30t)$, where t is the number of hours after the last high tide and d is the depth in metres of the sea-bed at the entrance.

Draw a graph to show the values of d at hourly intervals from the high tide at 1330 to the next (12 hours later), and hence give the times when a boat of draught 5 metres will be unable to enter the harbour.

38 Draw graphs of the two formulae in question 36 for $t = 0$ to $t = 12$.

23 Vectors

Notation   (Resultant)
Translation   (Problems)

● You need to know . . .

Vectors 1 to 8 (page 262)

1 Referring to Figure 23:1, describe the following vectors as column matrices, e.g. $\overrightarrow{AB} = \begin{pmatrix} 1 \\ 1 \end{pmatrix}$.

(a) $\overrightarrow{BC}$ (b) $\overrightarrow{CD}$ (c) $\overrightarrow{DE}$ (d) $\overrightarrow{EF}$ (e) $\overrightarrow{FG}$

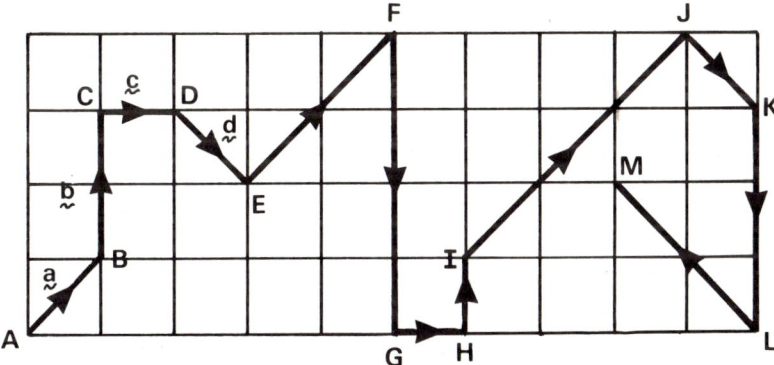

Fig. 23:1

2 Referring to Figure 23:1, state the following vectors in terms of $\underset{\sim}{a}$, $\underset{\sim}{b}$, $\underset{\sim}{c}$, or $\underset{\sim}{d}$.

(a) $\overrightarrow{EF}$ (b) $\overrightarrow{FG}$ (c) $\overrightarrow{GH}$ (d) $\overrightarrow{HI}$ (e) $\overrightarrow{IJ}$ (f) $\overrightarrow{JK}$ (g) $\overrightarrow{KL}$ (h) $\overrightarrow{LM}$

3 The magnitude, or length, of a vector is called its **modulus**. The symbol used is $|\overrightarrow{AB}|$ or $|\underset{\sim}{a}|$.

When the vector is not parallel to a grid line its modulus is found using Pythagoras' theorem.

Referring to Figure 23:1, find the length of:

(a) $\overrightarrow{BC}$ (b) $\overrightarrow{EF}$ (c) $\overrightarrow{IJ}$.

4 Calculate:

(a) $|\underset{\sim}{v}|$ where $\underset{\sim}{v} = \begin{pmatrix} 3 \\ 4 \end{pmatrix}$ (b) $|\underset{\sim}{w}|$ where $\underset{\sim}{w} = \begin{pmatrix} -2 \\ 5 \end{pmatrix}$.

5 Calculate the resultant of the following vectors, then check your answer by drawing.

(a) $\begin{pmatrix} 3 \\ 0 \end{pmatrix} + \begin{pmatrix} 2 \\ 0 \end{pmatrix}$ (b) $\begin{pmatrix} 1 \\ 0 \end{pmatrix} + \begin{pmatrix} 0 \\ 1 \end{pmatrix}$ (c) $\begin{pmatrix} 4 \\ -2 \end{pmatrix} + \begin{pmatrix} -3 \\ 1 \end{pmatrix}$ (d) $\begin{pmatrix} -2 \\ -1 \end{pmatrix} + \begin{pmatrix} -1 \\ -2 \end{pmatrix}$

6 Write as a column matrix the vector that translates the point $(3, -2)$ to $(-1, 2)$.

117

*7 (a) Copy Figure 23:2.

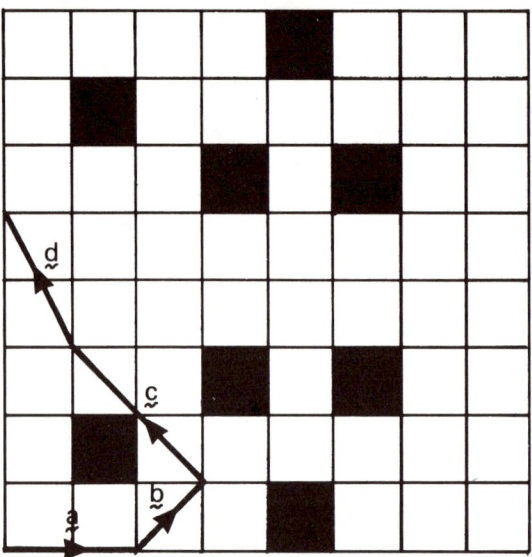

If you hit a black square (b), when doing part you have made a mistake!

Fig. 23:2

(b) Continue the vectors from the end of d as follows, marking an arrow on each one to show its direction.

$$\begin{pmatrix} 2 \\ 0 \end{pmatrix} \begin{pmatrix} 2 \\ 2 \end{pmatrix} \begin{pmatrix} 1 \\ -2 \end{pmatrix} \begin{pmatrix} -1 \\ -1 \end{pmatrix} \begin{pmatrix} 2 \\ -4 \end{pmatrix} \begin{pmatrix} 2 \\ 0 \end{pmatrix} \begin{pmatrix} -2 \\ 4 \end{pmatrix} \begin{pmatrix} 2 \\ 0 \end{pmatrix} \begin{pmatrix} -3 \\ 3 \end{pmatrix} \begin{pmatrix} 3 \\ 0 \end{pmatrix} \begin{pmatrix} -1 \\ -1 \end{pmatrix} \begin{pmatrix} 1 \\ 0 \end{pmatrix}$$

(c) Write its relationship to a, b, c, or d on each of the vectors you have drawn.

*8 $\overrightarrow{AB} = a$, $\overrightarrow{BC} = 2a$, $\overrightarrow{DB} = -a$, $\overrightarrow{CE} = b$.

$\overrightarrow{AB} = \begin{pmatrix} 2 \\ 0 \end{pmatrix}$ and $|\overrightarrow{AB}| = 2$ cm, $\overrightarrow{CE} = \begin{pmatrix} 0 \\ -2 \end{pmatrix}$ and $|\overrightarrow{AB}| = |\overrightarrow{CE}|$.

Sketch the following pairs of vectors:
(a) $\overrightarrow{AB}$ and $\overrightarrow{BC}$ (b) $\overrightarrow{AB}$ and $\overrightarrow{DB}$ (c) $\overrightarrow{BC}$ and $\overrightarrow{DB}$
(d) $\overrightarrow{AB}$ and $\overrightarrow{CE}$ (e) $\overrightarrow{BC}$ and $\overrightarrow{CE}$.

*9 If $a = \begin{pmatrix} 3 \\ 4 \end{pmatrix}$ and $b = \begin{pmatrix} 2 \\ -1 \end{pmatrix}$, find p and q when:

(a) $p = a + 2b$ (b) $q - a = 3b$.

H10 Refer to Figure 23:3. To express vector $\overrightarrow{AD}$ in terms of a and b find a way of getting from A to D by only travelling along vectors parallel to a and b.

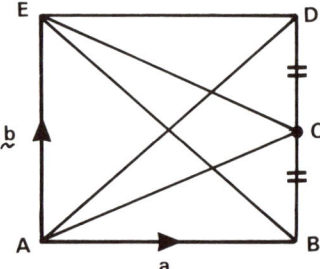

We could go along to B, then on to D.
so $\overrightarrow{AD} = \overrightarrow{AB} + \overrightarrow{BD} = a + b$.

Similarly $\overrightarrow{BE} = \overrightarrow{BA} + \overrightarrow{AE} = -a + b$.

Fig. 23:3

(a) In terms of b what is:
    (i) $\overrightarrow{BD}$ (ii) $\overrightarrow{DB}$ (iii) $\overrightarrow{BC}$ (iv) $\overrightarrow{CB}$ (v) $\overrightarrow{CD}$ (vi) $\overrightarrow{DC}$?

(b) Express in terms of a and b:
    (i) $\overrightarrow{DA}$ (ii) $\overrightarrow{EB}$ (iii) $\overrightarrow{AC}$ (iv) $\overrightarrow{CA}$ (v) $\overrightarrow{CE}$ (vi) $\overrightarrow{EC}$.

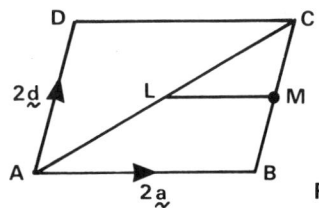

H11 In Figure 23:4, ABCD is a parallelogram. L and M are the mid-points of AC and CB.

Write in terms of $\underset{\sim}{a}$ and $\underset{\sim}{d}$:

(a) $\overrightarrow{CD}$ (b) $\overrightarrow{AC}$ (c) $\overrightarrow{LM}$ (d) $\overrightarrow{BL}$.

Fig. 23:4

H12 A plane flies due north at 400 km/h when a wind is blowing from the east at 40 km/h. Calculate its bearing and ground speed. Confirm your answer with a scale diagram.

H13 A 10 kg weight is supported by two strings angled at 50° and 30° to the vertical. Calculate the tension in each string. Confirm your answer with a scale diagram.

H14 On a still night a ship travelling at 20 knots sets a course of NNE ($022\frac{1}{2}°$), but the navigator reports it is travelling ENE ($067\frac{1}{2}°$) at 15 knots. Calculate the speed and direction of the current. Confirm your answer with a scale diagram.

---

15 In Figure 23:5, SR is parallel to OQ in triangle OTQ.

$$\overrightarrow{OP} = \underset{\sim}{x}, \quad \overrightarrow{PQ} = 3\underset{\sim}{x}, \quad \overrightarrow{OS} = \underset{\sim}{y}, \quad \overrightarrow{ST} = 3\underset{\sim}{y}$$

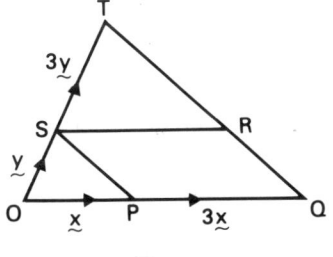

(a) Express in terms of $\underset{\sim}{x}$ and $\underset{\sim}{y}$:
(i) $\overrightarrow{SP}$ (ii) $\overrightarrow{TQ}$.

(b) Prove that SP is parallel to TQ.

(c) Express in terms of $\underset{\sim}{x}$ and $\underset{\sim}{y}$:
(i) $\overrightarrow{SR}$ (ii) $\overrightarrow{TR}$.

(d) Find the value of:
(i) $\dfrac{\text{area } \triangle SRT}{\text{area } \triangle OPS}$ (ii) $\dfrac{\text{area of PQRS}}{\text{area } \triangle OPS}$.

Not to scale   Fig. 23:5

(MEG)

16 In Figure 23:6, ABCD is a parallelogram in which $\overrightarrow{AB} = \underset{\sim}{p}$ and $\overrightarrow{BC} = \underset{\sim}{q}$. The point E in AD is such that $AE = \frac{1}{3}AD$.

(a) Express in terms of $\underset{\sim}{p}$ and $\underset{\sim}{q}$: (i) $\overrightarrow{AC}$ (ii) $\overrightarrow{BE}$.

(b) AC and BE intersect at F such that $\overrightarrow{BF} = k\overrightarrow{BE}$.
(i) Express $\overrightarrow{BF}$ in terms of $\underset{\sim}{p}$, $\underset{\sim}{q}$ and k.
(ii) Show that $\overrightarrow{AF} = (1 - k)\underset{\sim}{p} + \frac{1}{3}k\underset{\sim}{q}$.
(iii) Use this expression for $\overrightarrow{AF}$ and the fact that AFC is a straight line to find the value of k.

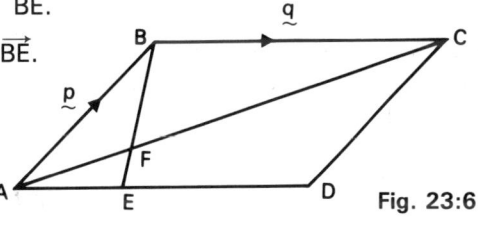

Fig. 23:6

(MEG)

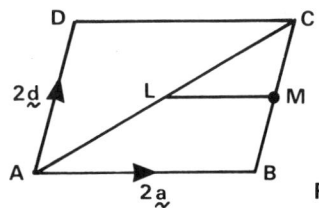

17 Figure 23:7 shows a regular hexagon where $\overrightarrow{OA} = \underset{\sim}{p}$ and $\overrightarrow{OE} = \underset{\sim}{q}$.

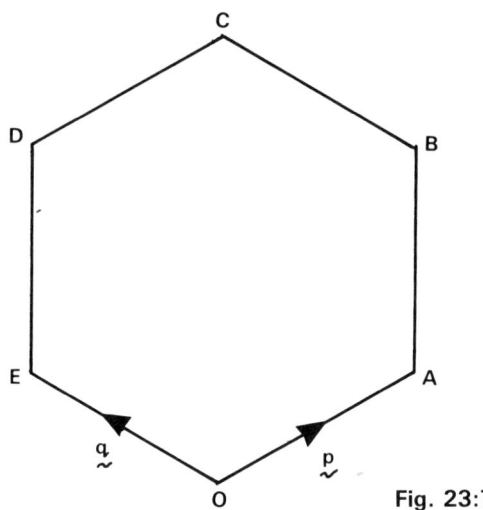

Fig. 23:7

(a) Express in terms of p and q:
  (i) $\overrightarrow{OD}$ (ii) $\overrightarrow{OB}$ (iii) $\overrightarrow{EA}$.

(b) Using your answer to (a) (i) and (ii) express the journey D to O to B in terms of p and q.

(c) What does this tell you about EA and DB?

(d) T is a point such that $\overrightarrow{OT} = 2p$, and R is a point such that $\overrightarrow{OR} = 2q$.
Express $\overrightarrow{RT}$ in terms of p and q.

(e) What does your answer to (d) tell you about RT and DB?

(f) O is the point (0, 0). A is the point ($\sqrt{3}$, 1) and E is the point ($-\sqrt{3}$, 1).
Write down $\overrightarrow{OA}$ and $\overrightarrow{OE}$ as column vectors.

(g) Calculate the co-ordinates of T, B, D and R.

(h) Calculate the length of RT. (SEG)

18 (a) OABC is a quadrilateral and the vectors $\overrightarrow{OA}$ and $\overrightarrow{OB}$ are equal to $\underset{\sim}{a}$ and $\underset{\sim}{b}$ respectively.
OPQR is the image of OABC under an enlargement with centre O and scale factor 2. Give the vectors $\overrightarrow{AB}$, $\overrightarrow{OP}$ and $\overrightarrow{PQ}$ in terms of a and b.

(b) XPYZ is the image of OPQR under the enlargement with centre P and scale factor 2. Give the vectors $\overrightarrow{PY}$, $\overrightarrow{OY}$ and $\overrightarrow{YB}$ in terms of a and b.

(c) By treating XPYZ as an enlargement of OABC, or otherwise, show that YB and ZC meet on OA. If they meet at T, use the fact that $\overrightarrow{OT} = \overrightarrow{OY} + k\overrightarrow{YB}$, where k is a number, to find $\overrightarrow{OT}$ in terms of a.

24 Transformations

Reflection   Translation
Rotation   (Inverse transformation)
Enlargement

● You need to know . . .

Transformations 1 to 4 (page 264)

1 In Figure 24:1, squares A3 and B2 have been shaded.
Lines p, q and r intersect at point M.

Which squares must be shaded to:
(a) reflect A3 in line q
(b) reflect B2 in line p
(c) reflect A3 in line r
(d) rotate B2 and A3 180° about M
(e) rotate B2 and A3 −90° (clockwise) about M

(f) translate B2 by the vector $\begin{pmatrix} 2 \\ -1 \end{pmatrix}$

(g) enlarge B2 by scale factor 2, centre the bottom left-hand corner of square A1.

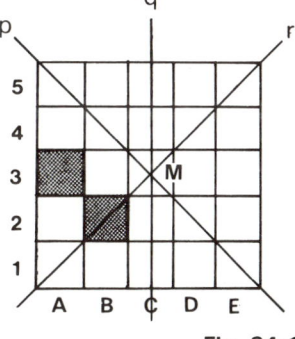

Fig. 24:1

2 (a) Draw axes, both from −8 to 12. Plot the triangle whose vertices are at (2, 2), (2, 4) and (4, 2).

(b) By drawing, or otherwise, state the co-ordinates of the vertices of the triangle when it is enlarged, centre the origin, by scale factor:
(i) 3   (ii) −1   (iii) $\frac{1}{2}$ (iv) −2.

3 Repeat question 2(a), then draw the triangle when it is enlarged as follows, stating the co-ordinates of the new vertices:
(a) scale factor 2, centre (2, 2)   (b) scale factor −1, centre (3, 1)
(c) scale factor $-\frac{1}{2}$, centre (4, 4).

4 PQRS is a rectangle with PQ = 3 cm and QR = 2 cm.

(a) Draw the rectangle, then enlarge it:
(i) by scale factor $1\frac{1}{2}$ from centre P   (ii) by scale factor $-\frac{1}{2}$ from centre P.

(b) Calculate the area of both transformed rectangles and state how many times as large as the original rectangle they are.

121

H5 Draw axes from −4 to 4 each. Plot the trapezium ABCD with corners at (−1, 1), (−1, 3), (−3, 3) and (−2, 1) respectively.

    (a) Reflect ABCD in the y-axis to give $A_1B_1C_1D_1$ then rotate $A_1B_1C_1D_1$ +270° about the origin to give $A_2B_2C_2D_2$.

    (b) Describe fully the single inverse transformation which maps $A_2B_2C_2D_2$ back onto ABCD.

---

6 Refer to Figure 24:2.

    (a) $A_1B_1C_1D_1$ is the image of ABCD under a single transformation. Write down the column vector which effects this transformation.

    (b) $A_2B_2C_2D_2$ is the image of ABCD under an anticlockwise rotation. Write down:
      (i) the angle of the rotation   (ii) the co-ordinates of the centre of the rotation.

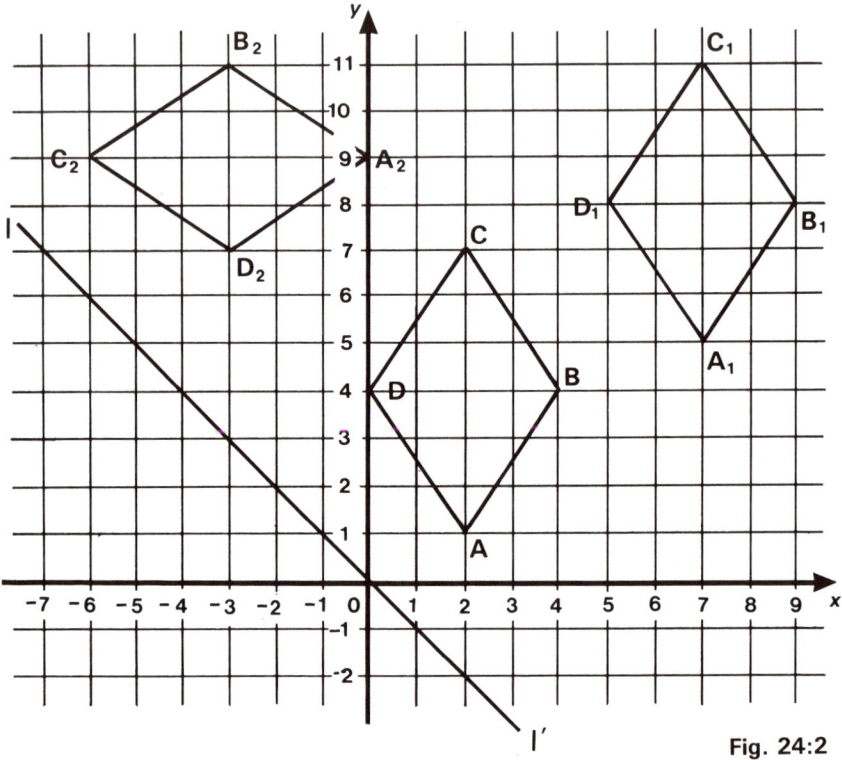

Fig. 24:2

    (c) The rhombus $A_3B_3C_3D_3$ (not shown in the figure) is the image of ABCD under a reflection in the line ll′. Write down: (i) the equation of line ll′, and   (ii) the co-ordinates of the point C_3.

7 (a) On squared paper draw rectangular axes O_x and O_y, using a scale of 1 cm to 1 unit. On these axes draw the graph of the line $y = 2x$.

(b) P is the point (5, 0). The operation of reflection in the line $y = 2x$ is denoted by **M**, and **R** is the operation of rotation anticlockwise through 90° about the origin. On your diagram mark the points **M**(P), **RM**(P), **MRM**(P), **RMRM**(P), together with their co-ordinates. Join these four points and name the figure formed.　(MEG)

8 In Figure 24:3, triangle ABC is isosceles, with AB = AC and ∠BAC = 100°.
The side AB is produced to D so that AD = BC.
Triangle A'B'C is the image of triangle ABC under an anticlockwise rotation about C.
(a) Calculate the angle of rotation.
(b) Write down the size of ∠B'A'C.
(c) Prove that A'B = BD, clearly stating your reasons.　(MEG)

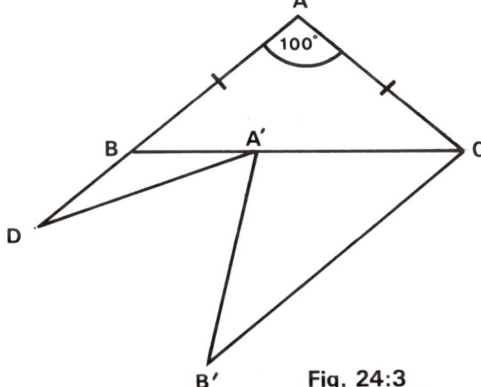

Fig. 24:3

9 Using a scale of 1 cm to represent 1 unit on each axis, draw the axes for $-3 \leqslant x \leqslant 9$ and $-3 \leqslant y \leqslant 9$. Also draw the line $y = x$.

The trapezium Q has vertices (3, 0), (4, 2), (6, 2) and (7, 0). Draw this trapezium on your diagram and label it Q.

Three transformations are defined:

M is reflection in the line $y = 0$.

T is translation with vector $\begin{pmatrix} 2 \\ 4 \end{pmatrix}$.

D is reflection in the line $y = x$.

On your diagram draw and label **M**(Q), **TM**(Q), and **DTM**(Q). The point P has co-ordinates (1, 3). Mark it on your diagram. Also label the point **DTM**(P). Hence describe the single transformation which maps Q onto **DTM**(Q).　(SEG)

25 Representation and interpretation of data

1 (a) Study Figure 25:1. Comment on how well it illustrates the data given in the table.

	1989	1990	1991	1992	1993
Value of car (£s)	10 000	6800	5600	5000	3600
Depreciation	0	3200	1200	600	1400
Repairs/service	0	250	280	300	440

(b) Design a pictogram to illustrate the increasing cost of repairs/service, using a spanner symbol to represent £100.

Depreciation

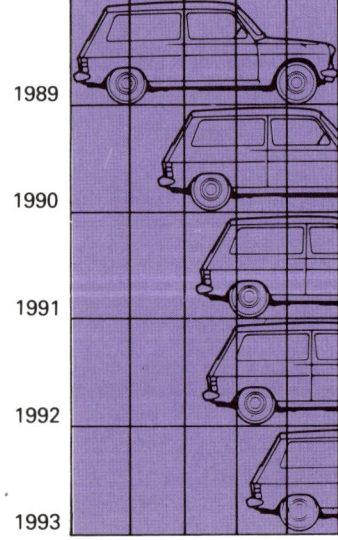

Fig. 25:1

2 Figure 25:2 shows a hospital patient's record sheet.

(a) What is the average human body temperature in °C?

(b) What units are used to measure pulse and respiration?

(c) What is the average human pulse rate?

(d) What is the average human respiration rate?

(e) Was the patient ill on the 15th February? If so, describe the symptoms. If not, say how you know.

(f) On what date did the patient's condition return to normal?

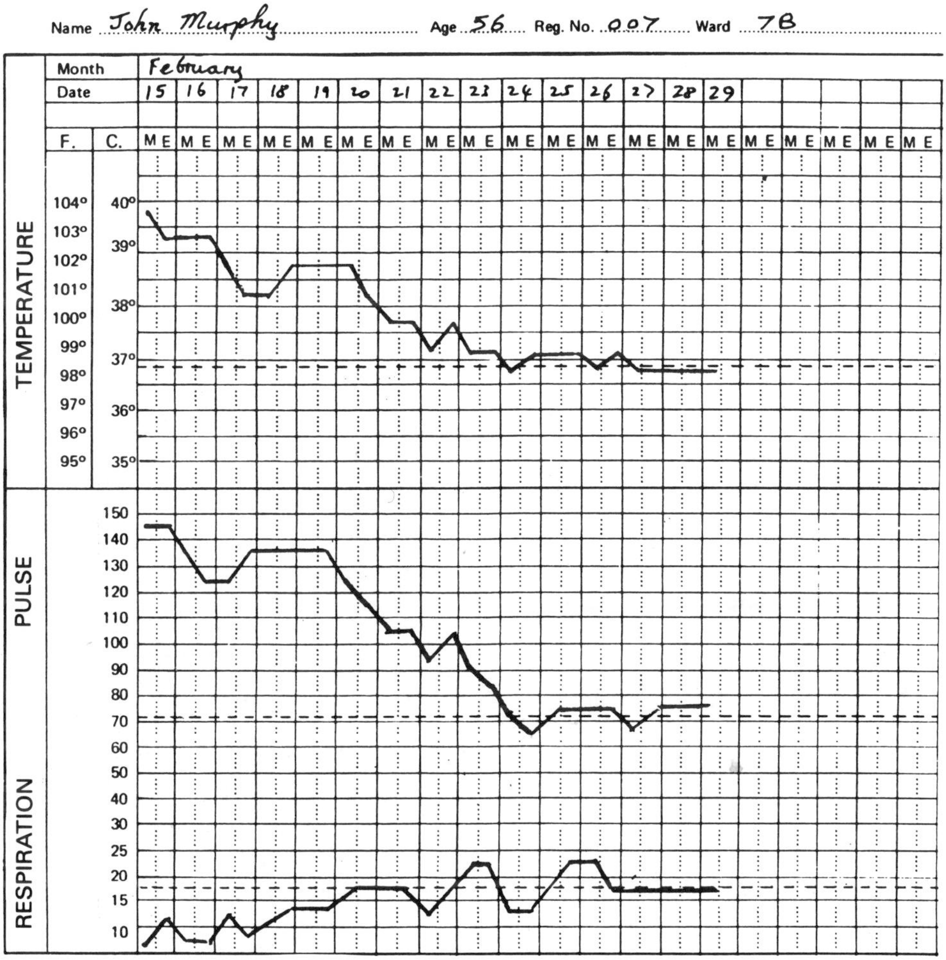

Name *John Murphy* .... Age *56* ...... Reg. No. *007* ...... Ward *7B*

Fig. 25:2

3

Marks	0	1	2	3	4	5	6	7	8	9	10
Frequency	1	4	8	5	7	3	1	1	0	1	0

(a) Draw a frequency polygon for the test results given in the table, using a continuous horizontal scale from 0 to 10.

(b) Comment on the results of the test.

4 A well-women clinic records the following heights and weights.

Height (cm)	150	158	166	151	152	151	159	166	155	157	152
Weight (kg)	47	53	55	40	65	48	51	56	42	71	49

Height (cm)	159	171	158	164	156	160	172	167	170	153	162
Weight (kg)	55	57	45	82	51	50	63	46	69	48	54

Height (cm)	174	172	172	157	164	178	174	176
Weight (kg)	62	50	75	52	56	67	47	80

(a) Draw up a table for the heights:

Class interval (h cm)	Tally	Frequency
150 ⩽ h < 155 155 ⩽ h < 160		

(b) Draw up a table for the weights, classes 40 ⩽ w < 45 etc.

(c) Illustrate the data with two frequency polygons, plotting at the mid-point of each class.

(d) Compare and comment on the shapes of the two graphs. Include a statement about their ranges.

5 What can you infer by comparing the two bar charts in Figure 25:3?

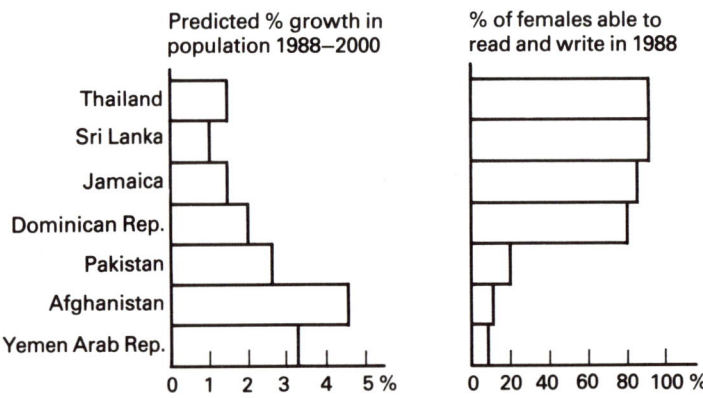

Fig. 25:3

6 Study Figure 25:4.

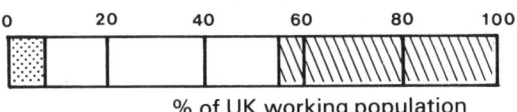

% of UK working population

▢ Raw material production

▢ Manufacture

▨ Services

Fig. 25:4

(a) What percentage of the UK working population are in each of the three categories?

(b) Redraw the chart:
 (i) as a bar chart with three separate bars
 (ii) as a pie chart.

7 (a) Chapter 6 question 7 gives a table showing stopping distances when driving. Draw a bar chart to illustrate this data, each bar to represent the total stopping distance divided into thinking and braking distances.

 (b) You are recommended to allow twice as far for braking in wet conditions. Redraw your bar chart to illustrate the effect that this has.

8

Mark (%)	0–9	10–19	20–29	30–39	40–49	50–59	60–69	70–79	80–89	90–100
No. of pupils	5	17	26	42	58	60	36	21	8	5

The table gives details of the marks obtained by a year-group in an examination.

(a) Draw a bar chart to illustrate the information, with a continuous horizontal scale running from 0 to 100.

(b) Draw a frequency polygon of the same data, using the mid-points of the classes.

9 (a) In 1985 an apprentice electrician's 'take-home' pay was £60 a week. His weekly budget was as follows:

Rent, food, heat and light £18
Clothes   £12
Entertainment   £16
Travel   £8
Savings and other items   £6

Draw a pie chart to represent his weekly budget.

(b) Figure 25:5 represents the 'average' family budget in 1985. The 'average family's' net income in 1985 was £6480.

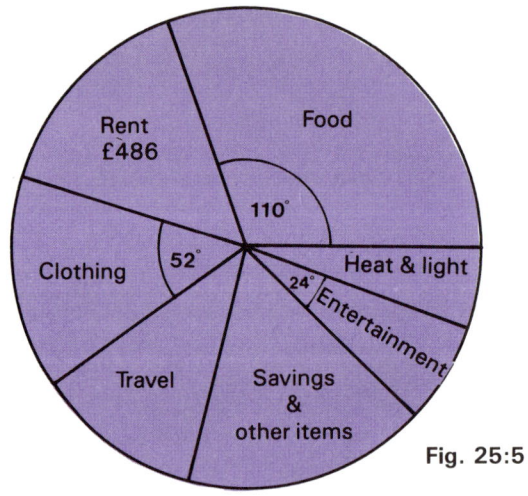

Fig. 25:5

Calculate:
  (i) how much was spent on food
 (ii) what angle is represented by rent
(iii) what percentage of the family's net income was spent on entertainment.

(c) By comparing the two pie charts, comment briefly on the major differences between the two budgets. (SEG)

10 For each pair of variables named below, state whether they are *positively correlated, negatively correlated* or *uncorrelated*.

(a) The heights and weights of adults.
(b) The weights and artistic abilities of 16 year olds.
(c) The height of a tree and its age.
(d) The volume of air remaining in a tyre and the time since the tyre valve was opened.
(WJEC)

11 What kind of correlation is shown by each scatter graph in Figures 25:6 to 25:8?

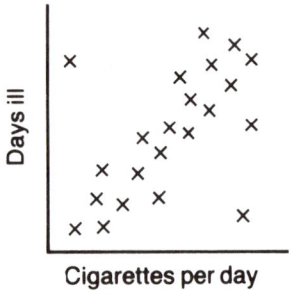

Fig. 25:6

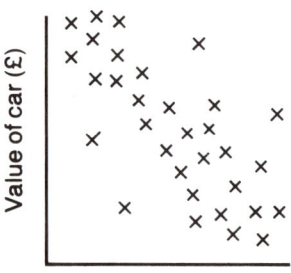

Fig. 25:7

Fig. 25:8

12 Illustrate your answers to question 10 with sketched scatter graphs.

13 (a) Draw a scatter graph for the data in question 4. Draw a line of best fit.

(b) A woman 150 cm tall should weigh between 44 and 52 kg. A woman 180 cm tall should weigh between 62 and 76 kg. Use this information to draw a band across your graph showing the healthy range.

14 The table shows fuel consumption by a car at various constant speeds.

Speed (m.p.h.)	20	30	45	50	55	60	65	70	75	80
Fuel consumption (m.p.g.)	52	56	58	50	47	41	35	31	27	24

Write brief comments on how fuel consumption changes with speed. Illustrate your comments with at least two different kinds of graph, one of which should be a scatter graph. Also explain why the car is unlikely to achieve 58 m.p.g. when it averages 45 m.p.h. on a 100 mile journey.

15 Figure 25:9 shows the cumulative frequency chart for average wages for men and women from a survey of 2000 people.

(a) Taking the lower quartile at 250, the middle quartile (the median) at 500, and the upper quartile at 750, find the median wages and the interquartile ranges for the men and for the women.

(b) Comment on the differences in the overall distribution of the men's and the women's wages.

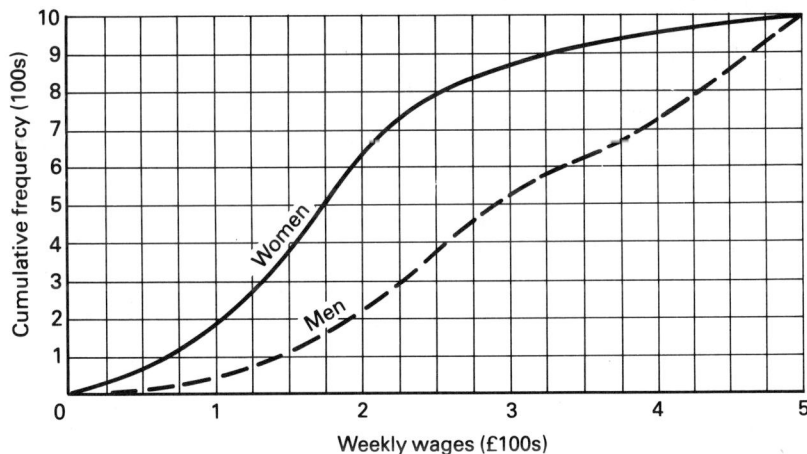

Fig. 25:9

16

Age in years	0–10	11–20	21–30	31–40	41–50
1921 population (100s)	25	19	15	19	16
1992 population (100s)	17	23	30	35	38

Age in years	51–60	61–70	71–80	81–90	91–100
1921 population (100s)	13	10	5	2	1
1992 population (100s)	37	29	19	12	9

The table gives the population of a town in 1921 and 1992.

(a) Redraw the table, giving each population figure as a percentage of the whole town to the nearest 1%.

(b) Redraw the table giving cumulative percentages.

(c) Draw the cumulative frequency graphs for the two populations.

(d) Comment on the differences between the two populations, suggesting possible reasons.

17

Maximum speed (m.p.h.)	0–29	30–39	40–44	45–49	50–54	55–59	60–80
Frequency	27	79	39	28	19	6	2

The table shows the speeds of cars passing a police check point in a 30 m.p.h. zone.

(a) Draw up a cumulative frequency table and graph. State the median, first and third quartiles, and the interquartile range.

(b) The police stop and warn drivers travelling at over 35 m.p.h. They prosecute those exceeding 47 m.p.h. Estimate how many there are in each category.

(c) The next day many drivers remember there was a speed check the day before at the same spot. Add to your graph another cumulative frequency curve to show any possible change in the speeds of 200 cars that day. Give reasons for any changes you make.

18

Price range (£1000s)	31–40	41–50	51–60	61–70	71–80	81–90	91–100
City A frequency	15	21	37	44	66	48	19
City B frequency	0	39	67	89	38	15	2

A newspaper gives the above table, stating it resulted from a random survey of 250 house prices in two cities.

(a) Draw on the same grid two frequency polygons to illustrate the data.

(b) Describe the main differences between the two samples.

(c) Do you believe the data? Why?

*19

Age range	0–9	10–19	20–39	40–59	60–69	70–79
Males	4	6	11	15	24	5
Females	10	12	21	18	39	19

The table shows the results of a survey into how many people watched Wimbledon Men's Finals on TV in 1993.

Draw bar charts to illustrate the data, then join the mid-points of the tops of the bars to form frequency polygons.

Comment on the differences in the two charts. Consider age and sex differences.

*20 The pie charts (Figure 25:10) show the percentages of the population in four age ranges in two countries.

Country A has a population of 2.5 million. Country B has a population of 8 million.

Copy and complete the table.

Age range	0–18	19–45	46–60	61 +
Country A (millions) Country B (millions)				

One country is in the Caribbean, one is in northern Europe. Which do you think is the Caribbean country? Why?

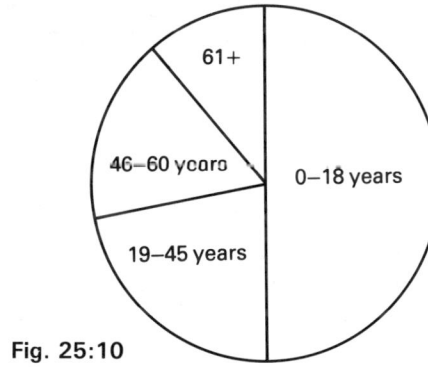

Country A

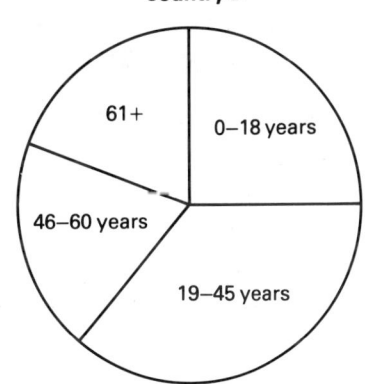

Country B

Fig. 25:10

*21 Two hundred oranges were weighed to the nearest 5 g. The weights were as shown in the table.

Weight (g)	70/75	80/85	90/95	100/105	110/115	120/125	130/135
Frequency	9	21	31	37	44	39	19

(a) Draw up a cumulative frequency table and draw the cumulative frequency curve.
(b) Use your curve to estimate the median weight.
(c) Approximately how many oranges weigh more than 105 g?
(d) On your curve show the lower and upper quartiles. State the interquartile range.

*22 The cumulative frequencies for the ages of the brides and bridegrooms at 100 weddings are given in the table.

Age	Under 20	Under 25	Under 30	Under 36	Under 40	Under 45
Brides	33	79	93	98	100	100
Grooms	7	42	81	91	96	100

(a) How many brides were under 25 years of age?

(b) How many grooms were not under 25 years of age?

(c) How many of these persons were under 20 years of age?

(d) How many more brides than grooms were under 30 years of age?

(e) On one set of axes draw the cumulative frequency curves for both the ages of the brides and for the ages of the grooms.

(f) Estimate from your graph (i) the median age for the brides, and (ii) how many more bridegrooms than brides were at least 27 years of age.

(SEG)

*23 State the interquartile ranges for the two sets of data in question 22. How do the different ranges illustrate the differences in the data?

H24 Figure 25:11 is a histogram. The area, not the height, of the bars represent the frequency (written in the bars in this figure). State what numbers should be placed at *a* and *b* on the frequency density axis, and at *c* and *d* on the two right-most bars.

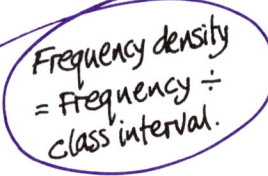

Frequency density = Frequency ÷ class interval.

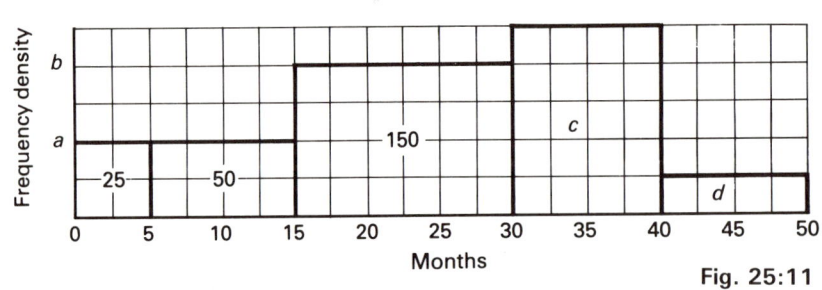

Fig. 25:11

H25 Represent the data in question 20 with two histograms. If you did not do this question you will need to first copy and complete the table.

Take 61+ as meaning 61–100, so use a horizontal continuous scale from 0 to 100.

State the unit area, and number the frequency density axes.

26 The histograms in Figure 25:12 show the pocket money paid to a sample of 11-year-olds and a sample of 14-year-olds.

(a) How many 11-year-olds and how many 14-year-olds were in the survey?

(b) How many of each age received over £3?

(c) Describe the differences between the distributions.

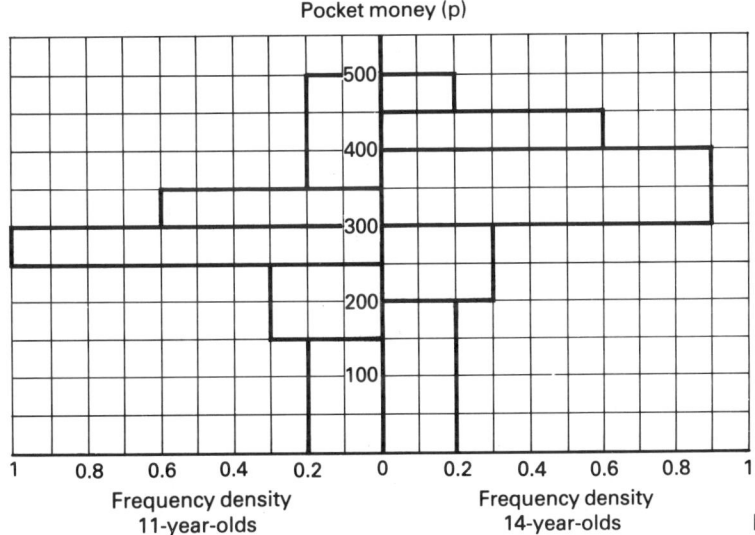

Fig. 25:12

27 The histograms in Figures 25:13 and 25:14 show the result of surveys into the number of people watching television on two Saturdays. Describe the differences between the distributions, suggesting possible reasons for them.

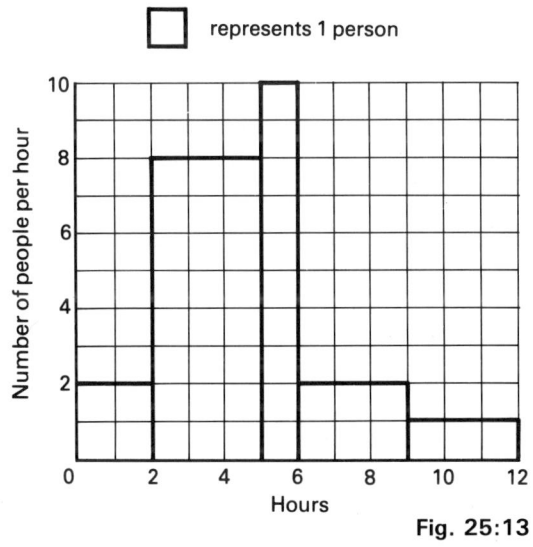

Fig. 25:13

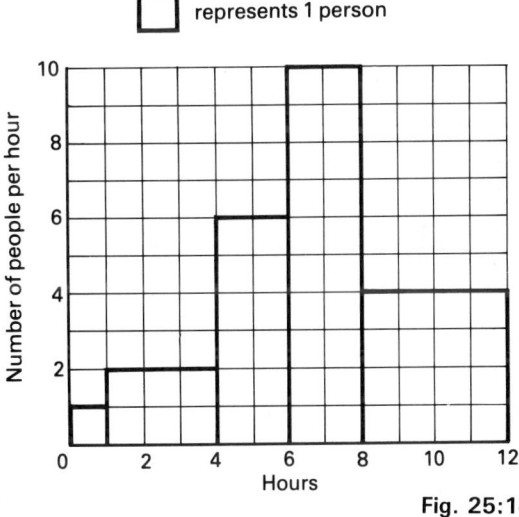

Fig. 25:14

28

Age (years)	16–19	20–24	25–34	35–44	45–60	Total
Thousands	876	192	56	18	8	1150

The table shows the number of full-time female students in Holland.

(a) What is the most and least possible number of total students if each frequency is rounded to the nearest 1000, with 500 rounding up?

(b) A histogram of the data is to be drawn using a scale of 2 cm to 5 years and 1 cm^2 to represent 10 000 students. Calculate the highest integral number required on the frequency density axis, and the heights of each column to the nearest 0.1 cm.

(c) Comment on the problems in representing the data with a graph.

29 Write a short article for the financial pages of a newspaper to go with Figure 25:15, a Carroll diagram/scatter graph. ('Lower Base Rates' refers to the Bank of England's official rate of interest. 'ERM' is the European Exchange Rate Mechanism).

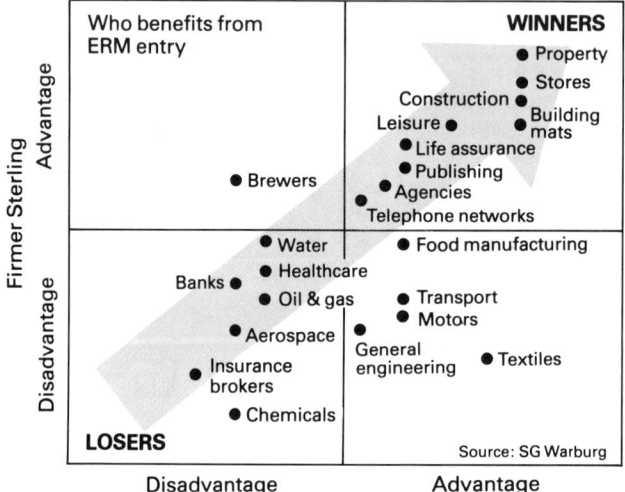

Fig. 25:15

Mean
Mode
Median

Range
(Standard deviation)

● You need to know . . .

Averages 1 to 4 (page 273)
Dispersion 1 (page 274)

1   'Lies, damned lies, and statistics.'   (Mark Twain)

Work out the mean, the mode, the median, and the range for each of the following maths tests, then comment on your answers.

Pupil code	a	b	c	d	e	f	g	h	i	j	k	l	m	n	o	p
Test 1	0	1	1	2	2	2	3	3	3	3	4	4	4	5	5	6
Test 2	0	3	3	3	3	3	3	3	3	3	3	3	3	3	3	6
Test 3	0	0	1	1	2	2	3	3	3	3	4	4	5	5	6	6
Test 4	0	0	0	0	2	3	3	3	3	3	3	4	6	6	6	6

2   What kind of data can never have a mean or a median, but can have a mode?

3   Bob says that he sleeps 8 hours a night on average. What do you think he means by this? How might he have arrived at the figure '8'?

4   Five researchers each ask ten people how much they earned last month. The replies are listed below. The researchers calculate the value of x for the headline 'Earnings average £x a month'. Calculate the three averages for each, then advise them which average they should use if:
(a) they are trying to be honest
(b) they are trying to make people seem badly paid.

Ann   £0, £0, £0, £0, £0, £0, £0, £0, £0, £10 000

Brenda   £980, £1020, £990, £1010, £1000, £995, £985, £1005, £999, £1016

Camille £0, £0, £0, £0, £0, £100, £1000, £2000, £3000, £3900

Davian   £0, £0, £500, £500, £500, £500, £2000, £2000, £2000, £2000

Erica   £10, £100, £100, £100, £200, £1000, £1000, £1000, £3200, £3290

5 Newspapers often talk about 'the average' without saying which average they are referring to.

Which average is used in the statements in Figure 26:1?

(a)

Average wage at Thumbles is £150 per week

Fact
Thumbles:
50 workers on £100 per week
1 owner on £2650 per week

(b)

On average, 15 out of 20 people use Zipsuds

Fact
Researchers for Zipzuds asked 160 people if they used Zipzuds. Every 20 people they recorded their findings:
Users of Zipzuds: 1/20, 0/20, 2/20, 15/20, 5/20, 3/20, 4/20, 15/20

(c)

Average Poll Tax in Bakerton is £112

Fact
In 1990, all 101 residents of Bakerton were asked what their Community Charge bill was. Fifty of them were still at school and paid nothing. Fifty paid the standard charge of £560 each, and one, a student, paid £112

Fig. 26:1

6 If the researchers in question 4 pooled their data, would any average give an honest impression of the wages earned?

7 The mean, the median, and the mode of seven numbers is 4. Write three possibilities for the numbers.

8

Number living in house	1	2	3	4	5	6	7	8
Number of houses	15	25	40	55	18	12	4	1
Total number of people	15	50						

The table shows the result of a council survey. Copy and complete the table, then find the mean, mode, median and range.

9 (a) Copy and complete the table for the answers given by a class to the number of cats and dogs they had at home.

Replies: 1, 0, 1, 3, 0, 2, 3, 1, 1, 6, 1, 7, 2, 1, 1, 2, 3, 0, 1, 2, 6, 2, 0, 3

No. of pets	0	1	2	3	4	5	6	7
No. of pupils								

(b) Find:
  (i) the number of pupils in the class
  (ii) the total number of cats and dogs in their homes
  (iii) the modal number of pets
  (iv) the mean number of pets per pupil
  (v) the median number of pets.

10 A new pupil joins the class in question 9. She has 4 pets at home. What is the effect on:
(a) the modal number of pets   (b) the mean number of pets per pupil
(c) the median number of pets?

11 Marks out of 100 for 60 pupils:
25, 34, 35, 36, 16, 7, 27, 38, 37, 37, 17, 49, 28, 39, 93, 46, 57, 68, 10, 22,
67, 45, 58, 71, 19, 25, 41, 42, 57, 76, 65, 49, 48, 43, 60, 75, 56, 31, 46, 50,
65, 79, 19, 19, 39, 45, 50, 67, 84, 23, 56, 57, 58, 69, 86, 28, 47, 44, 59, 66

Mark	Tally		Total pupils	Middle mark	Total marks
0–9	I		1	4.5	4.5
10–19	I			14.5	
20–29	II				
30–39	⅃⊢⊤ I				
etc.					
up to					
90–100					
		Totals			

(a) Copy and complete the grouped frequency table. We have tallied the first ten marks for you.

(b) Find:
  (i) the modal class
  (ii) the class containing the median mark
  (iii) an approximation for the mean mark
  (iv) the range of the marks
  (v) the interquartile range of the marks (from the 15th to the 45th).

(c) Do you think the teachers would be pleased with the results? Why?

12 The times taken by 100 pupils in a school cross-country are
17:18 (17 minutes 18 seconds), 15:22, 17:35, 16:07, 18:46, 15:51, 19:15, 16:33,
19:27, 16:34, 20:02, 16:42, 15:03, 13:52, 16:36, 13:59, 17:25, 16:02, 14:12,
13:35, 16:04, 14:54, 18:09, 15:25, 16:00, 14:42, 17:47, 15:36, 16:01, 14:44,
19:51, 16:39, 17:04, 15:33, 17:40, 15:17, 16:09, 15:07, 19:01, 16:08, 19:07,
16:29, 19:05, 16:14, 19:36, 16:37, 19:04, 16:12, 19:30, 16:35, 14:17, 13:50,
16:20, 15:35, 17:30, 15:07, 17:18, 15:00, 15:23, 14:26, 20:04, 16:44, 16:42,
14:48, 15:52, 14:39, 18:36, 15:49, 17:58, 15:25, 17:12, 15:35, 16:38, 15:10,
16:41, 15:13, 20:37, 16:49, 17:21, 15:57, 16:36, 13:59, 18:33, 15:47, 16:49,
15:18, 18:05, 16:24, 21:06, 16:50, 17:13, 15:36, 17:15, 15:43, 18:27, 15:44,
18:31, 15:46, 17:20, 15:56

Time taken	Tally		f	x	fx
13:30–13:59 14:00–14:29 etc.		*f means 'frequency'.* *x means 'middle mark'.*			

(a) Copy and complete the table.

(b) Find the modal class, the median class, and an approximation for the mean.

*13 The average wage at Saunders' Saw Mill is £100. The following table shows possible wages for the three workers at the mill.

Name	Wages A	Wages B	Wages C	Wages D	Wages E
Mrs Clarke	£100	£105	£10 000	£1000	£200
Mrs Rose	£100	£100	£100	£100	£50
Mr Bryden	£100	£95	£100	£0.50	£50

(a) What is the mean, median, mode and range for each set of wages, A to E?

(b) Is it true for each set that 'the average wage' is £100?

(c) For which sets of wages do you think it is misleading for Saunders' to advertise that their workers earn an average wage of £100?

*14 A shop conducts a survey to find how often its customers used the shop in a month. The results were recorded in the following table.

No. of times	1	2–4	5–7	8–10	11–15	16–20	21–25	26–30
Frequency	32	27	43	21	29	16	8	15

Find:
(a) the number of customers who answered the survey
(b) the modal class
(c) the class in which the median lies
(d) the mean number of times, correct to the nearest whole number, taking each frequency as that for the middle of the class.

15 Class 5A calculates their mean income is £2.75 a week. Class 5B calculates theirs as £3.25 a week. Class 5A has 25 pupils and class 5B has 35 pupils. What is the mean income of the combined classes?

16 The mean of five numbers is 18. Three of the numbers are equal and the other two have a mean of 10. Find the equal numbers.

17 Which five positive integers have a mode of 13, a median of 5 and a mean of 7?

18 Find the four positive integers with a mean of 11, a mode of 10, and a range of 20.

19 Sketch graphs to show the shape of the frequency against marks curve likely to be obtained when a maths test designed for average pupils is set to:
(a) a mixed ability class   (b) a class of pupils all very good at maths
(c) a class of pupils who find maths difficult.

H20 Many questions in this exercise have shown you that just knowing the three averages and the range often does not give a good picture of the data. Knowing the interquartile range helps, as this removes exceptional data at each extreme, but the most helpful guide to how 'normal' the data is, is to find the **standard deviation**. The symbol for this is σ. The reference notes on page 274 explain how to calculate σ although it is much easier to use a statistics mode calculator.

Figure 26:2 shows the normal curve.

We expect 68% of the scores to be within 1σ of the mean, about 95% to be within 2σ of the mean, and nearly 100% to be within 3σ of the mean, so you can easily see how normal the curve is, and there-fore how representative the calculated averages are likely to be.

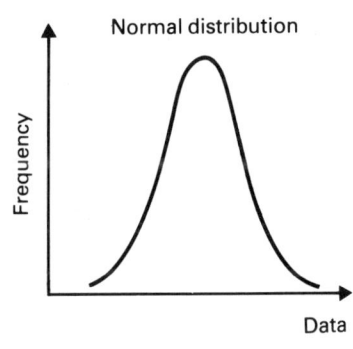

Normal distribution

Frequency

Data

Fig. 26:2

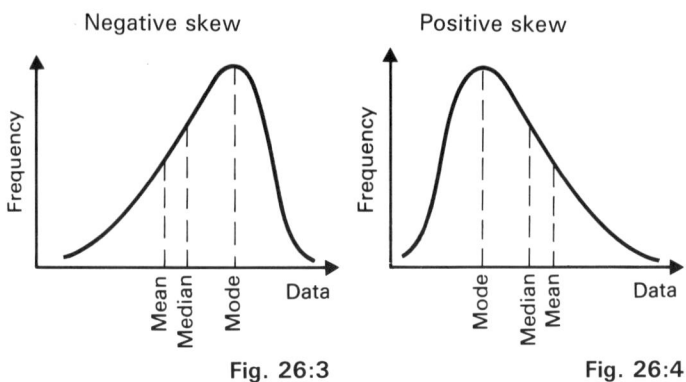

Negative skew

Positive skew

Fig. 26:3

Fig. 26:4

Figures 26:3 and 26:4 show skewed frequency curves.

When skewed to the left, more of the data is above the mean than below it.

When skewed to the right, more of the data is below the mean than above it.

Calculate the standard deviations for the data in questions 4 and 9. Are any of them normal distributions?

H21 Find the standard deviations for the data in question 11. What percentage of each data lie in the 1σ, 2σ and 3σ ranges? What do your answers tell you about the data?

---

22 Find σ for 1, 2, 2, 3. What is the effect on the mean and the standard deviation:
(a) when each piece of data is increased by 2
(b) when each piece of data doubles?

23 The data below shows the energy levels, in kilocalories per 100 g, of ten different snack foods, such as crisps and peanuts.

440, 520, 480, 560, 572, 550, 620, 680, 545, 490

(b) Calculate the mean and standard deviation of the energy levels of these snack foods.

(b) The energy levels, in kilocalories per 100 g, of ten different breakfast cereals had a mean of 350 kilocalories with a standard deviation of 28 kilocalories. Which of the two types of food show a great variation in energy level? Give a reason for your answer. (WJEC)

24 Karen is keen to improve her scores for darts. She always aims to hit the bull (at the centre of the darts board) and her last twelve scores were:

32, 25, 56, 52, 20, 60, 19, 36, 36, 20, 56, 20

(a) Calculate the mean and standard deviation of these scores.

(b) Karen decides to change her tactics and to aim to hit treble twenty (towards the top of the board). The mean of her next twelve scores is 33.75 and their standard deviation is 21.4. Karen is unfamiliar with statistical terms. What advice would you give her about her tactics and what explanation would you give? (MEG)

25 Gemma surveys her classmates to find their favourite number (from 1 to 100). She then calculates the mean to be 24.$\dot{6}$, the median to be 13, the mode to be 7, and the range to be 98. Comment on her decision to calculate these averages.

26 Miss Margaret said she would not be happy until every pupil's National Curriculum result was above average. Would she ever be happy? If not, what is the closest she can come to it?

27 Describe as far as you can the most likely wage structure of the following firms.

Firm	Mean wage	Modal wage	Median wage	Range	σ
A	£200	£200	£200	£0	£0
B	£200	£450	£50	£430	£204
C	£200	£200	£200	£250	£32
D	£200	£490	£10	£486	£237
E	£200	£5	£5	£975	£390
F	£200	£100	£150	£250	£105
G	£200	£200	£200	£400	£45

28 Which distributions in question 27 are normal, which are positively skewed and which are negatively skewed?

Surveys   **(Sampling)**
Questionnaires

●   ## You need to know . . .

Collecting data 1 to 3 (page 275)

There are no answers for this chapter in this book! Ask your teacher!

1. A school tuck-shop sells apples, choc bars, crisps, cola, buns, and sweets.

   Draw up a suitable data collection sheet to find out how many of each are sold in a week.

2. Suggest an issue of interest to you which could form the basis of a survey and for which data would need to be collected. Outline briefly how you would go about collecting the data.

3. You have been asked to observe a car park to find out how many drivers use the park and for how long they leave their cars. Design a simple observation sheet which could be used to collect the data.

4. Your council will not agree to a footbridge so that people can cross over a busy road safely. You want to convince them it is necessary. Design a suitable observation sheet for a team of researchers to use to gather data to back up your campaign.

5. Design a questionnaire to survey people's opinions about having a new industrial estate on the outskirts of their town.

6. A local shop has recently closed down and your teacher is considering resigning and setting up a business. What would help your teacher decide what sort of business it should be?

7. One of the things you might have suggested in question 6 is using surveys. Suggest three things you could investigate, then outline how you would go about one of them.

8. What can you find wrong with the following survey plans to find out what sort of shop would be best?

   (a) One afternoon phone up the first twenty people on random pages of the telephone directory who live near the shop and ask them what is their favourite kind of shop.

   (b) Make a list of every kind of shop in the area then suggest one that sells something completely different.

   (c) Spend every Saturday afternoon for a month outside the empty shop, stop everyone who passes, and ask them what they think the shop should sell.

   (d) Spend every evening for a week visiting homes in the area, asking every woman how much money she spends in local shops each week.

9 What is wrong with asking the following questions in a questionnaire? Suggest better questions.

(a) What should be done about the terrible hooligans on our streets at night?

(b) Taking everything into consideration would you say that the increase in the average age in our geographical area is likely to have a damaging effect on the ability of our social services to function with their accustomed efficiency?

(c) Do you smoke too much?

(d) What are your favourite television programmes?

(e) Have you been seriously ill in the past year?

10 Design a questionnaire which asks people to put in order of preference the things they would like to do on holiday, given a choice of six alternatives.

H11 Your school is conducting a survey to investigate pupil attitudes to 'compulsory school uniform' and you have been asked to choose a sample of pupils for the survey. State, with reasons, two factors which you consider to be important in your choice of sample so that the survey represents fairly the opinions of all the pupils in your school.

(WJEC)

H12 You are investigating peoples' feelings about closing down a local church. Explain how you might obtain a stratified sample that ensures the views of all sections of the community are represented in your survey.

H13 In your research for question 12 you might have used a completely random sample instead. How could you then have selected the people for your survey?

H14 Suggest some ways of selecting a sample for the survey in question 12 that would certainly not be either random or representative of the whole community.

H15 Jerome reckons that people who have more brothers than sisters are more likely to be able to spell correctly.

How would you go about seeing if his strange theory is correct? Include suitable data collection sheets, and say how you would choose a sample population. How could you present your conclusions in a clear and interesting way?

H16 The governors of a school decided that they wanted to change the colour of the school uniform.

Two pupils carried out a survey to find out what colour pupils would like. They asked their friends and people in their classes. Figure 27:1 is a page from their notebook.

(a) Design a better way to collect the information.

(b) When the pupils had completed their survey they realised that their results might not be very useful. Why may this be true? Suggest ways they could improve their survey.
(NEAB)

Black Red Grey Red Grey
Blue Blue Brown Black Green Black
Blue Red Black Green Blue Red Red
Blue

Fig. 27:1

143

17 Imagine that your headteacher decides that pupils should do more homework. (We only said 'imagine'!) Your friends suggest a strike, but you decide to carry out a survey and present a report to convince your headteacher that pupils already do enough homework. At the same time your headteacher carries out her own survey so that she can convince the parents that she is right in her view.

Suggest how both you and your headteacher might draw up questionnaires, organise surveys, and present a final report, to bias the views in your respective favour without actually giving any false information.

The probability scale
Single events

Mutually exclusive and independent probabilities by outcomes

● **You need to know . . .**

Probability theory 1 and 2 (page 276)

1 From a full pack of 52 playing cards, one card is drawn. What is the probability that the card is:
(a) a black suit (b) a diamond (c) a king (d) the ace of spades?

2 In one inner city the police calculate that householders have a 1 in 4 chance of being burgled during any one year. What is the probability that a householder will not be burgled during any one year?

3 Estimate, with reasons, the probability that:
(a) you will be late for a lesson next week
(b) there will be a general election in the next two years
(c) you will see a police car in the next 24 hours
(d) you will be absent from school through illness for at least two days next term.

4 Which of the following probabilities can be calculated without observation, experiment, or research into past records?

A Four people picked at random from a list of electors aged 18 to 60 being male.

B Six car drivers stopped at traffic lights being male.

C A fair die coming up 6 three times in a row.

D A biased die coming up 6 three times in a row.

5 Explain how a ball-point pen manufacturer could keep a check on the quality standard of her pens.

6 Explain why the chance of scoring a square number with a fair die numbered 1 to 6 is $\frac{1}{3}$, but the chance of a team winning a match, rather than drawing or losing it, is not $\frac{1}{3}$.

7 In tests 9 out of 10 people chose red as their favourite colour. If you ask ten people tomorrow what is their favourite colour, how many would you expect to answer 'Red'?

8 The table shows the report of Centrepoint night shelter in London on where the young people staying there came from.

North of England	30%	Scotland	15%
London	20%	South-east England	10%
Eire	15%	Northern Ireland	4%

What is the probability that someone in the night shelter who will not give their address comes from:
(a) Eire   (b) London or the South-east   (c) out of London
(d) none of the areas listed?

9 A coin tossed three times gives the following results in ten trials:
TTH, HHH, THH, HHT, HTH, HTT, HTH, HTH, HTT, THT

(a) Would you have any reason to be surprised at these results?

(b) After 100 trials about how many heads would you expect to have scored altogether?

(c) Which is more likely to happen in a trial of four throws: HHHH, HTHT, or TTHH?

10 By observing a certain set of traffic lights as I approach them each day I estimate the following probabilities:
P(amber) = 0.1, P(green) = 0.3, P(red with amber) = 0.1
(a) What is P(red)?   (b) What is P(green or amber)?

11 Explain why a cat being white and a cat being deaf are not-independent events.

12 Why is the following reasoning incorrect?

Two tossed coins can fall in three ways:
both heads, both tails, one of each.
Therefore, P(two heads) = $\frac{1}{3}$.

13 Three cards, marked with the number 7, 8 and 3 respectively, are placed in a box. Two cards are picked at random.

(a) Write down all possible numbers that result, e.g. a 7 and then a 3 gives 73.

(b) What is:
(i) P(78 is drawn)   (ii) P(an odd number is drawn)?

14 A small ferry can carry a maximum of 6 cars and a maximum of 2 coaches. Figure 28:1 is the possibility-space diagram showing all combinations of up to 6 cars and up to 2 coaches. However, there are the constraints that if 1 coach is aboard then only 4 cars can be fitted in, and if 2 coaches are aboard only 2 cars can be fitted in.

Copy the diagram and ring the dots that represent possible ferry loads. (The ferry does not sail if there is no vehicle aboard.)

Ferries at sea are inspected at random. Assuming that all possible loads are equally likely, what is the probability of the ferry holding:
(a) 2 coaches   (b) 3 or more cars   (c) more cars than coaches?

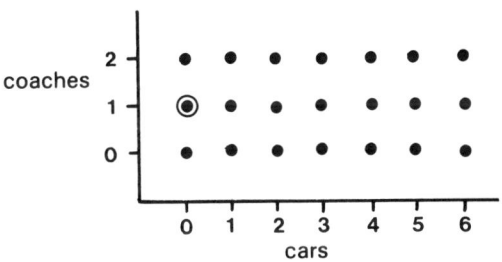

Fig. 28:1

15 Copy and complete the tree diagram (Figure 28:2). You will need 16 lines of your book for the final column. When you have completed the diagram, use it to find the probability of an equal number of heads and tails in four throws of a coin.

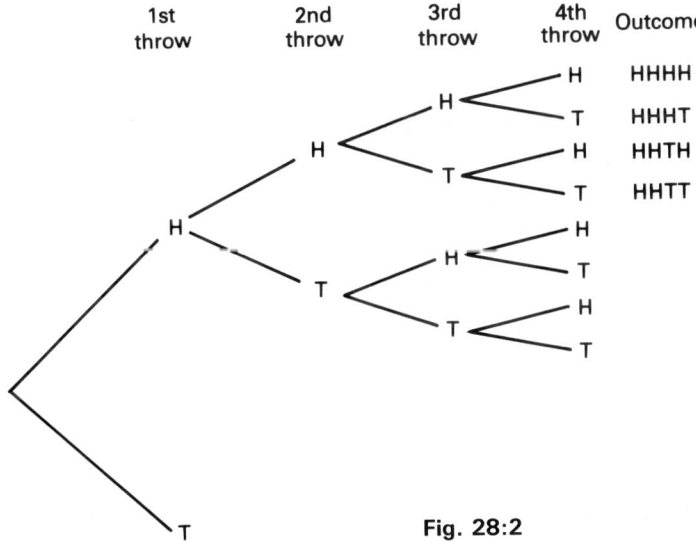

Fig. 28:2

16 Use the tree diagram you drew in question 15 to find the chance that a family of four children is made up of:
(a) all boys   (b) two boys and two girls   (c) at least one girl.

*17 Two fair dice are thrown. The possible scores are from 1 to 6 on each, and the total score is found by adding the two scores.

(a) Draw up a table to show all possible scores.

(b) Use your table to find the following total-scores probabilities.
(i) $P(3)$ (ii) $P(7)$ (iii) $P(16)$ (iv) P(more than 8) (v) P(less than 10)

18 A box contains two pairs of gloves, one pair of medium (M) size and one pair small (S).

(a) Two gloves are taken at random from the box. List all the possible results.

(b) What is the probability that a pair of gloves is picked?

(c) What is the smallest number of gloves you would need to pick to be certain of having picked a pair?

19 A six-sided die is biased so that a score of 6 is twice as likely as a score of 3, which in turn is twice as likely as a score of 2. The scores 1, 2, 4 and 5 are equally likely.

(a) Explain why the probability of a score of 1 is 0.1.

(b) Calculate the probability of each of the other scores.

(c) Explain why the total of the probability decimals must be 1.

20 The East London Squash and Badminton Club has 120 members. Of these 60 play squash, 72 play badminton and 20 play neither sport.

(a) How many members must play both sports?

(b) Copy and complete the Venn diagram in Figure 28:3 to show the numbers of members in each region.

(c) A member is chosen at random to be the club secretary. Find the probability that the person chosen:
(i) plays neither sport (ii) plays only squash.

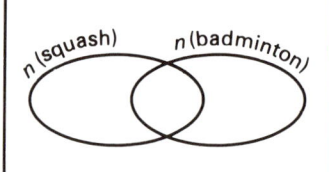

n (sports club)

n (squash) n (badminton)

Fig. 28:3

(NEAB)

21 In a game a £1 coin is hidden under one of three cups. You guess under which one it is. The operator (who knows) then turns over one of the other cups to show you it is not under that one. She then asks you if you want to change your guess. Should you, or does it not matter?

You will find it helpful to list all the possible situations, and what happens if you stick to your first choice or change. Like this:

You choose   Revealed cup   Third cup
    £1   0   0   Change, you lose

Note This should not be confused with the three-cup swindle when the operator uses sleight of hand to make you think an object was under a cup when it was not. You never win that game!

Mutually exclusive   **Independent**
(Not-independent)

● ## You need to know . . .

Probability calculations 1 to 3 (page 277)

1 After a national series of spot checks, traffic police consider that on checking a car the probability of finding a car with an illegally worn tyre is $\frac{1}{10}$, of a faulty light is $\frac{1}{12}$, and of a leaky exhaust is $\frac{1}{8}$.

What is the probability, on a random check, of a car having:
(a) a worn tyre and a faulty light   (b) a worn tyre and a leaky exhaust
(c) all three faults   (d) no faulty lights   (e) at least one fault?

2 In a laboratory making 'silicon chips', 1 in 5 chips on average are expected to be faulty. Three chips are tested at random.

What is the probability of finding:
(a) no faults   (b) none perfect   (c) at least one fault?

3 A coin is biased so that it is twice as likely to fall heads as tails. What is the probability of:
(a) three heads in a row   (b) two tails followed by a head?

4 Stan has a music collection on tape and compact disc. His collection can be summarised as:

	Rock	Heavy metal
Tape	15	5
CD	3	7

One evening Stan asked Fay, his little sister, to fetch him one tape and one CD from his collection. She does this, choosing at random.

(a) State as decimals the probability that:
    (i) the tape she chooses is rock   (ii) the CD is heavy metal.

(b) Draw a tree diagram to illustrate the four different combinations of recording media and music which Fay could have chosen. Write the probability of each choice on the tree, as simplified common fractions.

(c) Find the probability that both of the items chosen would be rock.

(d) Find the probability that at least one of the items chosen would be heavy metal.

(NEAB)

149

*5 Figure 29:1 represents a spinner at a fete. It is a wheel coloured in sections as shown.

What is the chance of the spinner stopping at:

(a) ■ (b) ▨ (c) ▧ or ☐ (d) ■ three times in a row (e) ☐ then ▨ ?

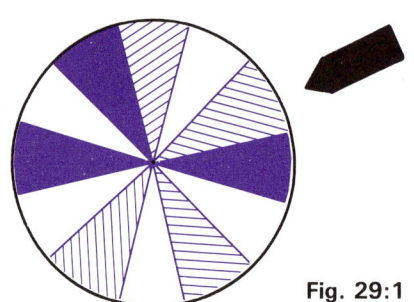

Fig. 29:1

*6 Three cards are dealt from a full pack. What is the probability of dealing:
(a) first an ace (b) first an ace, then a jack (c) first a heart, then a spade
(d) two aces, then a five of diamonds (e) three black cards
(f) all picture cards (ace, king, queen or jack)?

*7 A rat is trained to negotiate a maze set out in a series of blocks. The maze is represented in Figure 29:2. At each junction the rat can turn left (L), right (R) or go straight ahead (S). From past experience it is reckoned that the probabilities of these are:

$P(\text{L}) = \frac{1}{10}$ $P(\text{R}) = \frac{3}{10}$ $P(\text{S}) = \frac{3}{5}$

(a) The rat is in the maze as shown. If it takes the shortest route, what is the probability that it will emerge at:
(i) A (ii) B (iii) C (iv) D?

(b) What is the probability of the rat taking exactly two left turns at three consecutive junctions?

(c) What is the probability that if the rat starts at E it will go straight to A?

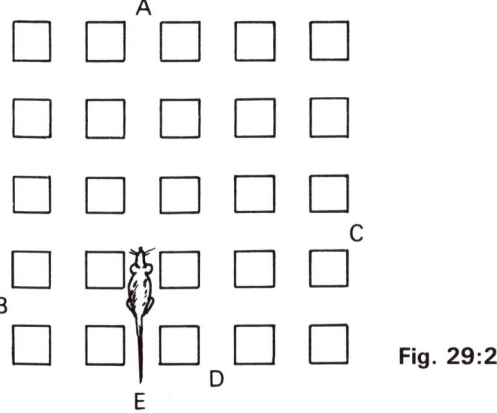

Fig. 29:2

H8 60% of the pupils taught by the Arrow School of Motoring pass their driving test at the first attempt. Each time a pupil retakes the test the chance of passing improves by 10%, i.e. a 70% chance of passing after one failure, etc.

    (a) Copy and complete the tree diagram in Figure 29:3 by writing in the missing probabilities.

    (b) Calculate the probabilities that a new Arrow pupil chosen at random will pass:
        (i) at the second attempt   (ii) at the third attempt.   (ULEAC)

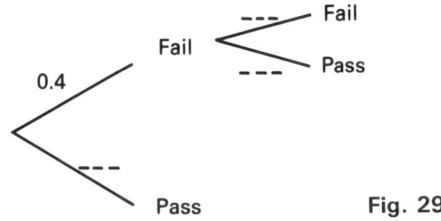

Fig. 29:3

H9 Explain why the 10% improvement quoted in question 8 cannot be entirely true.

H10 A computer is programmed to play chess. After each game that it wins it is a better player. Against an average player it is reckoned to have a 1 in 10 chance of winning if it has not won any previous games, a 1 in 5 chance of winning if it has won one game previously, and a 1 in 3 chance of winning if it has won two games previously.

Copy and complete the tree diagram (Figure 29:4) to show all the possible outcomes of the first three games the computer plays. Hence, or otherwise, calculate the probability of the computer winning:
(a) the first three games   (b) the first game, but losing the other two
(c) one of the first two games   (d) two of the first three games.

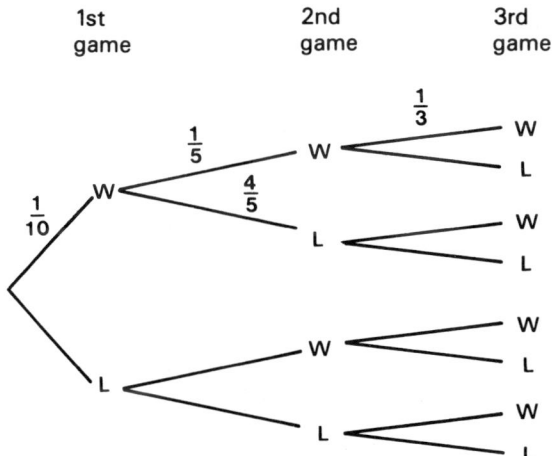

Fig. 29:4

H11   A bag contains four green balls and two amber balls.

(a)   Three balls are taken from the bag, one by one without replacement. Draw a tree diagram showing all possible outcomes, with their probabilities.

(b)   Calculate the probability that exactly two of the three balls drawn will be green.

(c)   What is the smallest number of balls you can take from the bag to have the maximum probability of getting exactly two green balls?

H12   A bag contains five red, one green, two white, and two blue balls. Calculate the chance of picking from the bag, each time without replacement:
(a)   two white balls   (b)   a white and a red ball   (c)   two red balls and one white ball
(d)   a green ball and two white balls   (e)   two balls of the same colour in two picks
(f)   exactly two balls of the same colour out of three picked.

H13   For a pack of cards:

$P(\text{red}) = \frac{26}{52},$ $P(\text{ace}) = \frac{4}{52},$ $P(\text{five}) = \frac{4}{52}$

Two cards are picked. Explain why:

(a) $P(\text{ace and a five})$ is $\frac{4}{52} \times \frac{4}{51} \times 2$ (b) $P(\text{two aces})$ is $\frac{4}{52} \times \frac{3}{51}$

(c) $P(\text{ace and a red})$ is not $\frac{4}{52} \times \frac{26}{51}$, and say what it is.

H14   Two cards are dealt from a shuffled pack. What is the probability of them being:
(a)   both hearts   (b)   one an ace and the other a heart
(c)   the first a black card and the second an ace
(d)   one a spade but the other not a spade?

H15   Before an election two political parties ask some of the 195 residents of a housing estate who they will vote for. The Conservative party asks a fifth of the residents and the Labour party asks a third of the residents.

(a)   What is the probability that any one resident will be asked by:
(i)   both parties   (ii)   neither party   (iii)   only one party?

(b)   Find (i) the least and (ii) the most possible number of residents that would **not** be asked.

H16   In a game of Bingo, Jeff has a card with the numbers 1, 7, 11, 23 and 24 on it, whilst Jane's card has 1, 7, 10, 15 and 22. The numbers are selected by drawing balls from a drum, the thirty balls being numbered from 1 to 30.

(a)   What is the probability that on the first draw Jeff will have one of his numbers called?

(b)   What is the probability that on the first draw neither Jeff nor Jane will have a number called?

(c)   What is the probability that Jane will have one of her numbers called in both of the first two draws?

(d)   Find the probability that Jane has a number called in the first draw but Jeff does not.

(e)   Find the probability that on the first draw Jeff has a number called, but Jane does not, then on the second draw Jane has a number called but Jeff does not.

17 Two people are selected at random from 10 married couples.

(a) How might the selection be made?

(b) Which of the following is most likely to be the outcome, or are they equally likely?
(i) two women   (ii) a man and a woman   (iii) two men

(c) A certain man is selected first. What is the probability his wife is selected second?

(d) What is the probability any married couple are selected?

18 A box contains various solids of different colours. One solid is picked at random from the box. The following colour probabilities are known:

P(white) = 0.65, P(red) = 0.2, P(purple) = 0.1

The following shape probabilities are known:

P(cube) = 0.35, P(sphere) = 0.4, P(cone) = 0.25

(a) How do you know there are only three shapes in the box?

(b) How do you know there is at least one colour in the box not described in the given probabilities?

(c) What is P(cube or cone)?

(d) P(a red shape or a cone) is 0.2 + 0.25. What does this tell you about the cones?

(e) Why cannot P(a white shape or a sphere) be 0.65 + 0.4?

(f) What does your answer to part (e) tell you about the spheres?

(g) Given that the box contains 20 objects, say how many there are of each shape.

(h) What is P(a cube and a cone) in two picks (without replacement)?

(i) What is P(exactly one cone in three picks without replacement)?

(j) Say whether the following probabilities involve outcomes that are:
A Mutually exclusive   **B** Not mutually exclusive   **C** Independent
D Not-independent
(i) P(a red shape or a purple shape in one pick)
(ii) P(a red shape and then a purple shape with replacement)
(iii) P(a red shape and then a purple shape without replacement)
(iv) P(a white shape or a sphere in one pick)
(v) P(a white shape twice in two picks)
(vi) P(a white shape and a sphere in two picks without replacement)

(k) How many objects would you need to pick from the box to have a fifty-fifty chance of having picked a cone?

19 Bag X contains three green balls and seven amber balls. Bag Y contains six green and five amber balls.

(a) If bag X has been chosen, what is the probability of a green ball being withdrawn?

(b) The experiment is repeated N times, where N is large.
  (i) How many times can you expect bag X to be chosen?
  (ii) How many times can you expect a green ball to be chosen from bag X?
  (iii) How many green balls can you expect to be drawn from both bags in N experiments?

(c) If a green ball is chosen, what is the probability that it came from bag X?

(NEAB)

20 (a) Robert buys two tickets in a raffle in which only 25 tickets were sold. There are two winning tickets. Find the probability that he wins:
  (i) both prizes   (ii) neither prize   (iii) exactly one prize.

(b) If Robert buys x tickets in the raffle, show that the probability of him winning both prizes is

$$\frac{x^2 - x}{600} .$$

By considering particular values of x, or otherwise, find the smallest number of tickets that Robert must buy so that he has at least an even chance of winning both prizes.

(WJEC)

21 (a) Copy and complete the following table, using the tree diagram in Chapter 28 question 15 or otherwise. Note that, for example, 2H1T means the number of different ways you can have 2 heads and 1 tail in the outcome, e.g. HHT, THH.

No. of throws	1		2			3			
Results	1H	1T	2H	1H1T	2T	3H	2H1T	1H2T	3T
No. of ways	1	1	1	2	1				

No. of throws	4				
Results	4H	3H1T	2H2T	1H3T	4T
No. of ways					

(b) The results are lines of Pascal's triangle:

```
1 throw                    1     1
2 throws                1     2     1
3 throws             1     3     3     1
4 throws          1     4     6     4     1
```

By continuing Pascal's triangle calculate the probability in six tosses of a coin of:
  (i) 6 heads   (ii) 3 heads and 3 tails   (iii) HHTTTT
  (iv) more heads than tails.

22 You will find reference to Pascal's triangle (see question 21) helpful in the last three parts of this question.

A game is played by two players, Paul and Ina. During the game the players try to score points. The game is won by the first player to score at least 4 points in total and at least 2 points more than the opponent.

When Paul plays Ina the probability of Paul winning each point is $\frac{2}{3}$.

List all the possible scores for a won game in order, starting with a total of 4 points and going up to a total of 8 points.

What is the probability of:
(a) Paul winning a game 4:0   (b) Ina winning a game 4:0
(c) Paul winning a game 4:1   (d) the game reaching 3:3
(e) Paul winning a game 5:3?

(WJEC)

23 (a) How many different three-figure numbers can you make by using the digits 1, 2, 3 without repetition?

(b) Repeat part (a) for four-figure numbers using 1, 2, 3 and 4.

(c) Check that the number of different n-figure numbers possible with n figures, without repetition, is $n!$ (called 'factorial n') where:

$$n! = n \times (n - 1) \times (n - 2) \times (n - 3) \times \cdots \times 3 \times 2 \times 1$$

You may have an $\boxed{x!}$ key on your calculator.

We can also think of the formula in this way: 'I have a choice of n digits for the first place, then $(n - 1)$ are left to choose from for the second place, making $n \times (n - 1)$ ways possible, then $(n - 2)$ for the third place, and so on until I only have 1 digit left.'

(d) We call the number of different arrangements of r objects chosen from n objects their **permutation**. It is written $^{n}P_{r}$ and is calculated by:

$$n \times (n - 1) \times (n - 2) \times (n - 3) \times \cdots \times (n - r + 1)$$

Show that this is the same as $\dfrac{n!}{(n - r)!}$.

Check that the formula calculates how many different numbers you can make using three digits chosen from 1, 2, 3, 4 and 5, without repeating a digit.

24 (a) You can only use the figures 1 and 2. How many different numbers can you make if they have:
   (i) two digits   (ii) three digits   (iii) four digits?

(b) Find a rule for the answers to part (a).

(c) How many different numbers can you make using 1 and 2 if they have:
   (i) five digits   (ii) six digits   (iii) seven digits?

25 (a) You have to select three letters from A, B, C, D and E. What **different** selections can you make? (ABC is not a different selection to BCA.)

(b) When we worked out ^nP_r in question 23 we counted selections like ABC and BCA as different. Now we are counting these as the same, so there are fewer total selections. We call these different selections **combinations**.

The number of combinations of n objects from r objects is written ^nC_r and is worked out by:

$$\frac{n!}{(n-r)! \times r!}$$

Compare this with the ^nP_r formula given in question 23.

Check the ^nC_r formula gives the correct answer for the number of different selections you made in part (a).

A knowledge of the ^nP_r and ^nC_r formulae can save a lot of time in harder probability questions, especially when the choices involve a lot of alternatives.

30 Circle geometry

(Perpendicular bisector
    of a chord)
(Tangent and radius)
(Angles in a circle)

(Alternative segment
    theorem)
(Intersecting chords)

● **You need to know . . .**

Circle geometry 1 to 6 (page 265)

Note: In the diagrams in this chapter the centre of a circle is called O and is marked with a large dot.

1 Link Figures 30:1 to 30:5 with the following facts:

A ⊥ bisector of a chord is a diameter

B ∠ at centre = 2∠ at circumference

C ∠s in same segment are equal

D ∠ in semicircle = 90°

E Opposite ∠s of cyclic quad. are supplementary.

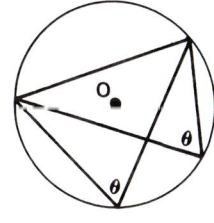

Fig. 30:1

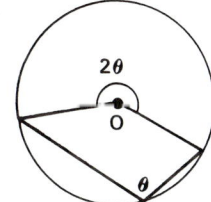

Fig. 30:2

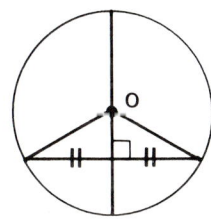

Fig. 30:3

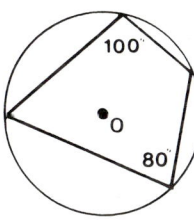

Fig. 30:4

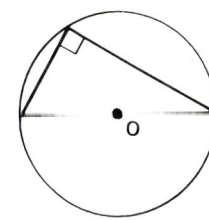

Fig. 30:5

2 Copy Figure 30:6 and show how to complete it to find the centre of the circle.

3 Calculate the shortest distance from the centre of a 10 cm-radius circle to a 16 cm chord. (Use Pythagoras' theorem.)

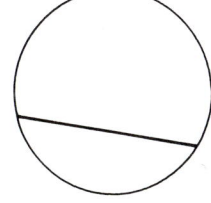

Fig. 30:6

4 In Figure 30:7, AB and CD are straight lines. E may, or may not, be the centre.

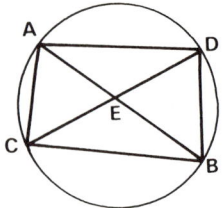

Fig. 30:7

(a) If E *is* the centre and ∠DEB = 80°, state the size of ∠BAD.

(b) If ∠BDC = 48°, state the size of ∠BAC.

(c) If AB is a diameter and ∠BAD = 42°, state the size of angle ABD.

(d) If ∠ACB = 100°, state the size of ∠ADB.

(e) If AD//CB and ∠BCD = 40°, state the size of ∠ABC.

5 Calculate the value of x in Figure 30:8.

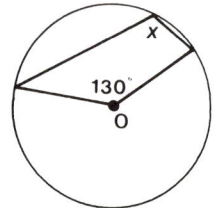

Fig. 30:8

6 For Figure 30:9:

(a) name two radii

(b) name two tangents

(c) state the size of:
(i) angle ASO   (ii) angle OTX
(iii) angle SOT   (iv) angle SOT reflex.

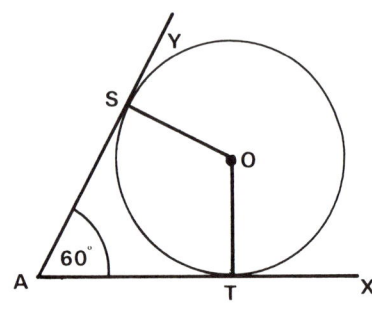

Fig. 30:9

7 (a) If in Figure 30:10 ∠RST = 110°, calculate the four angles of the cyclic quadrilateral.

(b) If ∠P = ∠R, calculate ∠RST.

(c) If PQ//SR (as well as PS//QR) explain why parallelogram PQRS must be a square or a rectangle.

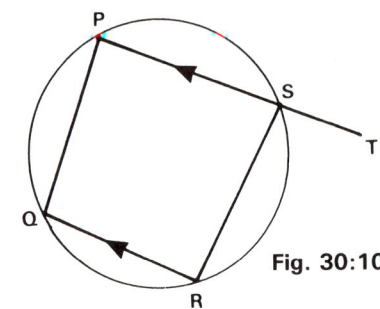

Fig. 30:10

8 Two chords AB and CD cross at E. If AE = AC, prove that DE = DB.

9 PR is a diameter of a circle and PQRS is a cyclic quadrilateral with ∠QPR = ∠RPS. Prove that ∠QRP = ∠SRP.

10 ABC is an equilateral triangle with its vertices on a circle. The bisector of ∠A meets the circle at D. Prove that AD is a diameter.

11 In cyclic quadrilateral ABCD, BA = BC and ∠ADB = 40°. Calculate ∠ABC.

12 Calculate the angle formed on a clock-face by joining the point representing 6 hours to the points representing 4 and 9 hours. (Use '∠s at centre and circumference'.)

13 In Figure 30:11, AX, BY and CZ are the altitudes of △ABC.

(a) Why must B, Z, Y and C be concyclic points (that is, points on the circumference of the same circle)?

(b) Why must B, Z, P and X be concyclic points?

(c) Name four other sets of concyclic points.

(d) Draw a large copy of Figure 30:15 and draw the six circles.

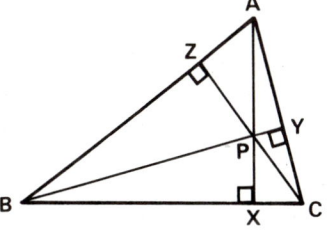

Fig. 30:11

14 For Figure 30:12:

(a) state the size of
(i) ∠ATC   (ii) ∠CBT.

(b) calculate
(i) ∠BTC   (ii) ∠BCT   (iii) ∠BDT.

Angle ATB will always equal angle BDT. This is called the **alternate segment theorem**. See Figure 30:13.

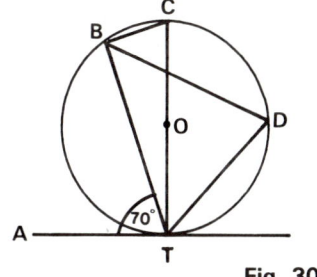

Fig. 30:12

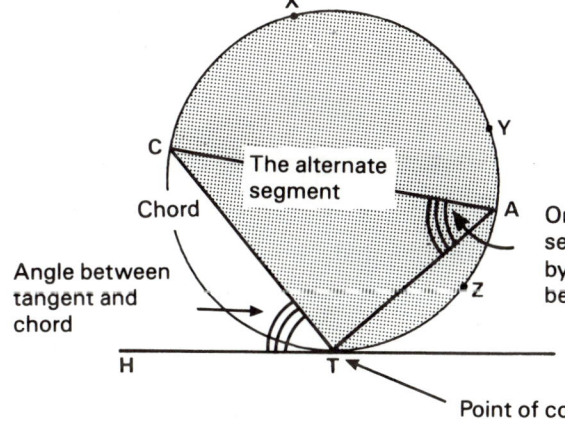

C
Chord
Angle between tangent and chord
H   T
The alternate segment

One of the angles in the alternate segment. You could draw many more by joining C and T to any point between C and T like X, Y or Z.

Tangent

Point of contact   Fig. 30:13

15 In Figure 30:14:

p = t (alternate segment theorem, tangent TAM, chord AC).

Write in a similar way the angle-pairs starting with:
(a) v   (b) r   (c) u.

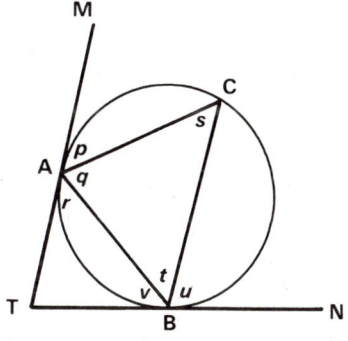

Fig. 30:14

16 Prove that angle QPT in Figure 30:15 is a right angle for all values of ∠VTU. (**Hints:** Let ∠VTU = $\theta°$; then find angles TPV, PVT, PTV, PQV and QPV in terms of θ.)

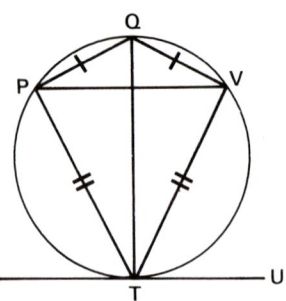

Fig. 30:15

17 In Figure 30:16, ADE is straight, CBT is a tangent, but EB is not. Prove AC//BE.

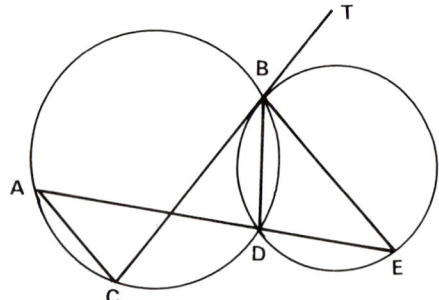

Fig. 30:16

18 In Figure 30:17, XZ is parallel to RS. Prove that YT bisects ∠RTS.

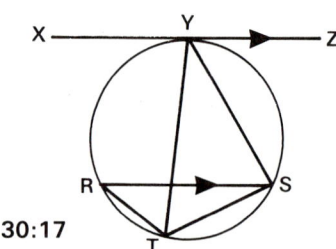

Fig. 30:17

19 TC is a tangent to a circle at C. CB is a chord such that CB = CT. Line BT crosses the circle at A. Prove that △CAT is isosceles.

20 TA and TB are tangents touching a circle at A and B. Chord BC is drawn parallel to TA. Prove that △ABC is isosceles.

21 AB and CD are two chords of a circle, intersecting at P, which is not the centre of the circle. Explain why △APC must be similar to △DPB and show that AP × PB = CP × PD.

This is called the **intersecting chord theorem**. The product AP × PB is sometimes called 'the rectangle contained by the two parts (AP and PB) of the line AB'. Can you see why?

22 If, for the circle in question 21, AB = 8 cm, AP:PB = 1:3, and CP = 3 cm, calculate PD.

23 PQ is the diameter of a 19 cm-radius circle. ABC is a chord cutting PQ at B. PB = 32 cm and CB = 24 cm. Calculate the length of chord AC, verifying your answer with a scale drawing.

24 Two circles intersect at M and N. P is a point on MN. APB and CPD are straight lines, A and B being points on the circumference of one circle, C and D being points on the other. Prove that ACBD is a cyclic quadrilateral.

31 Matrices

(Matrix multiplication)
(Matrix transformations)

● **You need to know . . .**

Matrices 1 and 2 (page 267)

1 $A = \begin{pmatrix} 3 \\ 2 \end{pmatrix}$ $B = (4 \quad 1)$ $C = \begin{pmatrix} -1 \\ -3 \end{pmatrix}$ $D = (-2 \quad 1)$ $E = \begin{pmatrix} 1 & 3 \\ 2 & 4 \end{pmatrix}$ $F = \begin{pmatrix} 0 & 1 \\ 2 & 0 \end{pmatrix}$

$G = \begin{pmatrix} -1 & 3 \\ 0 & -2 \end{pmatrix}$

Calculate where possible:

(a) A + C   (b) B + D   (c) C + D   (d) 3A   (e) $\frac{1}{2}$D   (f) 2G   (g) BC
(h) CE   (i) DE   (j) DG   (k) FC   (l) EF   (m) FE   (n) FG   (o) GF.

2 (a) On squared paper draw eight sets of axes, each from −6 to 6.

   On each grid plot the triangle represented by the matrix $\begin{pmatrix} 1 & 4 & 5 \\ 1 & 2 & 4 \end{pmatrix}$.

(b) Write the eight possible matrices $\begin{pmatrix} a & b \\ c & d \end{pmatrix}$ where either $a = d = 0$, and $b, c \in \{1, -1\}$;
   or $b = c = 0$, and $a, d \in \{1, -1\}$.

E means 'are members of the set of'.

(c) Apply each matrix in turn to the triangle's matrix and plot the resulting transformation, one on each pair of axes. Describe each transformation.

3 By investigating its effect on the triangle given in question 2, or otherwise, find the transformation effected by the matrix:

(a) $\begin{pmatrix} 1 & 2 \\ 0 & 1 \end{pmatrix}$ (b) $\begin{pmatrix} 1 & 0 \\ -2 & 1 \end{pmatrix}$ (c) $\begin{pmatrix} 2 & 0 \\ 0 & 2 \end{pmatrix}$ (d) $\begin{pmatrix} -2 & 0 \\ 0 & -2 \end{pmatrix}$ (e) $\begin{pmatrix} 2 & 0 \\ 0 & 1 \end{pmatrix}$

(f) $\begin{pmatrix} 1 & 0 \\ 0 & 2 \end{pmatrix}$.

4 The **determinant** of matrix $\begin{pmatrix} a & b \\ c & d \end{pmatrix}$ is $ad - bc$. This determinant gives the change in area when matrix $\begin{pmatrix} a & b \\ c & d \end{pmatrix}$ is used to transform a shape.

Example $\begin{pmatrix} 2 & 1 \\ 1 & 2 \end{pmatrix}$ has a determinant of $4 - 1 = 3$.

So $\begin{pmatrix} 2 & 1 \\ 1 & 2 \end{pmatrix}$ increases the area of a shape three times.

Plot the rectangle $\begin{pmatrix} 0 & 3 & 3 & 0 \\ 2 & 2 & 0 & 0 \end{pmatrix}$. Transform it using the matrix $\begin{pmatrix} 2 & 1 \\ 1 & 2 \end{pmatrix}$ and plot the resulting quadrilateral. Check the area of the quadrilateral is three times that of the rectangle.

5 State the effect on the area when a shape is transformed by using the matrix:

(a) $\begin{pmatrix} 3 & 1 \\ 2 & 1 \end{pmatrix}$　　(b) $\begin{pmatrix} 2 & 1 \\ 1 & 3 \end{pmatrix}$　　(c) $\begin{pmatrix} 1 & 2 \\ 2 & 2 \end{pmatrix}$　　(d) $\begin{pmatrix} 0 & 1 \\ -1 & 0 \end{pmatrix}$.

6 In Figure 31:1, O is the origin, A is the point (2, 0) and B is (2, 1). OAD and OCE are triangles onto which the triangle OAB is mapped by certain transformations.

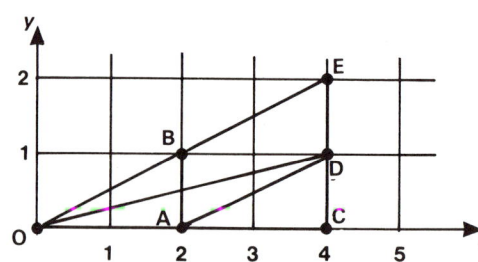

Fig. 31:1

(a) Name the type of transformation T_1 which maps OAB onto OAD.

(b) State the transformation T_2 which maps OAB onto OCE.

(c) Find the matrix which determines transformation T_1 and write it down.

(d) Find the matrix which determines transformation T_2 and write it down.

(e) What is the area of the triangle onto which OAB is mapped by the transformation $T_2 T_1$? Find the matrix of this transformation. (WJEC)

7 On squared paper, using a scale of 1 cm to represent 1 unit on each axis, draw axes to show values of x from 0 to 10 and values of y from -10 to 10.

Draw and label the rectangle with vertices at O (0, 0), A (10, 0), B (10, 5) and C (0, 5).

(a) The points O, P, Q and R are the images of O, A, B and C respectively under the transformation represented by the matrix $\frac{1}{5}\begin{pmatrix} 3 & 4 \\ -1 & 7 \end{pmatrix}$.

Find, draw, and label the quadrilateral OPQR on your diagram. Draw, and label, the line s which is invariant under this transformation and write down the equation of the line.

(b) Find the matrix representing the transformation which maps the points O, P, Q and R onto O, A, B and C respectively. Hence deduce the 2×2 matrix representing the single transformation which maps the points O, P, Q and R onto O, L (2, 0), M (2, 1) and N (0, 1) respectively. (SEG)

8 Draw x- and y-axes on graph paper with values on the x-axis from -8 to 8 and on the y-axis from -12 to 8, with a scale of 1 cm to 1 unit on each axis.

The triangle T_0 has vertices O (0, 0), P (2, 0) and Q (2, 1). This triangle is to be transformed to triangle T_1, by the matrix A where $A = \begin{pmatrix} 1 & -1 \\ 1 & 1 \end{pmatrix}$. Similarly, A will transform T_1 to T_2, and so on.

(a) Draw T_0 on your diagram. Calculate the co-ordinates of the vertices of T_1 and T_2, and draw these two triangles on your diagram.

(b) Either by noticing the geometric effect of A on these triangles, or by calculation, draw triangles T_3, T_4 and T_5 on your diagram.

(c) The transformation represented by A is a combination of a rotation and another transformation. State the angle of rotation, and describe precisely the other transformation.

(d) The matrix A^n will transform T_0 to T_n. What is the smallest value of n (excluding $n = 0$) for which A^n represents a simple enlargement with a positive scale factor? For your value of n, give the co-ordinates of the vertices of T_n, and also express A^n as a 2×2 matrix. (MEG)

9 The transformation T of the plane is represented by the matrix M where $M = \begin{pmatrix} 1 & 1 \\ -1 & 1 \end{pmatrix}$.

The images of the points O (0, 0), A (5, 0) and B (0, 3) under this transformation are O, A' and B' respectively.

(a) Obtain the co-ordinates of A' and B', and show triangles OAB and OA'B' on a grid with axes from -6 to 6 each.

(b) Transformation T is a combination of a rotation and an enlargement.
   (i) Write the size of the angle of rotation.
   (ii) Calculate the scale factor of the enlargement.

(c) Find positive integers n and k such that:
   $M^n = k \begin{pmatrix} 1 & 0 \\ 0 & 1 \end{pmatrix}$. (MEG)

10 The co-ordinates of the vertices of two squares are:
A (2, 1), B (4, 1), C (4, 3), D (2, 3)
A' (-4, -1), B' (-6, -1), C' (-6, -3), D' (-4, -3)

The square ABCD is mapped onto the square A'B'C'D' by a rotation about the origin followed by a translation. Describe these two transformations in terms of matrices. (NEAB)

Defining regions   **Maximising**

● **You need to know . . .**

Linear programming 1 (page 280)

---

1 (a) Draw axes, y from 0 to 4, x from 0 to 6. Shade outside the region where:
$x \geqslant 1$, $y > 1$ (dotted line), $x + y \geqslant 3$, $y \leqslant -\frac{2}{3}x + 4$.

(b) Give the co-ordinates of four points in this region where x and y are integral.

(c) What are the best values for x and y if $y - 2x$ has to be as large as possible?

2 (a) Draw axes, y from 0 to 8, x from 0 to 6. Shade outside the region where:
$y < 8$, $x + y \geqslant 6$, $y \leqslant 4x$, $y > 2x - 4$, $3y + 8x \leqslant 48$.

(b) Find the integral values for x and y when $y + 4x$ is a maximum.

3 A firm makes two kinds of jewel box. The standard model (s) takes 6 hours to make, the deluxe model (d) takes 10 hours. Up to 10 workers may be employed, 8 hours per day for 5 days a week. Each box uses £20 of materials and the firm can afford to spend up to £1000 a week on materials. The profit to be made is £50 on a standard model and £75 on a deluxe model.

(a) How many of each model should the firm make per week to make a maximum profit if sales of each are expected to be the same?

(b) If sales indicate that it is better to make twice as many standard as deluxe models, what is now the best manufacturing schedule?

Hints: **1** Show that the time information gives
$6s + 10d \leqslant 400 \rightarrow 3s + 5d \leqslant 200$.

2 Show that the cost information gives
$20s + 20d \leqslant 1000 \rightarrow s + d \leqslant 50$.

3 Draw the graphs $3s + 5d = 200$ and $s + d = 50$. Use d for the horizontal axis.

4 Shade the unacceptable region.

5 Show that the profit line for situation (a) is a line of the family $s = -\frac{3}{2}d + a$ and that this gives $s = 25$ and $d = 25$ as the best solution.

6 Show that situation (b) means that the solution lies on $s = 2d$ and that this gives $s = 32$ and $d = 16$ as the best solution.

4 A distributor receives an order from two customers for new cars. Customer 1 needs 35 cars and customer 2 needs 30 cars. The cars are at two depots, 60 at depot A and 25 at depot B. The table shows the distances of the depots from the customers:

	Customer 1	Customer 2
Depot A	40 km	25 km
Depot B	20 km	30 km

The cost of transporting the cars is directly proportional to their distance from the customer. How should the distributor arrange to despatch the order at least cost?

Hints: 1 Let customer 1 receive x cars from depot A and let customer 2 receive y cars from depot A.

2 Show that $x \leqslant 35$, $y \leqslant 30$, $x + y \leqslant 60$, $x + y \geqslant 40$.

3 Show that the solution is on one of the family of lines $y = 4x + a$ and that this gives the solution:
Depot A sends 10 cars to customer 1, and 30 to customer 2.
Depot B sends all 25 cars to customer 2.

Note: You may think that the solution was obvious from the start! Linear programming at GCSE level is a bit like that!

5 A newspaper page of area $800\,cm^2$ contains articles, pictures and advertisements. The policy of the paper is that each page must contain all three, that articles must take at least half the page, advertisements at least an eighth of the page, and that no picture must be smaller than $150\,cm^2$. The ratio of articles to pictures must be at least $2:1$.

What is the maximum possible area of the articles if:
(a) the advertisement area is kept to a minimum
(b) pictures take up the maximum possible area?

6 A bus company can offer a school two sizes of coach, a 48-seater and a 64-seater. At least 384 pupils are to be transported as cheaply as possible. The 48-seater costs £50 to hire, the 64-seater deluxe model costs £80. The company has seven drivers available and five 64-seater coaches. What coaches should the school ask the company to send?

7 Gemma is a pharmacist. She has to mix two drugs, A and B, each of which contains three active ingredients, X, Y and Z, as shown in the table:

mg per g			
	X	Y	Z
A	80	15	2
B	40	50	8

The final mix must contain less than 60 mg of X, not more than 35 mg of Y and at least 4 mg of Z.

Find how many grams of A and B Gemma must mix to fulfil these conditions, given that her scales cannot weigh to a greater accuracy than 0.25 g.

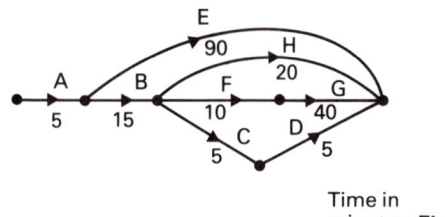

Time in
minutes **Fig. 33:4**

2 Figure 33:4 shows a planning network for a roast dinner. The job has been broken down into:

Preparation
A Prepare meat
B Prepare vegetables
C Lay table
D Make gravy

Cooking
E Cook meat
F Par-boil potatoes
G Roast potatoes
H Boil greens

(a) The meal is to be ready at 1 p.m. What is the latest time the cook can begin work?

(b) Which jobs lie on the critical path?

(c) Write the latest possible starting times for jobs H to B.

(d) Write the jobs to be done in the order they must commence.

(e) By using a hotter oven, the cooking time for the meat can be reduced to 60 minutes. If the cook starts work at 11:25, what is the soonest lunch can be served and what jobs make up the critical path?

(f) The vegetables preparation time can be divided into:
B_1 5 minutes for the potatoes
B_2 10 minutes for the greens.

Draw a new planning network to show how the critical time with the hotter oven could be reduced to 65 minutes, show the 'before zero hour' flags for each job, then list the latest starting times for each of the jobs H to A to have the meal ready by 1 p.m.

(g) Write the jobs to be done in the order they must commence.

3 The activities involved when two workers replace a broken window, and the times taken for each activity, are given in the table.

Activity	Duration in minutes	Preceding activity
A Remove broken pane	20	—
B Measure size of pane	15	—
C Purchase glass and putty	20	B
D Put putty in frame	10	A, C
E Put in new pane of glass	5	D
F Putty outside and smooth	10	E
G Sweep up broken glass	5	A
H Clean up	5	F, G

Using a critical-path diagram, find out how the two workers should share the activities to complete the replacement of the broken pane in the minimum possible time. State this minimum time.
(ULEAC)

4 Val and Winston have set up as gardeners. Today they are working at Golden Court. Figure 33:5 shows a rough plan of the garden.

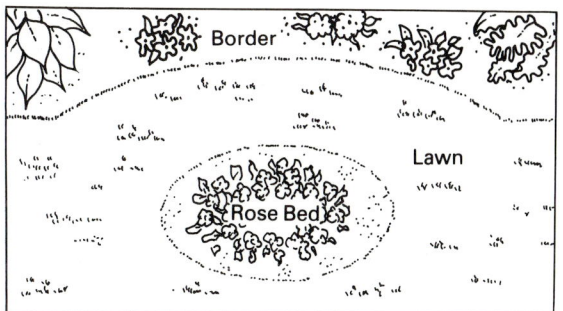

Fig. 33:5

The jobs they have to do are:

Travel: A Cycle to Golden Court   B Cycle back home
Lawn: C Mow   D Trim edges   E Fertilise
Roses: F Summer prune   G Weed   H Spray   I Fertilise
Border: J Dig   K Fertilise   L Plant
Finish: M Clean tools   N Brush up   O Get paid

Figure 33:6 shows the diagram they draw to show which jobs must be done in order and which may be done simultaneously, together with a planned time for each.

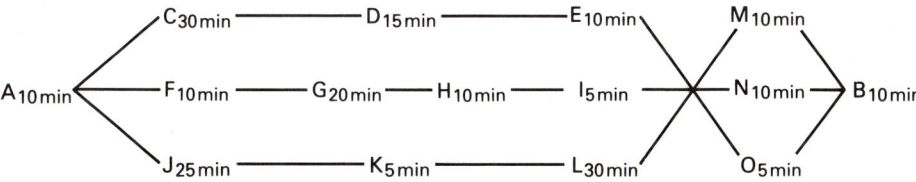

Key A$_{10min}$ means job A takes 10 minutes   Fig. 33:6

(a) Assuming their estimated times are correct, draw up a planning network. Try to avoid either person being idle, and assume that they do not have enough tools to share any job.

(b) State the critical path and its time, listing who does which job.

(c) Alison sees their plan and suggests that she could come too. The total time for all the jobs is 205 minutes, so in theory the three of them should be able to do the work in about 70 minutes. Show why this is not possible, and state how much time would be saved by having three workers instead of two.

Papers

Paper 1

1 Calculate, using a calculator and giving your answers correct to two decimal places:

(a) $348 - 6.3(3.46 - 8.03)$ (b) $\dfrac{236 \times 14.8}{6.4 \times 5.09}$ (c) $\dfrac{(7.34 + 6.74)}{(4.89 - 2.93)}$.

2 A drawing is made to a scale of 1 cm to 5 m. Find the actual distance represented by a line on the drawing of length:
(a) 6 cm   (b) 4.7 cm   (c) 32 mm.

3 Find the time taken in hours and minutes for a car to travel 180 miles at 50 m.p.h.

4 A family is going to buy a new television set. They have a choice of two methods of payment:

CASH   A single payment of £480.

HIRE PURCHASE   A deposit of £96 followed by 36 monthly payments of £16.

(a) If they choose hire purchase:
   (i) After paying their deposit how much more will they have left to pay?
   (ii) How much more do they have to pay above the cash price?
   (iii) Express the extra amount as a percentage of the cash price.

(b) The family could also rent the set at £14.50 per month. How much would this method cost over:
(i) 2 years   (ii) 3 years?

*5 (a) Copy and complete the table of values for $y = 2x^2$.

x	-3	-2	-1	0	1	2	3
y	18			0	2		

(b) Draw a graph of the results using a horizontal scale of 2 cm to 1 unit (x-axis) and a vertical scale of 1 cm to 1 unit (y-axis).

(c) On the same set of axes draw the graph of $y = -x + 6$.

(d) From the graph read off the values of x where the lines $y = 2x^2$ and $y = -x + 6$ cross.

6 Figure P1:1 shows part of an ornament formed from a cylinder 10 mm in diameter and 30 mm long. One end is surmounted by a hemisphere. The other end has a right cone-shaped piece removed. The height of the cone is 12 mm and it has a 10 mm-diameter base.

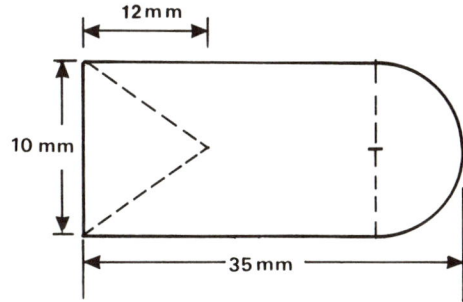

Fig. P1:1

Calculate:

(a) the slant height of the cone
(b) the volume of the ornament correct to four significant figures
(c) the total surface area of the ornament to the nearest 10 mm²
(d) the weight of 40 pieces, correct to the nearest gram, given that the weight of the material is 8.4 grams per cm³.

(Use the calculator value of π.)

Volumes: sphere = $\frac{4}{3}\pi r^3$, cylinder = $\pi r^2 h$, cone = $\frac{1}{3}\pi r^2 h$.
Surface areas: sphere = $4\pi r^2$, cylinder = $2\pi rh$, cone = πrl where *l* is the slant height.

Paper 2

1 Write in standard form the answer to:
(a) $3 \times 10^{-1} \times 3 \times 10^3$ (b) $6 \times 10^3 \times 7 \times 10^{-1}$ (c) $4 \times 10^2 + 7 \times 10^{-1}$.

2 A coat costing £85 and a jacket costing £56 are offered in a sale at a reduction of 25%.

(a) Find the sale price of each item.

(b) The sale price is then reduced by a further 15%. Find the new price of each item, correct to the nearest penny.

(c) Express the final price as a percentage of the original price, correct to two decimal places.

3 Solve:
(a) $5a - 3(a - 1) = 17$ (b) $2(5 - 3a) + 7(2a - 1) = 19$.

4 Eva is 2 years older than Mark who is 1 year older than Saskia. The sum of their ages is 49 years. How old are they?

5 The charges made by British Gas in 1992 were 47p per unit plus a standing charge of £9.90 per quarter. At the end of the autumn quarter my meter reading was 8437. At the end of the previous quarter it was 8251.

(a) Find the total cost of the bill for the autumn quarter.
(b) A neighbour paid a bill of £70.53. How many units had she used?

*6 A distribution service charges the following rates for the number of leaflets delivered:

No. of leaflets	100	200	400	600	800
Cost	£13.50	£15	£18	£21	£24

On a piece of A4 graph paper draw a graph of these charges. Use a vertical scale (Cost) of 1 cm to £1 and a horizontal scale (No. of leaflets) of 2 cm to 100 leaflets.

Use the graph to find:
(a) the fixed charge
(b) the cost of delivering 280 leaflets
(c) approximately how many leaflets you can have delivered for £20.

A rival company has no fixed charge but simply a rate of £3.50 per 100 leaflets delivered. On the grid you have already drawn plot a graph of their rates. Use this new line to determine:

(d) the cost of delivering 280 leaflets
(e) approximately how many leaflets you can have delivered for £20
(f) the number of leaflets for which the charge is the same for both firms.

7 During a week in July, 21.4 mm of rain fell on a flat roof measuring 6 m by 1.2 m. The rain was collected in a cylindrical butt of 70 cm diameter and height 1.2 m.

Calculate:
(a) the area, in cm², of the roof
(b) the volume, in cm³, of the rain which fell onto the roof
(c) the base area of the butt, in cm² (use the calculator value of π)
(d) the height to which the water in the butt rose, to the nearest cm
(e) how many litres of rain, to the nearest litre, fell on the roof.

Paper 3

1 Which of the following are prime numbers?
(a) 91   (b) 1003   (c) 1807   (d) 2003

2 A bicycle which cost £320 is sold for £240. What is the percentage loss?

3 Change 72 km/h to m/s.

4 The world record in 1992 for the 5000 metres was slightly under 13 minutes. Assuming a steady pace, how long did it take to cover 100 metres?

5 In 1992 a firm gave its employees a pay increase of 5%. In 1993 it gave an increase of 4%.

(a) In 1991 Michael earned £8600. Calculate his salary in: (i) 1992 (ii) 1993.

(b) In 1992 Vera received a £400 increase on her 1991 salary. What did she earn in 1991?

(c) In 1993 Richard earned £13 311.48. How much did he earn in: (i) 1992 (ii) 1991?

*6 The length of a rectangular field is 84 metres and the ratio of its length to its breadth is 8 : 5. Find:

(a) its breadth   (b) its area in m²

(c) the cost of fencing the field at £1.26 per metre.

A similar-shaped field has a length of 144 metres. Find:

(d) its breadth   (e) its area in hectares.

1 hectare = 10 000 m²

7 I belong to a bridge club. The membership fee is £12 per year and every time I play bridge I pay £1.50.

(a) How much does it cost me altogether to play bridge
   (i) 10 times a year   (ii) 40 times a year?

(b) Explain why the total cost £C of playing bridge n times per year is given by
$$C = 1.5n + 12.$$

(c) Draw the graph of $C = 1.5n + 12$ using a horizontal scale, n, of 2 cm to 10 times and a vertical scale of 2 cm to £10.

Non-members pay £3 every time they play bridge.

(d) On the same grid draw the graph of $C = 3n$.

(e) How many times a year can a non-member play bridge before it becomes cheaper to join the bridge club?

(NEAB)

8 Solve simultaneously: $3x - 2y = 12$ and $\dfrac{5x}{2} + 3y = -4.$

Paper 4

1 Estimate, giving your answer correct to 2 significant figures:

(a) 47.6×69.37 (b) $\dfrac{683 \times 2.14}{392}$ (c) $\dfrac{4812}{0.042}.$

2 Using ruler and compasses, construct a triangle ABC such that AC = 8 cm, AB = 9 cm and CB = 7 cm.

(a) Measure the perpendicular distance of B from the line AC.

(b) Calculate the area of the triangle ABC.

(c) Bisect perpendicularly the sides of the triangle to find the centre of the circumcircle. Draw the circumcircle and measure its radius.

3 (a) Taking 5 miles as being equivalent to 8 kilometres, calculate the kilometre equivalent of:
  (i) 70 m.p.h.   (ii) 30 m.p.h.

 (b) Travelling on a motorway Vera Palmer averaged 68 m.p.h. for $2\frac{1}{2}$ hours. Find the distance she travelled in:
  (i) miles (ii) kilometres.

 (c) How long would it have taken Vera to do the same journey at 50 m.p.h.?

 (d) What is the difference in minutes between the times taken in (b) and (c)?

 (e) If Vera's car did 36 miles per gallon and petrol cost £1.98 per gallon, how much did the journey cost her?

*4 The Wells family received bills for gas, electricity and telephone during the same week! Gas is subject to a standing charge of £9.90 and each unit used costs 45.9p; electricity is subject to a standing charge of £11.10 and each unit used costs 8.6p; telephones are subject to a standing charge of £18.46 and each unit used costs 4.2p. VAT at 17.5% is added to telephone bills only.

 (a) Find the bill for using:
  (i) 136 units of gas   (ii) 1210 units of electricity   (iii) 432 telephone units.

 (b) What is the sum of the bills presented?

5 Using a scale of 2 cm to represent 1 unit on the x-axis and 1 cm to represent 1 unit on the y-axis, draw an x-axis from 0 to 6 and a y-axis from 0 to 16.

Copy and complete the table below which gives the values of the function $y = 16 - \dfrac{9}{x}$.

x	$\frac{3}{4}$	1	2	3	4	5	6
y	4					14.2	

On the axes draw the graph of $y = 16 - \dfrac{9}{x}$.

Using the same axes draw the graph of $y = 3x + 4$.

At the points of intersection of the two lines, the equations are equal. Write down the equation in x which satisfies these two points and show that it simplifies to $x^2 - 4x + 3 = 0$.

Use your graph to solve the equation $x^2 - 4x + 3 = 0$.

6 The area of a circle may be calculated by the formulae $A = \pi r^2$ or $A = \frac{1}{4}\pi d^2$, where r is the radius and d the diameter.

 (a) Show that the second formula may be obtained from the first.

 (b) Calculate the area of a 4 cm-radius circle, taking π as 3.14.

 (c) Draw a 4 cm-radius circle on 5 mm-squared paper, and verify your answer to part (b) by counting squares.

Paper 5

1 Evaluate, giving your answer in standard form:
   (a) $(3.4 \times 10^2) \times (2.1 \times 10^3)$ (b) $(34.6 \times 10^2) \times (16.4 \times 10^{-3})$.

2 Change 240 km/h to m/s.

3 Solve:
   (a) $7a - (4 + a) = 0$ (b) $7 - 3a = 5 - 2a$ (c) $10x - (2x + 3) = 21$.

4 A school clothing shop has as part of its stock 150 jumpers, 65 skirts and 180 pairs of training shoes.

   (a) The jumpers are bought for £22 each and are sold for £30.36 each. Find the profit and express it as a percentage of the cost price.

   (b) The skirts are sold for 35% profit. What is the selling price of a skirt which cost the shop £38.40?

   (c) The training shoes are sold for £39 thus earning the shop 25% profit. Find the cost to the shop.

   (d) Find the total cost of all the items to the shop.

   (e) Find the total amount of money received if all the items are sold.

   (f) Express the profit as a percentage of the cost price.

   Last year the shop bought jumpers at £18.80 and skirts at £28.60. These became shop-soiled.

   (g) The jumpers were sold for £14.10. What was the percentage loss on each jumper?

   (h) The skirts were sold at 35% loss. What was the selling price?

*5 A family of four travelled to France, exchanging £500 into francs at a rate of exchange of £1 = 9.54 francs. They spent their first night in an hotel, which charged 140 francs per person, before travelling to Strasbourg which was to be their centre for the holiday.

   While there they bought a jacket for 420 francs, presents for 375 francs, petrol for 660 francs and food for 1842 francs.

   On the return journey they stayed for one night at the same hotel as on the outward journey.

   (a) How many francs did they receive when they changed their £500?

   (b) What was the pound sterling equivalent of:
       (i) the hotel bill on the outward journey   (ii) the jacket which they bought
       (iii) the petrol they used   (iv) the food they bought?

   (c) Calculate the number of francs they had left when they boarded the boat for the return journey.

   (d) On the boat they exchanged their remaining francs for pounds at a rate of exchange of £1 = 9.66 francs. How much did they receive?

6 (a) Figure P5:1 shows a speed–time graph for the motion of a car lasting 30 seconds.

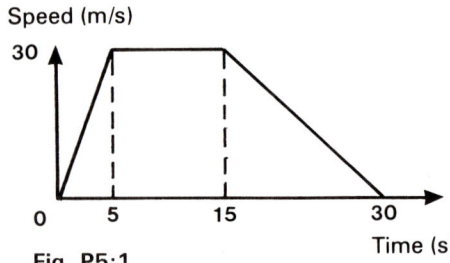

Fig. P5:1

    (i) Find the rate of acceleration in the first five seconds.

    (ii) Find the rate of deceleration during the last 15 seconds.

    (iii) Find the rate of acceleration 10 seconds from the start.

    (iv) Find the total distance travelled.

(b) The table gives the motion of a particle undergoing acceleration in a straight line:

Time (s)	0	1	2	3	4	5
Speed (m/s)	0	4.5	8	10.5	12	12.5

    (i) Draw a speed–time graph to represent this motion using a time scale of 2 cm to 1 unit and a speed scale of 1 cm to 1 unit.

    (ii) Using the points you have plotted draw straight lines between them and use the trapezoidal method to estimate the distance travelled by the particle in 5 seconds.

    (iii) By drawing a tangent to your graph, estimate the acceleration of the particle after 2 seconds.

7 (a) Draw the graph of $y = \dfrac{6}{x}$ for values of x between 1 and 6 inclusive, taking a scale of 2 cm to 1 unit on each axis.

(b) On the same axes draw the graph of $y = \frac{3}{2}(5 - x)$.

(c) Write down the equation in x which is solved where the graphs cross and show that it simplifies to $x^2 - 5x + 4 = 0$.

(d) Use your graph to solve the equation $x^2 - 5x + 4 = 0$.

Paper 6

1 Two partners in a firm shared their profits of £30 000 in the ratio 3:2. How much did they each receive?

2 Find the average speed in km/h of a car which covers 408 km in 4 h 15 min.

3   A group of 34 people travelled by train from Birmingham to Weston-Super-Mare for the day. The train times were:

Birmingham New Street   0915 | 2111 ↑
Weston-Super-Mare   1119 ↓   1856 |

The cost per person for a group return ticket was £16.40. A normal day-return ticket cost £18.56. For the journey find:
(a) how long the morning train took   (b) how long the evening train took
(c) the total cost for the group at the cheap rate
(d) the total amount saved by not buying normal day-return tickets.

4   Solve simultaneously: $2x + 3y = 5$ and $x = 2y - 8$.

5   Solve:
(a) $3(x - 2) = x + 5$ (b) $2(3 + 2x) = 3(2x - 4)$ (c) $2(x - 4) + 1 = 3(2x - 5)$.

*6   Design a flow chart that could be used as the instructions to find and repair a suspected puncture in a cycle tyre (or any other process which you are familiar with). Include at least one question box (to which the answer must be 'yes' or 'no').

7   Which of the following are rational?
(a) 3.3   (b) $\sqrt{2}$ (c) 5.$\dot{5}$ (d) $8 \div 0.8$ (e) 2π

8   Two parts of a child's toy are made from a piece of wood with a square cross-section. One piece is a cube, the other is a cuboid, as shown in Figure P6:1. The volume of the cuboid is $75 \, \text{cm}^3$ greater than the cube.

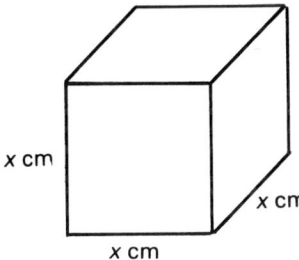
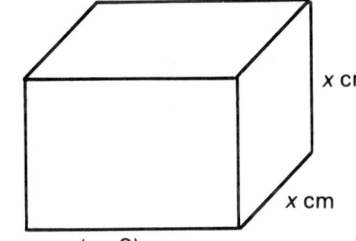

x cm

x cm

x cm

x cm

x cm

$(x+3)$ cm

Fig. P6:1

(a) Write down an expression in x for:
(i) the volume of the cube   (ii) the volume of the cuboid.

(b) Use these expressions and the difference in volume to write down an equation in x and show that this simplifies to $x^2 - 25 = 0$.

(c) Find the dimensions of the blocks.

(d) Write down an expression, in terms of x, for the total surface area of the two blocks.

(e) Hence, or otherwise, find the total surface area of the two blocks.

(f) How many pairs of blocks could be painted from 1 litre of paint if one litre covers $8 \, \text{m}^2$?

(g) How many metres of wood, cross-section x^2, would be needed to make the blocks in part (f)? Give your answer correct to the nearest centimetre.

P

Paper 7

1   A club buys 46 jumpers for £391. How much would they have to pay for 70 jumpers?

2   A car travels between London and Birmingham, a distance of 110 miles, at an average speed of 65 m.p.h.

(a) Find the time taken to the nearest second.

A second car does the journey at an average speed of 50 m.p.h.

(b) How long does this car take? Give your answer to the nearest second.
(c) How much time, in minutes and seconds, does the first car save?

3   Refer to Figure P7:1.

(a) If AB = 6 cm,   AM = 3 cm and AC = 5 cm, find AN.
(b) If AB = 9 cm,   AM = 6 cm,   AN = 4 cm and BC = 6 cm, find:
   (i) AC   (ii) MN.

(It is sometimes easier to understand this sort of question if you separate the diagram into two triangles, i.e. ABC and AMN.)

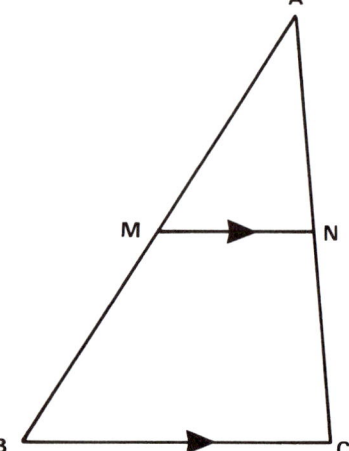

Fig. P7:1

4

Width (cm)	TRIUMPH WORK SURFACES Price List (£)						
	Length (cm)						
	50	75	100	125	140	165	180
30	1.20	1.80	2.40	3.00	3.40	3.90	4.40
45	1.80	2.70	3.60	4.50	5.10	5.80	6.40
60	2.40	3.60	4.80	6.00	6.80	7.20	8.80

(a) Using the table, find the cost of a piece 125 cm by 45 cm.

(b) Find the cost per square metre of a piece 30 cm wide and one metre long.

(c) A bench measures 135 cm by 50 cm.
   (i) What size surface would be needed if the bench was to be covered in one piece?
   (ii) What percentage of the sheet would be wasted?
   (iii) If the bench was to be covered exactly without waste, what size pieces would you buy?
   (iv) How much would you save using this method rather than covering the bench in one piece?

*5 (a) Calculate 4.89×34.2, giving your answer correct to:
(i) 2 significant figures (ii) 2 decimal places (iii) to the nearest 10.

(b) Write your answer to part (a) (i) in standard form.

*6 A map is drawn to a scale of $1 : 25\,000$.
(a) What distance, in metres, does 1 cm on the map represent?
(b) What distance, in km, does 8.7 cm on the map represent?
(c) A wood measures 3 cm by 1.5 cm on the map. What is its true area in hectares?

*7 (a) Add 12 hours 48 minutes 36 seconds to 18 hours 31 minutes 29 seconds.
(b) Subtract 12 hours 48 minutes 36 seconds from 18 hours 31 minutes 29 seconds.

*8 If $a = \frac{2}{3}$ and $b = \frac{3}{4}$ find:
(a) $a + b$ (b) $b - a$ (c) $b \div a$ (d) ab.

9 (a) Figure P7:2 shows a picture frame. The length of the picture is 2 cm longer than the width (x cm). The surround is 2 cm wide. The total area of the picture and its surround is 168 cm². Find the dimensions of the picture.

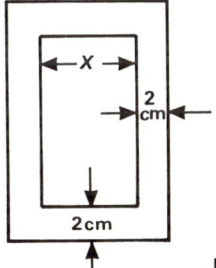

Fig. P7:2

(b) Factorise:
(i) $ax + ay + bx + by$
(ii) $am - an - bm + bn$
(iii) $6a + 4b - 9ac - 6bc$.

10 Four functions, e, f, g and h are defined so that

$$e(x) = 2x, \quad f(x) = 3x - 3, \quad g(x) = x^2, \quad h(x) = \frac{6}{x}.$$

(a) Find the values of:
(i) $e(3)$ (ii) $f(-3)$ (iii) $g(3)$ (iv) $h(-3)$.

(b) Find the value of x if $f(x) = 18$.

(c) Solve $f(e(x)) = 9$.

(d) Find the values of x for which:
(i) $e(x) = f(x)$ (ii) $g(x) = e(x)$.

(e) Find expressions for the inverse for each of the functions of x; that is, find:
$e^{-1}(x)$, $f^{-1}(x)$, $g^{-1}(x)$, $h^{-1}(x)$.

Paper 8

1 Divide 800 in the ratio $3 : 5 : 8$.

2 A lamp costing £24.60 has VAT added at 17.5%. What is the selling price?

3 Find the average speed of a cyclist who covers 31.5 miles in 1 hour 45 minutes.

4 Refer to Figure P8:1.

(a) Write down the equations of lines a, b and c.

(b) Write down the values of x and y where:
  (i) line a intersects line b
  (ii) line b intersects line c
  (iii) line a intersects line c.

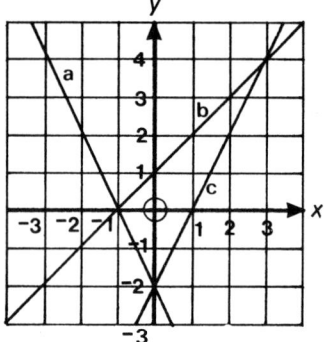

Fig. P8:1

5 Using axes from -1 to 5 draw the graphs of $y = x - 1$ and $y = -2x + 5$. What are the values of x and y at the crossing point of the lines?

*6 (a) Taking values of x from -3 to $+3$, make a table of values for $y = 2x^2$.

(b) Taking 2 cm to 1 unit on the x-axis and 1 cm to 1 unit on the y-axis, draw the graph of $y = 2x^2$.

(c) Using the same axes, draw the graph of the straight line $y = -x + 3$.

(d) From your graph read off and write down the values of x where the graphs cross.

(e) Solve $2x^2 + x - 3 = 0$.

(f) What is the connection between the equation in part (e) and your graphs?

7 ABCO is a rhombus and OCD is an equilateral triangle. $\overrightarrow{OC} = p$, $\overrightarrow{DO} = q$.

(a) Express in terms of p and q:
  (i) $\overrightarrow{OB}$ (ii) $\overrightarrow{DC}$ (iii) $\overrightarrow{AC}$.

(b) Show the journey D to O to A to B to C is equivalent to $\overrightarrow{DC}$.

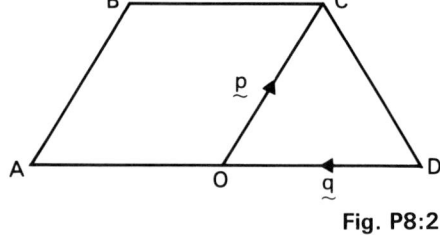

Fig. P8:2

(c) F is the point such that $\overrightarrow{OF} = -p$ and E is the point such that $\overrightarrow{OE} = -(p + q)$. Express $\overrightarrow{EF}$ in terms of p and q, as simply as possible. What does your answer tell you about EF?

(d) If O is the origin and D = (2, 0) write down the co-ordinates of:
  (i) C   (ii) B.

8 (a) A shop sold x bottles of lemonade at 30p each and y packets of nuts at 42p each. The total number of bottles and packets sold was 36 and the amount received from their sale was £13.20. Write down one equation in x and y to find the total number sold and another equation to find the total cost. Hence find the number of bottles sold and the number of packets sold.

(b) Solve the equation $4x^2 - 3x - 5 = 0$ correct to 2 decimal places.

Paper 9

1 A family bought the following items in a German shop:

200 g	Dutch cheese	2.58DM
1000 g	frozen chips	0.99DM
500 g	tomato ketchup	2.29DM
250 ml	vegetable oil	2.48DM
10	eggs	2.75DM
1000 g	peaches	1.49DM

(a) How much was their bill?

(b) How much change did they receive from a DM20 note?

(c) Using an exchange rate of £1 = DM2.92, calculate the equivalent sterling price for each item.

2 Calculate the gradient of the line joining:

(a) (3, 0) to (4, 3)   (b) (1, 6) to (−1, 2).

3 Taking values of x from 1 to 8, make a table of values for the equation $y = \dfrac{8}{x}$.

On axes from −3 to +8, using 1 cm to 1 unit on both axes, plot the graph of $y = \dfrac{8}{x}$.

On the same axes draw the graph of $y = 3x - 2$.

Write down the value of x where the lines intersect.

*4 A motorway is 115 miles long and runs from north to south. A lorry with a heavy load joins the southern end of the motorway at noon and travels at an average speed of 30 m.p.h. After 2 hours the driver stops at a service station for 40 minutes before continuing at the same speed as before to the end of the motorway.

A car joins the motorway at 2 p.m. and travels the whole length of the motorway from south to north at an average speed of 69 m.p.h.

A motorbike joins the motorway 35 miles from the northern end at 1 p.m. and travels south at an average speed of 60 m.p.h.

Draw on the same axes graphical representations of these journeys, taking 3 cm to 1 hour on the horizontal axis and 2 cm to 10 miles on the vertical axis.

At what time does:

(a) the car reach the end of the motorway

(b) the lorry reach the end of the motorway

(c) the motorbike reach the end of the motorway

(d) the motorbike pass the car (approximately), and about how far is it then from the southern end of the motorway?

5 Using a scale of 1 cm to represent 1 unit on each axis, draw x- and y-axes, taking values of x from -7 to $+10$ and values of y from -9 to $+13$. (Use A4 graph paper.)

Draw and label the triangle T with vertices at (3, 1), (3, 4) and (1, 4).

(a) The single transformation A maps the triangle T onto the triangle A(T) with vertices at (9, 3), (9, 12) and (3, 12). Draw and label the triangle A(T). What is the transformation?

(b) The single transformation B maps the triangle T onto the triangle B(T) with vertices at $(-6, -2)$, $(-6, -8)$ and $(-2, -8)$. Draw and label the triangle B(T). What is this transformation?

(c) The transformation R is an anticlockwise rotation of $90°$ about the origin. Draw and label the triangle R(T).

(d) The transformation C is the translation $\begin{pmatrix} -4 \\ -5 \end{pmatrix}$. Draw and label the triangle C(T).

(e) Draw the triangle RC(T).

(f) The single transformation D is represented by the matrix $\begin{pmatrix} 0 & 1 \\ 1 & 0 \end{pmatrix}$. Draw and label the triangle D(T). Describe fully the transformation D.

Paper 10

1 The distance between Gloucester and Birmingham is 82 kilometres. How far apart will they be, in cm, on a map of scale 1 : 500 000?

2 The total cost of a holiday in France is 2744 francs when there are 9.54 francs to the pound. How much does the holiday cost in pounds?

3 A regular octagon has sides 4 cm long. What is the length of a side of the regular hexagon which has the same perimeter as the octagon?

4 Write in algebraic form each of the following:

(a) Three times the sum of b and c.

(b) Three times the product b and c.

(c) The square of c added to the square of d.

(d) The total cost of 10 bananas at b pence each and 6 pears at p pence each.

(e) Your answer to part (d) expressed in £.

(f) The number of pupils on a school roll if f pupils are absent and p pupils are present.

5 Make y the subject of the following:
(a) $p = y - c$ (b) $Ty + b = R$ (c) $q = py - 4$.

*6 Simplify:
(a) $y^2 \times y^3$ (b) $a^4 \times a^{-3}$ (c) $a^6 \div a^3$ (d) $a^6 \div a^{-3}$ (e) $3a^3 \times 4a^{-1}$
(f) $24a^4 \div 6a^{-1}$.

*7 (a) (i) $-4 + (-6)$ (ii) $-8 \div (-2)$ (iii) $+12 - (-6)$ (iv) $-6 \times (-2)$

(b) If $a = -2$, $b = 3$ and $c = 4$, evaluate:
(i) $3a - b$ (ii) $4bc - 3a$ (iii) $a - b + 2c$ (iv) $c - a - b$
(v) $a^2 - 3c$ (vi) $2b^2 - a^2$.

(c) Multiply out the brackets and simplify:
(i) $2y + 2(y - 4)$ (ii) $3y(y - 4) - 2(y - 3)$ (iii) $3(y - 2) - (y - 2)$.

8 A concrete base for a garage has a perimeter of 23 metres.

(a) Letting x be the width, write down in terms of x an expression for the area of the base.

(b) The base has an area of $30\,\text{m}^2$. Show that $x^2 - 11.5x + 30 = 0$.

(c) Solve the quadratic equation in part (b) to find the dimensions of the base.

Fig. P10:1

9 Simplify: $\dfrac{4x - 2}{3} - \dfrac{2x + 3}{2}$.

10 Solve:

(a) $\dfrac{x + 3}{2} = \dfrac{x - 4}{5}$ (b) $\dfrac{3}{x + 2} - \dfrac{2}{x - 1} = 0$ (c) $\dfrac{x - 4}{3x} = \dfrac{x + 3}{3x + 7}$.

Paper 11

1 Work these out, using a calculator and giving your answers correct to 2 decimal places:

(a) $\dfrac{120}{6.4 + 18.3}$ (b) $\dfrac{14.3}{5.6} + \dfrac{6.42}{3.5}$ (c) $\dfrac{1}{6.4} - \dfrac{1}{14.3}$ (d) $\dfrac{1}{3.48} \times \dfrac{1}{0.37}$

(e) $\dfrac{1}{4} + \dfrac{3}{5} - \dfrac{3}{8}$.

2 Solve: $2(x + 5) - 3(2x - 4) = 2(2 - 3x)$.

3 The sum of three consecutive numbers is 27. Write down an equation to show this, using x to represent the smallest number. Solve your equation to find the three numbers.

4 One number is five times bigger than another. The sum of the numbers is 162. Find the numbers.

P

*5 Three boys weigh themselves. Mike is 8 kg less than John who is 6 kg less than Philip.

(a) By using x to represent Mike's weight, write down expressions for John's and Philip's weight.

(b) If their total weight is 202 kg, find the weight of each of them.

*6 (a) If the perimeter of the triangle in Figure P11:1 is 58 cm, find the lengths of the sides.

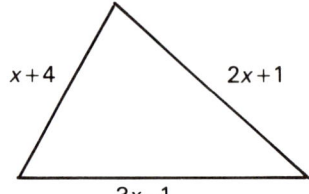
Fig. P11:1

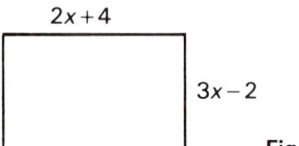
Fig. P11:2

(b) If the perimeter of the rectangle in Figure P11:2 is 124 cm, find the lengths of the sides.

(c) The result of multiplying a number by 3 is the same as adding 12 to it. What is the number?

7 (a) The distance a boulder falls from rest is proportional to the square of the time taken. If the boulder falls 125 metres in 5 seconds, how far will it fall in 10 seconds? How long will it take to fall 64 metres?

(b) Two variables A and B are such that $\dfrac{a}{A} + \dfrac{b}{B} = 2$, where a and b are constants.

If $A = 2$ when $B = 14$ and $A = -9$ when $B = 3$ calculate:
(i) the values of a and b (ii) the value of A when $B = -2$.

8 (a) Factorise $2x^2 - 7x - 4$.

(b) Solve the inequality $2x^2 - 7x - 4 \geqslant 0$.

9 A, B, C and D are four points such that A $= (-3, 2)$, C $= (6, -3)$, $\overrightarrow{AB} = \begin{pmatrix} 5 \\ 4 \end{pmatrix}$ and $\overrightarrow{CD} = \begin{pmatrix} -5 \\ -4 \end{pmatrix}$.

Calculate:
(a) the co-ordinates of B and D   (b) the vectors $\overrightarrow{BC}$ and $\overrightarrow{AD}$.

What can you conclude about:
(c) vectors $\overrightarrow{BC}$ and $\overrightarrow{AD}$ (d) vectors $\overrightarrow{AB}$ and $\overrightarrow{DC}$ (e) quadrilateral ABCD?

Paper 12

1 A prize of £900 is shared so that Colin receives three times as much as Sarah, who receives twice as much as Liz. How much do they each receive?

2 A record is sold for £5.04, thereby making a profit of 40%. What was the cost price?

3 A car costing £6800 depreciates in value by 26% during the first year and then by 15% of its value at the beginning of the year each year thereafter. Calculate:
  (a) its value after 3 years
  (b) the percentage loss after 3 years, to 2 significant figures
  (c) the average annual depreciation over the first 3 years
  (d) the number of complete years which will elapse before the value of the car is less than £2700.

4 Find the angles marked with letters in Figure P12:1.

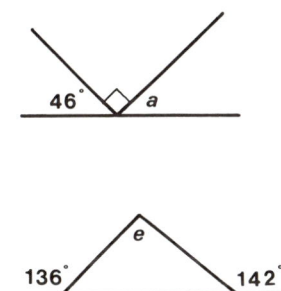
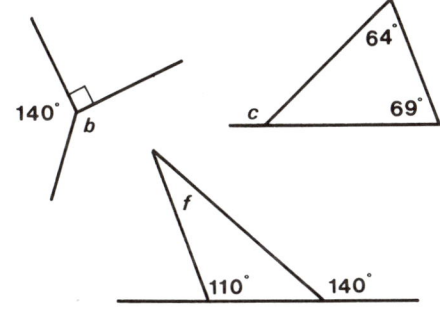
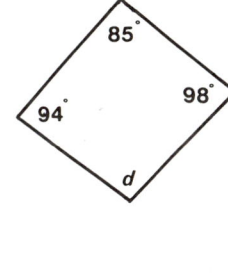

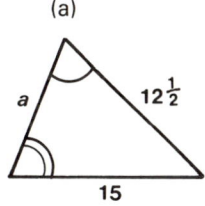

Fig. P12:1

5 In the pairs of similar figures shown in Figure P12:2, find the sides marked with letters.

(a)
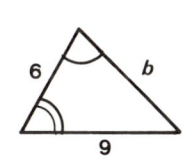

(b)
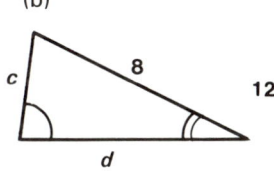
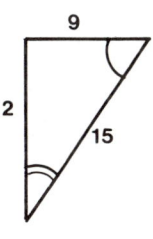

(c)
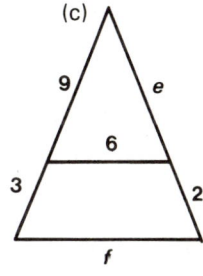

Fig. P12:2

*6 Draw a flow chart to read the mass and volume of 10 substances, calculate each density, and then print out a table of results. Your flow chart should start by printing out the headings:
Mass   Volume   Density

7 Decide which of the pairs of triangles in Figure P12:3 are congruent and which conditions of congruence they satisfy.

(a)   (b)   (c)

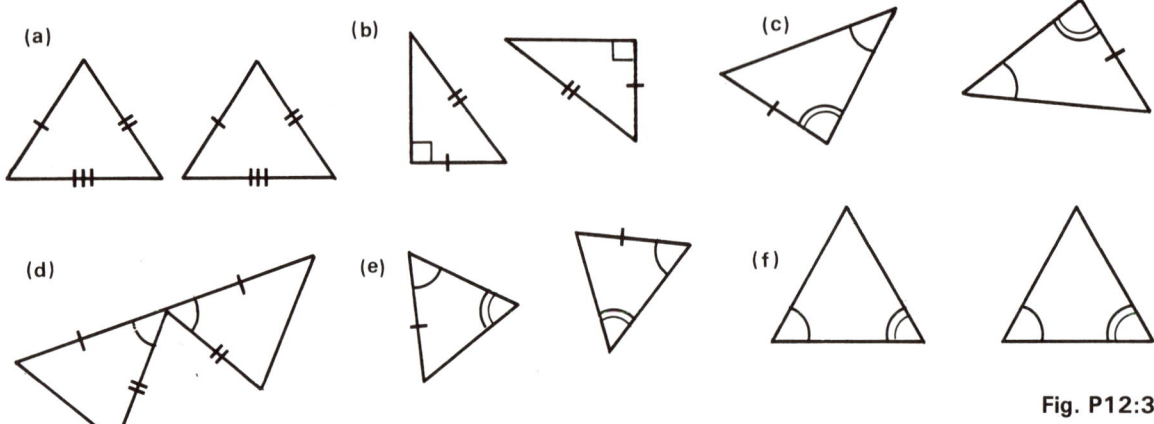

(d)   (e)   (f)

Fig. P12:3

8 P is proportional to y^2. If $P = 9$ when $y = 7\frac{1}{2}$, calculate:
(a) the value of P when $y = 10$ (b) the value of y when $P = 4$.

Paper 13

1 While on a 12-day holiday in Japan a family spent a total of £768.

(a) What was their average spending per day?

(b) Using an exchange rate of £1 = 235 Yen, find how many Yen they spent during their holiday.

2 An adult train fare is £12 more than a child's fare. If the combined fare is £40, find the cost of each fare.

3 Find the angles marked with letters in Figure P13:1.

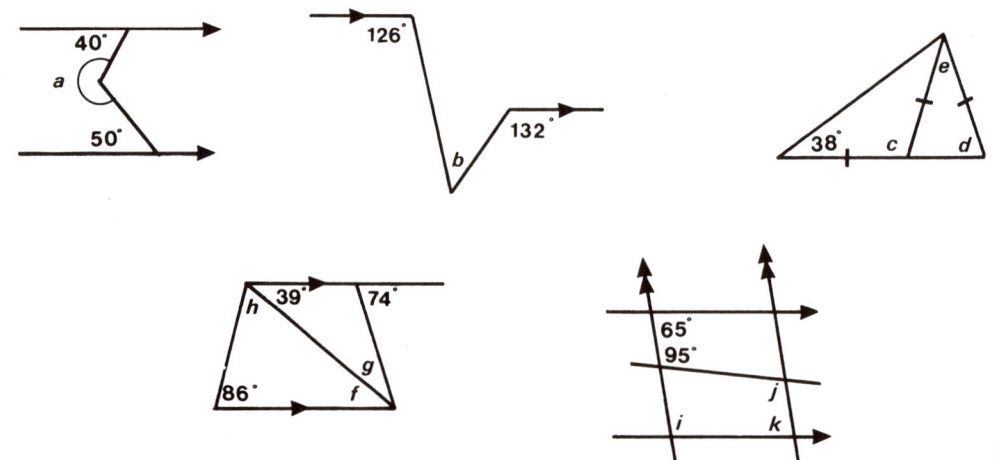

Fig. P13:1

4 (a) Calculate the number of sides of a regular polygon:
(i) whose exterior angles are each 40° (ii) whose interior angles are each 156°.

(b) ABCD is a quadrilateral in which ∠A = 96°, ∠B = 94°, ∠C = 48° and ∠DBC = 79°. Calculate ∠BDC and ∠BDA.

(c) A pentagon has angles $m°$, $(m + 10)°$, $(m + 22)°$, $(m + 48)°$ and $(m - 5)°$. Find m.

*5 Mary has some square patches and some regular hexagonal patches, each with sides of 10 cm. She wants to use them to make a patchwork quilt. Make a sketch of each shape.

(a) What is the size of each interior angle of:
(i) a square patch (ii) a hexagonal patch?

(b) Mary finds that she cannot fit the pieces together to make a quilt. Why not?

(c) (i) What other single regular-shaped patch would she need to enable her to make the quilt?
(ii) What is the size of each interior angle of the new patch?

6 A wooden post consists of a cylinder, 20 cm in diameter and 274 cm long, with the last 24 cm being tapered to a point. It is shown in Figure P13:2.

(a) Calculate:
(i) the slant height of the conical point
(ii) the cost of painting the post to the nearest penny if it costs 1p to paint 35 cm².

(b) Calculate the weight of the post, in kg, if 1 cm³ of wood weighs 1.4 grams.

(c) A similar post has a diameter of 15 cm. Calculate
(i) its length (use the properties of similar figures)
(ii) the cost of painting it
(iii) its weight in kg.

(The volume of a cone $= \frac{1}{3}\pi r^2 h$, curved surface area of a cone $= \pi r l$. Use the calculator value of π.)

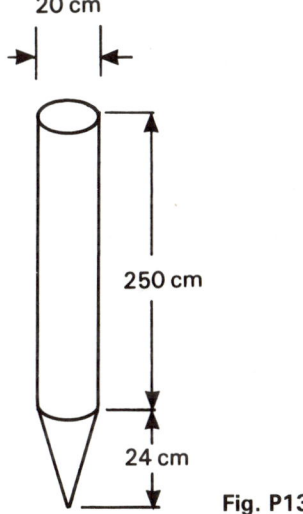

20 cm

250 cm

24 cm

Fig. P13:2

Paper 14

1 Work these out, using a calculator and giving your answer correct to two decimal places:

(a) $\sqrt{\dfrac{4.374}{2.18}}$ (b) $\sqrt{\dfrac{3.74}{18.4}}$ (c) $\sqrt{\dfrac{16.47 + 11.34}{15.67}}$.

2 Draw an example of each of the following quadrilaterals: square, rectangle, parallelogram, rhombus, kite, isosceles trapezium, trapezium. Mark all lines of equal length and all parallel lines. Under each figure list the number of lines of symmetry and the order of rotational symmetry.

3 The 32 members of a club buy 125 balls at £1.49 each. They share the cost equally between them. The treasurer uses his calculator to work out the cost to each member. He gets an answer of 44 pence.

Without using a calculator, use a rough estimate to check whether his answer is correct. You must show all your working. (SEG)

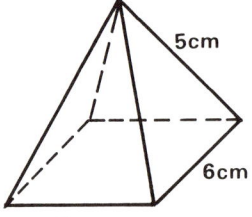

4 Draw the net of the square-based pyramid in Figure P14:1.

Fig. P14:1

*5 A model of a building is made to a scale of 1:50. The dimensions of the building are length 22 m, width 12 m and height 34 m. Find the dimensions of the model, in cm. If the total area of the windows in the model is 6500 cm², find in m² the total area of the windows in the actual building.

*6 A pilot takes a family on a pleasure flight. He sets out to fly the following route: to Alcester 70 miles N 30° E, then to Breen 50 miles on a bearing of 150°, then to Cappel 50 miles due south and then back to the airfield. Taking a scale of 1 cm to 10 miles make a plan of the route.

Use your drawing to find:
(a) the distance and bearing of the airfield from Breen
(b) the distance and bearing of the airfield from Cappel.

7 Calculate the sizes of the lettered angles in Figure P14:2. O indicates the centre of the circle.

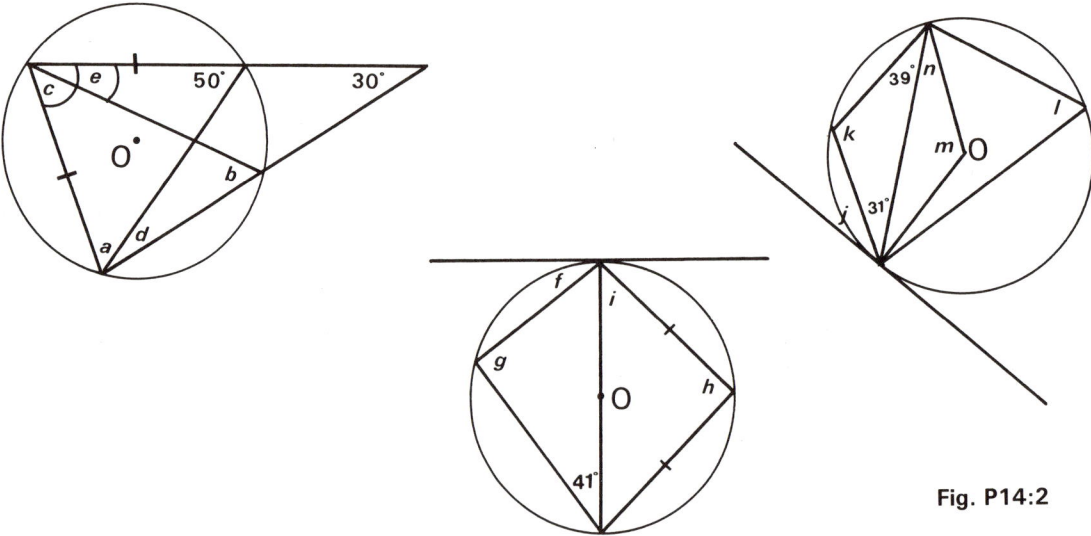

Fig. P14:2

Paper 15

1 A motorist travelled 1500 miles in a month at an average of 7.5 miles per litre. How much did she spend on petrol if one litre cost 50p?

2 Figure P15:1 represents the plan of a garden. Find its area in m².

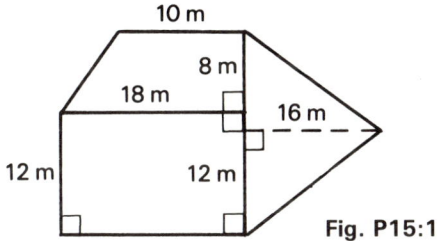

Fig. P15:1

3 Figure P15:2 represents a template cut out of a piece of card. Calculate:
(a) its area
(b) its perimeter.

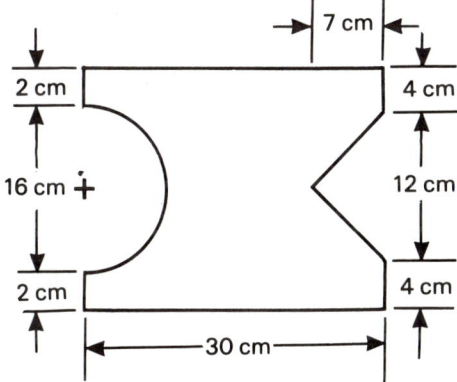

Fig. P15:2

*4 A cylindrical barrel has a radius of 25 cm and a height of 115 cm.

(a) Calculate its volume in cm³ to the nearest whole number.

(b) A cylindrical tin has a radius of 6 cm and a height of 14 cm. Calculate its volume in cm³ to 2 decimal places.

(c) If the contents of a full barrel is poured into tins, how many tins will be needed?

(d) Calculate the total area of metal used to make all of the tins, in m² to 4 significant figures. (Ignore the seams. Include the lids.)

5 (a) Construct the triangle ABC in which BC = 12 cm, AB = 10 cm and angle BAC = 60°.

   (b) Construct the bisector of the angle ABC and let this bisector meet AC at D.

   (c) Construct the bisector of the angle ACB and let this bisector meet AB at E.

   (d) Mark the point where BD meets CE and call this point O.

   (e) From O construct a line perpendicular to BC to meet BC at F.

   (f) Using OF as a radius, centre O, draw a circle.

   (g) Extend OF to P so that OP = 10 cm. From P construct a tangent to the circle, marking the point of contact as T.

   (h) Measure and write down the distances:
       (i) CD   (ii) AE   (iii) OF   (iv) PT.

6 Draw the graph of $y = \dfrac{12}{x} + x - 5$ for values of x from 1 to 7.

   (a) By drawing a suitable straight line on the graph, calculate an estimate of the gradient of the curve when $x = 2$.

   (b) By taking three strips of equal width and using the trapezium rule, calculate an estimate of the area of the region between the curve and the x-axis from $x = 4$ to $x = 7$.

   (c) By drawing a suitable line, use the graph to find estimates of the solutions of the equation
   $8 - x = \dfrac{12}{x} + x - 5$.

   (ULEAC)

Paper 16

1 Find the value of $3x^2$ if:
   (a) $x = 3$ (b) $x = -2$ (c) $x = \frac{1}{3}$.

2 Solve simultaneously: $2x - 3y = 5$ and $3x + 2y = -12$.

3 Solve the equation:
   $3(2 + x) - (x - 2) - 20 = 0$.

*4 Find the lengths of p and q in Figure P16:1 and hence the area of the triangle ABC.

5 Solve the equation: $4 = x(5 - x)$.

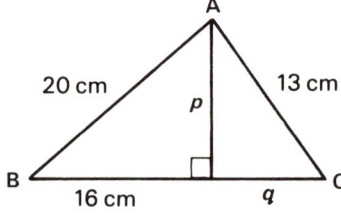

Fig. P16:1

6 Construct a triangle ABC in which AB = 8 cm, BC = 10 cm and angle ABC = 60°.

   (a) Construct the perpendicular bisector of AC.

   (b) Construct the locus of the point P such that PB = PC.

   (c) Mark O, the point at which the perpendicular bisector of AC meets the locus of the point P.

   (d) Using O as the centre construct a circle radius OC.

   (e) Produce OB to X so that OX = 12 cm. From X draw a tangent to touch the circle at T.

   (f) Measure:
      (i) AC   (ii) OC   (iii) XT.

7 In Figure P16:2, AC is a diameter and AD bisects the angle BAC. Calculate:
   (a) BC   (b) ∠BAC   (c) CD.

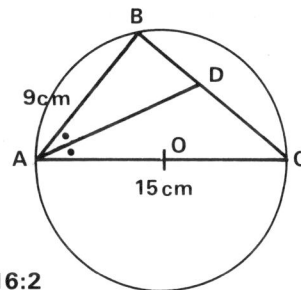

Fig. P16:2

8 (a) In Figure P16:3, OA and OB are radii of concentric circles. Calculate:
      (i) the area ABCD correct to 2 decimal places
      (ii) the difference in the lengths of arcs BC and AD correct to 3 decimal places.

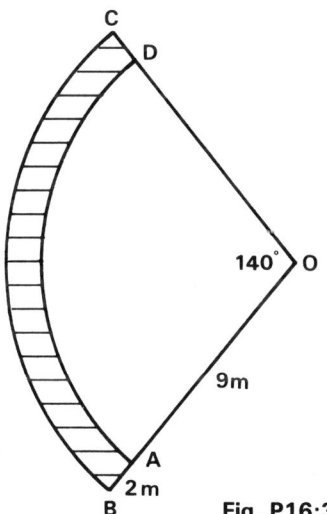

Fig. P16:3

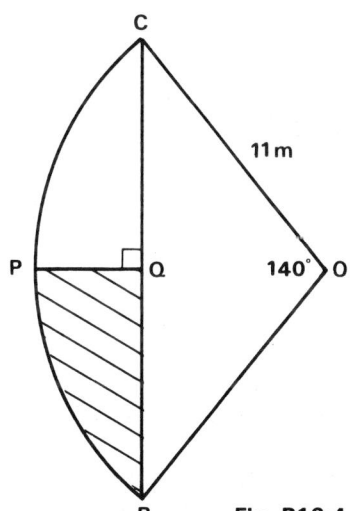

Fig. P16:4

   (b) In Figure P16:4, OB is the radius of a circle and PQ is the perpendicular bisector of chord BC. Giving all your answers correct to 2 decimal places, calculate:
      (i) BC   (ii) the area of sector OBPC   (iii) the area of triangle OBC
      (iv) the hatched area BPQ.

Paper 17

1 Work these out, using a calculator and giving your answer correct to 3 significant figures.

(a) $\left(\dfrac{2.46}{1.64}\right)^4$ (b) $6.47 - 1.64^2$ (c) $\dfrac{1}{3.4^2} - \dfrac{1}{4.7^2}$ (d) $\left(\dfrac{49.87^2 - 17.36^2}{3.7}\right)^2$.

2 Using one set of axes from 0 to 5, sketch the lines:
(a) $x = 0$ (b) $y = 0$ (c) $x + y = 5$ (d) $y + x = 3$.

3 In Figure P17:1 CBE is a straight line.
Calculate:
(a) BC   (b) AE   (c) BE.

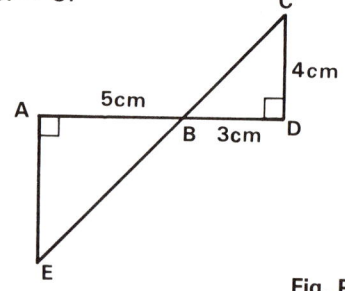

Fig. P17:1

4 (a) Solve:
(i) $3p - 6 - 2p + 8 + 4p = 0$ (ii) $4(p - 6) = 2p - 9$.

(b) Peter is x years of age and his mother is 24 years older. If the sum of their ages is 56 find Peter's age.

(c) $A = 4y^2 - x^2$ and $B = 4(2y + x)$
(i) If $y = 5$ and $x = 3$, calculate A and B.
(ii) If $x = 4$ and $B = 48$, calculate y and A.

*5 In Figure P17:2, the faces EFGH, DHGC, ABCD and AEFB are rectangles of length 3.2 m.

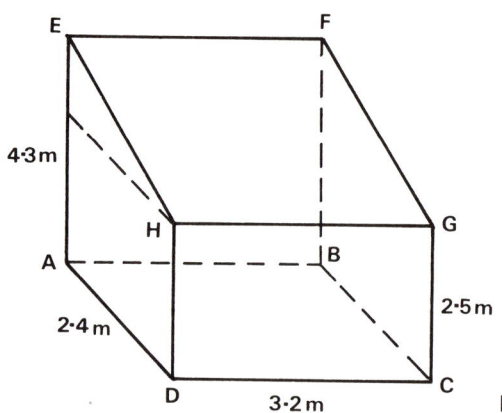

Fig. P17:2

Using Pythagoras' theorem calculate the length of:
(a) the diagonal CH   (b) the diagonal AC   (c) the diagonal EC   (d) the length EH.
(e) Calculate the total surface area of the figure.   (f) Calculate the volume of the figure.

Using a scale of 1 cm to 1 m draw a net of the solid.

6 (a) In Figure P17:3, AB is a chord 8 cm long with a perpendicular distance of 3 cm from the centre of the circle O. Sector OCD has angle COD = 60°. Calculate:
(i) the radius of the circle
(ii) the length CD
(iii) the length of arc CD
(iv) the area of sector OCD.

(b) In Figure P17:4, calculate:
(i) x to 4 significant figures
(ii) h to 2 decimal places.

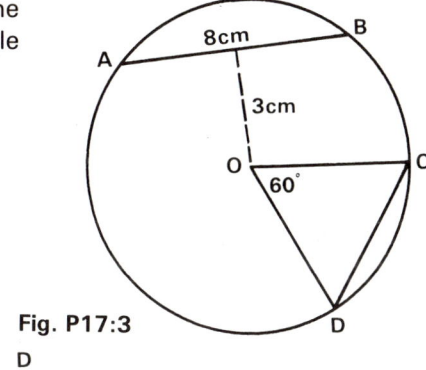

Fig. P17:3

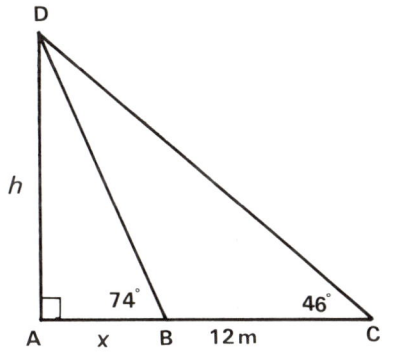

Fig. P17:4

7 Simplify:
(a) $8^{\frac{1}{3}}$ (b) $27^{\frac{2}{3}}$ (c) 4^{-1} (d) $8^{-\frac{2}{3}}$.

8 (a) If $\underset{\sim}{p} = \begin{pmatrix} 2 \\ 3 \end{pmatrix}$ and $\underset{\sim}{q} = \begin{pmatrix} 4 \\ -2 \end{pmatrix}$ find:
(i) $2\underset{\sim}{p} + \underset{\sim}{q}$ (ii) $\underset{\sim}{p} - 3\underset{\sim}{q}$.

(b) If A is (3, 2) and B is (6, 6):
(i) write $\overrightarrow{AB}$ as a vector   (ii) find $|\overrightarrow{AB}|$.

9 m varies directly as p^2 and inversely as n.
$m = 3$ when $p = 3$ and $n = 6$.

Find n when $m = 10$ and $p = 5$.

10 Solve the equation $2x^2 - 3x - 4 = 0$ correct to 2 decimal places.

Paper 18

For questions 1–20 you are asked to select the correct answer from the four given. Be careful, as the three wrong answers are worked out by making common mistakes; try to ignore the given answers until you have done your own working.

1 $4 \times (8 - 3) =$ A 8   B 20   C 44   D 25

2 $72.37 + 4.9 =$

    A 72.46 **B** 72.86 **C** 77.27 **D** 121.37

3 $-8 - 3 =$

    A 11 **B** −5 **C** −11 **D** 5

4 4% of £3.50 is

    A 14p **B** £1.40 **C** £1.25 **D** 13p

5 The next two numbers in the sequence 2, 3, 5, 8, 13 are:

    A 18, 24 **b** 20, 29 **C** 20, 27 **D** 21, 34

6 50% of people asked said they ate brown bread. Which of the following must be true?

    A 100 people were asked.
    B Half the people ate brown bread.
    C 50% ate white bread.
    D 50 people ate brown bread.

7 If $x = \sqrt{8}$, then $x^2 =$

    A 8 **B** 64 **C** 16 **D** 4

8 The mean of 4, 7, 10, 5, 4, is

    A 5 **B** 4 **C** 6 **D** 30

9 The median of 4, 7, 10, 5, 4 is

    A 5 **B** 4 **C** 6 **D** 30

10 The mode of 4, 7, 10, 5, 4 is

    A 5 **B** 4 **C** 6 **D** 30

11 Which of the letters of STOP have line symmetry?

    A S, T **B** S, T, O **C** O, P **D** T, O

12 $46 - 9 - 8 =$

    A 45 **B** 29 **C** 63 **D** 47

13 Which is the nearest approximate value of $\sqrt{160}$?

    A 17 **B** 13 **C** 40 **D** 4

14 0.034 km expressed in metres is

    A 34 **B** 3.4 **C** 340 **D** 0.34

15 If $4x - 3 = 9$, then x is

    A 1.5 **B** 6 **C** 12 **D** 3

16 If $a = 3$ and $b = 2$, then $2a^2b$ equals

    A 24 **B** 72 **C** 36 **D** 32

17 The sum of 1.5 kg, 250 g, $2\frac{1}{2}$ kg and 0.75 kg is

    A 5 kg **B** 4.25 kg **C** 277.25 kg **D** 7.25 kg

18 The area of triangle ABC in Figure P18:1 is

    A 12 cm² **B** 40 cm² **C** 24 cm² **D** 25 cm²

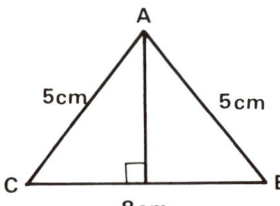

Fig. P18:1

19 Length AB in Figure P18:2 is

A 10 cm **B** 13 cm **C** 169 cm
D 17 cm

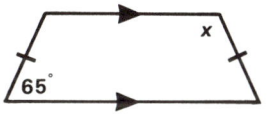

A

5 cm

12 cm

B

Fig. P18:2

20 Angle *x* in Figure P18:3 is

A 115° **B** 180° **C** 65° **D** 125°

x

65°

Fig. P18:3

21 (a) A family's budget for a month was as follows:

Mortgage	£180	Heat, light, etc.	£120	Food	£140
Savings	£80	Travel, car	£160	Clothes, etc.	£40

Using a 4 cm-radius circle, draw a pie chart to represent their budget.

(b) Figure P18:4 represents a survey of a group of school-leavers' plans for the following year. 28 pupils didn't know what they were going to do.

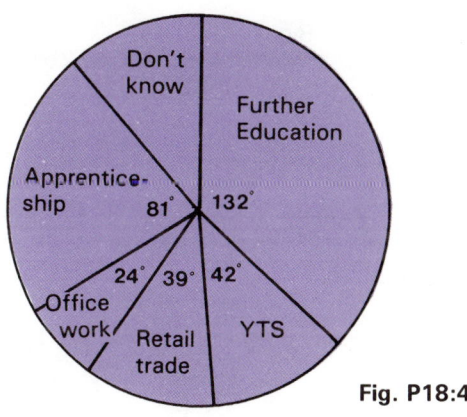

Fig. P18:4

Calculate:
(i) the number of pupils in each sector   (ii) how many pupils were questioned.

22 A motorist buys 5 gallons of petrol and 2 pints of oil for £14.66. A second motorist buys 3 gallons of the same grade petrol and 1 pint of oil for £8.44. Find the prices of petrol per gallon and of oil per pint.

23 Figure P18:5 represents a field. Calculate:
- (a) the lengths AB and BC in metres, to the nearest 0.1 metre
- (b) angles BAD, BCD and ABC in degrees to the nearest 0.1°
- (c) the area of the quadrilateral ABCD in hectares correct to 4 significant figures (1 hectare = 10 000 m²).

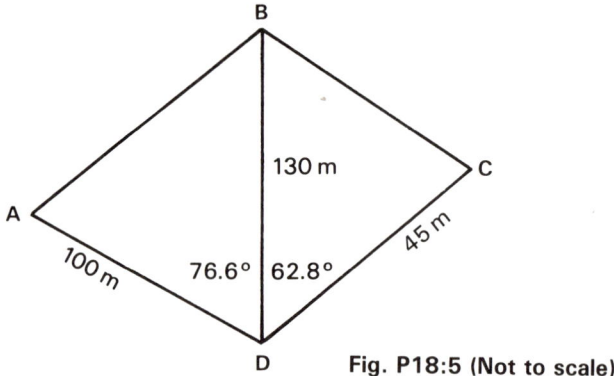

Fig. P18:5 (Not to scale)

Paper 19

Questions 1–20 are multiple choice, as in Paper 18.

1 A rectangle measures 50 cm by 18 cm. A square of equal area has sides of

    A 25 cm   **B** 45 cm   **C** 90 cm
    D 30 cm

2 In Figure P19:1, the value of x is

    A 85°   **B** 136°   **C** 125°   **D** 75°

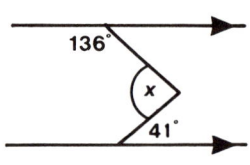

Fig. P19:1

3 In Figure P19:2, angle x is

    A 72°   **B** 64°   **C** 128°   **D** 98°

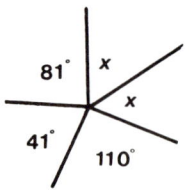

Fig. P19:2

4 Given that $ac + d = p$ then $c =$

    A $\dfrac{p - d}{a}$ **B** $p - d - a$ **C** $\dfrac{p}{ad}$ **D** $\dfrac{p}{d} - a$

5 In Figure P19:3, the value of x is   **A** 43°   **B** 32°   **C** 46°   **D** 51°

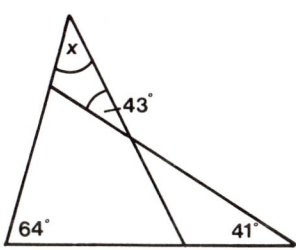

Fig. P19:3

6 The exterior angle of a 12-sided regular polygon is   **A** 36°   **B** 45°   **C** 30°   **D** 42°

7 The factors of $x^2 + 4x - 12$ are   **A** $(x + 12)(x - 1)$ **B** $(x + 6)(x - 2)$
                                      C $(x - 6)(x + 2)$ **D** $(x - 4)(x + 8)$

8 In Figure P19:4, $\sin x = $ **A** $\dfrac{3}{\sqrt{34}}$ **B** $\dfrac{3}{4}$ **C** $\dfrac{5}{3}$ **D** $\dfrac{4}{3}$

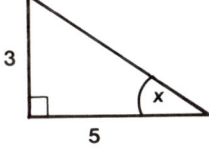

Fig. P19:4

9 The area, in cm², of a circle with diameter 19 cm is approximately   **A** 300   **B** 1200   **C** 120   **D** 60

10 $(0.08)^2 = $ **A** 0.16   **B** 0.64   **C** 0.016   **D** 0.0064

11 The ratio of the heights of two similar cylinders is 1 : 3. The ratio of their volumes is   **A** 1 : 6   **B** 1 : 9   **C** 1 : 27   **D** 4 : 9

12 Given that $x = 4$, $y = 2$, $a = 2$ and $b = -1$, then the value of $x^a + y^b$ is   **A** 6   **B** 16.5   **C** 7.5   **D** 14

13 In Figure P19:5, the equation of the line is   **A** $y = -3x + 3$ **B** $y = -x + 3$
                                       C $y = x + 3$ **D** $y = -3x - 3$

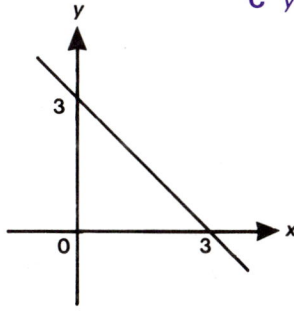

Fig. P19:5

14 Which of the following has the greatest value?   **A** $\frac{2}{3} + \frac{1}{4}$ **B** $\frac{2}{3} \times \frac{1}{4}$ **C** $\frac{2}{3} - \frac{1}{4}$
                                                                   D $\frac{2}{3} \div \frac{1}{4}$

15 In Figure P19:6, the height AB is

A $\sqrt{153}$ cm B 5 cm C 4 cm
D 9 cm

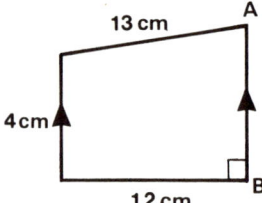

Fig. P19:6

16 A cylinder has a radius of 7 cm and a height of 10 cm. Its volume is

A 385 cm³ B 440 cm³
C 1540 cm³ D 880 cm³

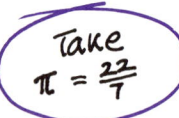

Take
$\pi = \frac{22}{7}$

17 If $x = 4.0 \times 10^{-2}$ and $y = 6.0 \times 10^4$, then xy equals

A 2.4×10^2 B 2.4×10^3
C 2.4×10^{-8} D 2.4×10^{-9}

18 £500 increased by 100% is

A £500 B £1000 C £600 D £550

19 The mean of 4, 3, 9, 3, 1 is

A 3 B 4 C 5 D 9

20 A rhombus has sides of length 13 cm and a diagonal of length 24 cm. The area of the rhombus in cm² is

A 240 B 169 C 156 D 120

21 Using a circle of 5 cm radius, make a copy of Figure P19:7 and mark on it the sizes of all angles.

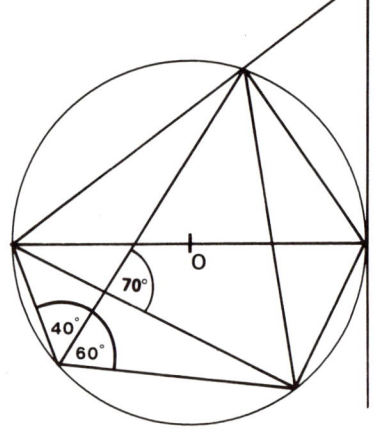

Fig. P19:7

Paper 20

1 Telephone charges are 4.2p per unit plus a standing charge of £18.46 per quarter. VAT at 17.5% is added to this total. Find the payment due if the number of units used is:
(a) 432   (b) 744.

2 The hour-hand of a clock is 12.8 cm long and the minute-hand 17.4 cm long. Calculate the distance between the tips of the hands at 9 o'clock.

3 For Figure P20:1, calculate:
   (a) the angles BAC, BCA and ABC
   (b) the length AC.

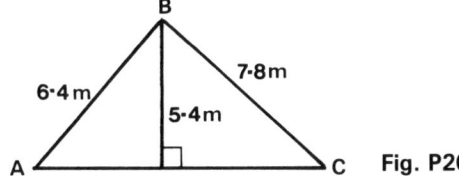

Fig. P20:1

4 Find the angles *a*, *b* and *c* in Figure P20:2.

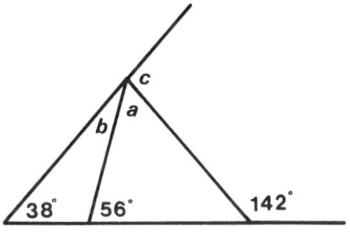

Fig. P20:2

5

Marks	3	4	5	6	7	8	9	10
No. of pupils	2	4	5	10	14	8	5	2

The table shows the result of a test. Find:
(a) the mean mark   (b) the modal mark   (c) the median mark.

*6 A class set out to find the speed of vehicles on a road. When a vehicle passed a lamp-post a signal was given to pupils 100 yards away to start their stop-watches to time the vehicles over the distance. They had worked out that the formula they needed to make a conversion graph was:

$$\text{Time over 100 yards} = \frac{60 \times 60 \times 100}{1760 \times \text{speed in m.p.h.}} \text{ seconds}$$

Cancel the fraction, reducing its denominator to 11.

Copy the table and complete it for speeds between 10 and 120 m.p.h.

Speed over 100 yards

Speed (m.p.h.)	10	12	15	20	30	40	60	80	100	120
Time (s)	20.5				6.8					

Using a scale of 1 cm to 1 second on the *x*-axis and 1 cm to 10 m.p.h. on the *y*-axis draw a graph of your results. Use your graph to convert the following times over 100 yards to m.p.h.:
(a) 7 s   (b) 12 s   (c) 3 s.

7 A function f is defined by $f : x \rightarrow x^2 - 5$.

   (a) Evaluate:
      (i) $f(3)$ (ii) $f(2)$ (iii) $f(3^{-1})$ (iv) $f(7^{\frac{1}{2}})$.

   (b) If $f(p) = \frac{4}{9}$, find the two possible values of *p*.

8 Simplify: $\dfrac{3a - 2}{5} - \dfrac{2a - 3}{4}$.

9 Solve: $\dfrac{1}{x + 2} - \dfrac{1}{x} = \dfrac{2}{x - 4}$.

10 A group of explorers above a ravine want to measure its width and depth. They stand at a point (A) directly opposite a tree (T) on the other side of the ravine. From where they stand the angle of depression of a rock (R) at the bottom of the ravine, directly beneath the tree, is 53°. They measure out 100 metres at right angles to line TA to a point (B), from where they find that the angle to the tree is now 33°. Find:
(a) the width of the ravine
(b) the depth of the ravine.

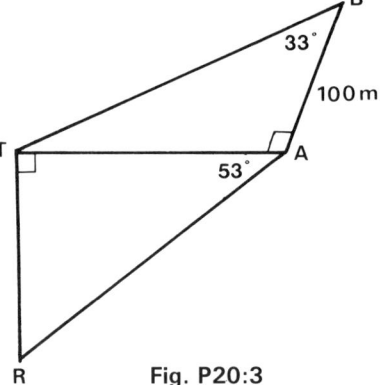

Fig. P20:3

11 A garden contains 50 trees and shrubs. Their heights were measured, in metres to the nearest 10 cm, to be:

12.1, 3.6, 6.2, 6.6, 8.9, 6.1, 4.8, 4.9, 7.5, 11.2, 5.1, 6.3, 9.2, 4.2, 5.8, 7.2, 6.5, 4.3, 5.1, 8.4, 8.1, 9.0, 7.9, 9.2, 7.1, 0.5, 4.9, 5.7, 2.8, 5.1, 8.3, 8.0, 6.1, 9.1, 4.1, 8.4, 10.2, 8.1, 8.3, 4.7, 6.4, 7.8, 6.1, 7.8, 5.1, 7.3, 1.1, 6.5, 4.4, 5.9

(a) Make a grouped frequency distribution table of their heights using classes of 0.0–, 2.0–, 4.0– etc.

(b) Use the table to construct a cumulative frequency curve, using a scale of 2 cm to 2 m on one axis, and 2 cm to represent 10 trees/shrubs on the other axis.

(c) By using your diagram, or otherwise, estimate the median and the quartiles.

(d) Estimate the percentage of trees and shrubs less than 6.5 m high.

(e) Calculate the standard deviation.

Reference notes: Number

Basic arithmetic

The number in brackets after each heading indicates the chapter in Book 5 which contains questions on the topic. An H after the subsection heading number indicates that the topic is only needed at Levels 9 and 10.

Basic arithmetic 1   Approximation (1)

Many amounts used in life are approximations to (not exactly) the true amount. This may be because there is no exact amount (e.g. the length of a line), or to make the number easier to read or remember (e.g. a football crowd of 21 000).

Approximations may be expressed in many ways. For example, 15.79 is 16 **to the nearest whole number**; 31 215 is 31 200 **to the nearest hundred**; 7.68 cm is 7.7 cm **to the nearest mm**; £10 ÷ 3 is £3.33 **to the nearest penny**.

You will often need to approximate calculator answers. How approximate you should make them usually depends on the information supplied. For example, if a question is based on an average speed correct to the nearest km/h, and a distance correct to the nearest km, then it is silly to give a time for the journey correct to the nearest second, and even sillier to give an answer like 1.245 367 hours.

Two special approximations are used in mathematics:

Decimal places (d.p.)

This states the number to a given number of figures after the decimal point. Clearly it is of no use when there are no figures after the point!

Examples $7.0145 \rightarrow 7.015$ to 3 d.p.
$7.0145 \rightarrow 7.01$ to 2 d.p.
$7.98 \rightarrow 8.0$ to 1 d.p. (the 'key' 8 makes $9 \rightarrow 10$)

Significant figures (s.f.)

All figures are counted, except zeros between the decimal point and the first non-zero digit, and place-value zeros before the point.

Examples $126.87 \rightarrow 130$ to 2 s.f. (The zero is not a significant figure, but it is needed to show the empty units' column, otherwise $126.87 \rightarrow 13$, which is silly.)
$0.001\ 34 \rightarrow 0.0013$ to 2 s.f.
$0.0598 \rightarrow 0.060$ to 2 s.f.

It is very important to check that your approximated answer *is* approximately the same size as the original number. (Students have been known to state that 1236.8 is approximately 12 to 2 s.f.)

Basic arithmetic 2   Range of error (1)

Given a number, n, correct to the nearest value, x, the range of error is $n \pm \frac{1}{2}x$.

Example   A number given as 15 to the nearest 5 could be anything from $15 - 2.5$ to $15 + 2.5$, that is 12.5 to 17.5.

Example   A weight given as 14.5 kg to the nearest 10 g could be anything from $14.500 - 0.005$ kg to $14.500 + 0.005$ kg, that is 14.495 to 14.505 kg.

Example   If the dimensions of a rectangle are given as 14.5 m by 13 m correct to the nearest 10 cm, then the length could be from 14.45 m to 14.55 m and the width from 12.95 m to 13.05 m.

The area could be between 14.45×12.95 m^2 to 14.55×13.05 m^2, that is from 187.1275 m^2 to 189.8775 m^2, a range of error of 2.75 m^2.

Note   Although you have been taught to round numbers up when the key figure is 5, this is only a convention. 16.5 is exactly midway between 16 and 17, so 16 could be as big as 16.5 and still be called 16 to the nearest whole number.

Basic arithmetic 3   Standard form (1)

Numbers above 1

$$1.2346 \quad 08$$

Fig. N1

This shows the way most scientific calculators display the answer to $123\,456 \times 1000$. Because the answer ($123\,456\,000$) is too long for the display it has been switched to standard form. (The calculator we used cuts off all figures after the first five, so the number has also been rounded to 5 s.f. Yours may shown more, or fewer, figures.)

The 08 at the right is called the **exponent**. It tells you that 1.2346 is 8 columns too small. Moving the figures up the 8 columns gives the answer as $123\,460\,000$, which is the most accurate this calculator can achieve. You may find it easier to think of the 08 as meaning that there are 8 figures between the first figure and the decimal point.

When we handwrite standard form we use the form $A \times 10^n$, where A is between 1 and 10 and n is an integer.

$$123\,456\,000 \to 1.234\,56 \times 10^8 \quad (\text{i.e. } 1.234\,56 \times 100\,000\,000)$$
$$318 \to 3.18 \times 10^2 \quad (\text{i.e. } 3.18 \times 100)$$

A calculator would not normally use standard form for numbers like 318, but using the $\boxed{\text{EXP}}$ or $\boxed{\text{EE}}$ key you can type in 318 as 3.18 $\boxed{\text{EXP}}$ 2 to give the display 3.18   02. When you type $\boxed{=}$ the the calculator will probably switch it back to 318. If it does not, try typing $\boxed{\times}$ 1 $\boxed{=}$ instead. (Many calculators have an optional SCI (scientific notation) mode which displays all numbers in standard form.)

Computers also use standard form for very large numbers (how large depends on your computer), but they show the exponent by an E, without leaving a gap, so that 1.234   06 becomes 1.234E6.

Numbers below 1

$$7.8 \quad -02 \quad \text{Fig. N2}$$

The calculator display here shows 0.078 in standard form. The -02 tells us that the 7.8 should be shifted two columns to the right to give the true value. You may think of the -2 as meaning there are two leading zeros, including the one before the decimal point.

This would be handwritten as 7.8×10^{-2}.

10^{-2} is the index way of writing the fraction $\dfrac{1}{10^2} = \dfrac{1}{100}$.

Basic arithmetic 4   Prime numbers (1)

Primes have only two different factors. For example, 19 is prime because its factors are 1 and 19; 9 is not prime because it has three factors, 1, 3 and 9.

Example   Write 162 as a product of prime factors.

$162 \;\rightarrow\; 2 \times 81 \;\rightarrow\; 2 \times 3 \times 27 \;\rightarrow\; 2 \times 3 \times 3 \times 9 \;\rightarrow\; 2 \times 3 \times 3 \times 3 \times 3$

This may be written:

```
2)  162
3)   81
3)   27
3)    9
3)    3
      1
```

Basic arithmetic 5   Highest common factor (HCF) (1)

A factor divides exactly into a number. The HCF is the highest factor that divides exactly into a set of numbers. For large numbers a prime factor method is useful.

Examples   The HCF of {12, 15, 18} is 3. (This can be done by just thinking about it.)

To find the HCF of 168 and 180:

$168 \;\rightarrow\; ②\times②\times 2 \times ③ \times 7$

$180 \;\rightarrow\; ②\times②\times ③ \times 3 \times 5$

$\text{HCF} = 2 \times 2 \times 3 = \underline{\underline{12}}$

N

Basic arithmetic 6   Lowest common multiple (LCM) (1)

A multiple is made by multiplying by an integer. The LCM is the lowest number that is a multiple of each member of a given set of numbers.

Example   The LCM of {6, 8, 12} is 24. (This can be done by just thinking about it.)

To find the LCM of 18, 30 and 36:
As 36 is a multiple of 18, we need not think about the 18. All multiples of 30 end in a zero, therefore the answer is a multiple of 36 that ends in a zero and also divides exactly by 30. The answer is 180.

For large numbers a prime factor method is useful.

Example   To find the LCM of {18, 24, 64}:
18 → 2 × 3 × 3
24 → 2 × 2 × 2 × 3
64 → 2 × 2 × 2 × 2 × 2 × 2
The prime factors of the LCM will consist of 2s and 3s. We need two 3s for 18 and six 2s for 64.
Hence the LCM is 2 × 2 × 2 × 2 × 2 × 2 × 3 × 3 = 576.

Computation

Computation 1   Divisibility (2)

A number divides exactly by:	2	3	5	6	9	10
if its digit-sum is:	any	3; 6; 9	any	3; 6; 9	9	any
and its last digit is:	even	any	0; 5	even	any	0

A number divides exactly by 4 if its last two digits divide exactly by 4.

A number divides exactly by 8 if its last three digits divide exactly by 8.

Computation 2   Directed numbers (2)

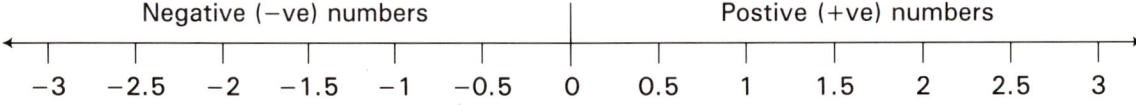

Negative (−ve) numbers   Postive (+ve) numbers

−3   −2.5   −2   −1.5   −1   −0.5   0   0.5   1   1.5   2   2.5   3

Positive (plus) numbers need no signs; negative (minus) numbers need − signs.

Like signs multiply to make a plus:
$$- -3 \;\to\; +3 \qquad -3 \times -2 \;\to\; +6$$

Unlike signs multiply to make a minus:
$$- +3 \;\to\; -3 \qquad + -3 \;\to\; -3 \qquad -3 \times +2 \;\to\; -6 \qquad +3 \times -2 \;\to\; -6$$

Computation 3 Decimal fraction arithmetic (2)

When adding or subtracting, be sure to keep the units' digits in a vertical line.

When multiplying, ignore zeros at the beginning or end of the number; multiply the resulting integers, then replace all omitted 'end' zeros; finally replace the decimal point so that there are as many figures after it as there were after the points in the original question.

Example $0.381 \times 10\,700 \rightarrow 381 \times 107 \rightarrow 40\,767 \rightarrow 4\,076\,700 \rightarrow 4076.700 \rightarrow \underline{4076.7}$

Written working for 381×107:

$$\begin{array}{r} 381 \\ \times\,107 \\ \hline 26_567 \\ 38\,100 \\ \hline 40\,767 \\ \hline \scriptstyle 1 \end{array}$$

Show the carry figures clearly.

When dividing, multiply both numbers by the power of 10 needed to change the divisor (the number you are dividing by,) into an integer, then divide as usual. No further change in the position of the point is required.

Example $18.324 \div 0.09 \rightarrow \dfrac{18.324}{0.09} \xrightarrow{\text{multiply top and bottom by 100}} \dfrac{1832.4}{9} \rightarrow \underline{203.6}$

Written working for $1832.4 \div 9$: $\quad 9\overline{)1\,8\,3^32^5.4}\quad = 2\,0\,3.6$

Show the carry figures clearly.

Computation 4 Common fraction arithmetic (2)

Common fraction to decimal fraction

Example $\frac{3}{5} \rightarrow 3 \div 5 \rightarrow 5\overline{)3.0}\;\;^{0.6} \rightarrow \underline{0.6}$

Decimal fraction to common fraction

Example $0.375 \xrightarrow[\substack{\text{last figure is in the} \\ \text{thousandths' column}}]{} \dfrac{375}{1000} \rightarrow \dfrac{\cancel{375}^{3}}{\cancel{1000}_{8}} \rightarrow \underline{\frac{3}{8}}$

Addition and subtraction

It is best to deal with the whole numbers first.

Examples $\quad 6\frac{3}{8} + 1\frac{1}{4} \rightarrow 7\frac{3}{8} + \frac{2}{8} \rightarrow \underline{7\frac{5}{8}}$

$\qquad\qquad 4\frac{1}{3} - 2\frac{2}{5} \rightarrow 2\frac{5}{15} - \frac{6}{15} \rightarrow 2 - \frac{1}{15} \rightarrow \underline{1\frac{14}{15}}$

Multiplication of a fraction by an integer

Examples $\quad \dfrac{^2\cancel{8}}{_1\cancel{12}} \times \cancel{9}^3 \rightarrow \underline{6}$

$\qquad\qquad 2\frac{1}{6} \times 3 \rightarrow \dfrac{13 \times \cancel{3}^1}{\cancel{6}_2} \rightarrow \dfrac{13}{2} \rightarrow \underline{6\frac{1}{2}}$

Fraction multiplied by fraction

Change all mixed numbers to improper (top-heavy) fractions first.

Example $3\frac{3}{4} \times 1\frac{1}{5} \rightarrow \dfrac{^3\cancel{15}}{_2\cancel{4}} \times \dfrac{\cancel{6}^3}{\cancel{5}_1} \rightarrow \dfrac{9}{2} \rightarrow 4\frac{1}{2}$

Fraction divided by integer

To divide by n, multiply instead by its reciprocal $\left(\dfrac{1}{n}\right)$.

Example $\frac{3}{4} \div 4 \rightarrow \frac{3}{4} \times \frac{1}{4} \rightarrow \frac{3}{16}$

Division by a fraction

To divide by a fraction, multiply instead by its inverse.

Examples $3 \div \frac{1}{2} \rightarrow 3 \times \dfrac{2}{1} \rightarrow \underline{\underline{6}}$

$2\frac{1}{2} \div 1\frac{2}{3} \rightarrow \dfrac{5}{2} \div \dfrac{5}{3} \rightarrow \dfrac{^1\cancel{5}}{2} \times \dfrac{3}{\cancel{5}_1} \rightarrow \dfrac{3}{2} \rightarrow \underline{\underline{1\frac{1}{2}}}$

Alternatively buy a calculator with a fraction capability!

Percentages

Percentages 1   Fraction to a percentage (3)

Multiply the fraction by 100%.

Note that because $100\% = \frac{100}{100} = 1$ we do not increase the fraction when we multiply it by 100%.

Example $\dfrac{11}{15} \rightarrow 11 \times \dfrac{^{20}\cancel{100}}{_3\cancel{15}}\% \rightarrow \dfrac{220\%}{3} \rightarrow 73\frac{1}{3}\%$

Percentages 2   Percentage of an amount (3)

Example To find 35% of £45:

35% is another way of writing $\dfrac{35}{100}$.

Hence 35% of £45 $\rightarrow \dfrac{35}{100} \times £45 \rightarrow \dfrac{^7\cancel{35}}{_{420}\cancel{100}} \times £\cancel{45}^9 \rightarrow £15.75$

Percentages 3   One amount as a percentage of another (3)

Write it as a fraction, then use the method of *Percentages 1*.

Example To find 32 as a percentage of 128:

Write this as $\dfrac{32}{128}$, then $\dfrac{32}{128} \rightarrow \dfrac{32}{128} \times 100\% \rightarrow \dfrac{^{1}\cancel{32}}{_{4}\cancel{128}} \times 100\% \rightarrow 25\%$.

Remember One amount as a percentage of another amount is the first over the second times 100%.

Percentages 4   Percentage changes (3)

A change can be an increase, a decrease, a profit, a loss, etc.

Change % = change over original × 100%.

Example   To find the percentage loss if a book bought for £20 is re-sold for £18:
The change in the cost is £2.
The original cost was £20.

Hence the percentage loss is $\dfrac{2}{20} \times 100\% = 10\%$.

Percentages 5   Increase and decrease by a percentage (3)

It is possible just to find the increase, then add it on, but a better method to use is:

To increase by $r\%$, multiply by $\dfrac{100 + r}{100}$.

To decrease by $r\%$, multiply by $\dfrac{100 - r}{100}$.

Examples   To increase by 12%, you multiply by $\dfrac{100 + 12}{100} \rightarrow \dfrac{112}{100}$ or 1.12

To decrease by 12%, you multiply by $\dfrac{100 - 12}{100} \rightarrow \dfrac{88}{100}$ or 0.88

Percentages 6   Calculator percentage key (3)

Unfortunately different calculators do not always use the same method, but one of the following two methods will probably work:

To find 8% of £60:   key: 60 ☒ 8 ▩ or 60 ☒ 8 ▩ ▣

To increase £60 by 8%:   key: 60 ⊞ 8 ▩ or 60 ⊞ 8 ▩ ▣ or 60 ☒ 8 ▩ ⊞

To decrease £60 by 8%:   key: 60 ⊟ 8 ▩ or 60 ⊟ 8 ▩ ▣ or 60 ☒ 8 ▩ ⊟

Percentages 7   Percentage changes (inverse calculations) (3)

Sometimes you known the amount resulting from a percentage change and have to find the amount before the change took place. This would occur in a shop where the price given includes 17.5% VAT and the customer wants to known what the price was before VAT.

A shopkeeper once told me that the VAT on a £100 television was £17.50, because 17.5% of £100 was £17.50. He was wrong, because the VAT was 17.5% of the 'before-VAT' price. This is what he should have done:

To increase by 17.5%, multiply by 1.175,
so 1.175 × before-VAT price = £100
→ before-VAT price = £100 ÷ 1.175 = £85.11
The VAT was therefore £14.89.

Example   Selling price £28, profit 12%, find the cost price.

The temptation is to work out 12% of £28, then take this away from the £28, but this is not correct, for the 12% profit is reckoned on the cost price. The correct method is:

To increase by 12% multiply by $\dfrac{112}{100}$ or 1.12

Then 1.12 × cost price = £28 → cost price = £28 ÷ 1.12 = £25.

Example   Selling price £21, loss 40%, find the cost price.

To reduce by 40% multiply by $\dfrac{100 - 40}{100}$ → $\dfrac{60}{100}$ or 0.60

Then 0.60 × cost price = £21 → cost price = £21 ÷ 0.60 = £35.

Percentages 8   Paying interest (7)

Interest is paid to someone who lends money. If you lend money to the government through your Post Office Savings Bank then the government will pay you interest.

If the interest rate is $5\frac{1}{2}$% per annum (p.a.) then the government will pay you £5.50 a year for every £100 you lend them.

If the interest is *not* added to the loan, it is called **simple interest**.

If the interest *is* added to the loan, it is called **compound interest**. Most everyday-life interest is compound interest.

Examples   £100 loaned at 5% p.a. simple interest for 3 years gives £5 each year, making £15 interest altogether.

£100 loaned at 5% p.a. compound interest for 3 years gives £5 interest the first year, but 5% of £105 = £5.25 the second year, and 5% of £110.25 = £5.51 the third year, a total of £15.76 interest.

There are two formulae which you can use, but they are not essential:

Simple interest = $\dfrac{P \times R \times T}{100}$ Compound interest = $P\left(1 + \dfrac{R}{100}\right)^{T} - P$

where P is the principal (the initial amount lent)
    R is the interest rate p.a.
    T is the number of years for which the principal is lent.

Using a calculator

(a) Simple interest

To find the interest on £50 at 8%:

Key: 50 ☒ 8 %

or   50 ☒ 8 %=

or   50 ☒ 0.08 =

(b) Compound interest

To find the total amount on £100 at 8% compound interest for 5 years:

To increase by 8%, multiply by 1.08.

      ☒

1.08 ☒ **or** 100 =====

      K

Note   ☒☒ or ☒K switches on the constant multiplier function.
Some calculators only need one press of ☒.
Some calculators multiply by the second number keyed in. If yours does, exchange the 1.08 and 100 in the above key sequence.

Using the formula

$$\text{Compound interest} = P\left(1 + \frac{R}{100}\right)^T - P$$

to find the compound interest on £700 for 15 years at $8\frac{1}{2}$%:

$$P = £700; \quad 1 + \frac{R}{100} = 1.085; \quad T = 15$$

Key: 700 ☒ 1.085 y^x 15 ☐ 700 =

giving £1679.82 interest!

To depreciate by 8% each year, repeatedly multiply by 0.92.

Ratio

Ratio 1   The meaning of ratio (4)

A ratio states the connection between two quantities. For example, the ratio of weight of cheese to number of eggs for a cheese omelette could be 20 g cheese to every one egg.

Ratios are often expressed with a colon (:), e.g. 2:3 (say '2 to 3'). In this case, both numbers must be in the same units, so we could not write the omelette example in this way, but we could say that the ratio of the weights of flour to margarine for plain scones is 4:1. That is, you use 4 times as much flour as margarine, e.g. 400 g flour and 100 g margarine.

Ratio 2   Simplifying ratios (4)

Ratios may be simplified by dividing by a common factor, as we do with fractions.

Example   2 litres water to 12 cl Jeyes Fluid
           → 200:12 (both units are now cl)
           → 50:3 (dividing both 200 and 12 by 4).

Ratio 3　Expressing a ratio in the form $n:1$ and $1:n$ (4)

Ratios are easier to use if one of the quantities is 1. For example, a ratio $1:2\frac{1}{4}$ clearly shows that the second amount is $2\frac{1}{4}$ times the first; this is not so obvious when the same ratio is written as $4:9$.

Example　Express $17:6$ in the ratio　(a) $n:1$　(b) $1:n$.

(a) $17:6 \xrightarrow{\text{divide both by 6}} 2\frac{5}{6}:1$

(b) $17:6 \xrightarrow{\text{divide both by 17}} 1:\frac{6}{17}$

Ratio 4　Given one amount, how to find the other (4)

Example　A 200 g jar of coffee granules makes about 120 cups of coffee.
The ratio of coffee to cups is 200 to 120 $\to$ 5 to 3.
Therefore a 250 g jar should make about $250 \times \frac{3}{5}$ cups $=$ 150 cups.

For 50 cups we need about $50 \times \frac{5}{3} \simeq 85$ g of coffee.

Note how the ratio $5:3$ became $\frac{3}{5}$ or $\frac{5}{3}$, depending on whether the required answer is to be bigger or smaller than the given amount.

Ratio 5　Given the total, how to find each (divide in a ratio) (4)

Example　Concrete for a path should consist of 1 part cement, 2 parts sand, and 3 parts coarse aggregate. If $3\,\text{m}^3$ of dry mix is required, what volume of each material should be purchased?

Cement : sand : aggregate $= 1:2:3$

Total $= 1 + 2 + 3 = 6$ parts

$3\,\text{m}^3$ in 6 parts $\to \frac{1}{2}\,\text{m}^3$ per part

$\therefore$ use $1 \times \frac{1}{2} = \frac{1}{2}\,\text{m}^3$ cement
$2 \times \frac{1}{2} = 1\,\text{m}^3$ sand
$3 \times \frac{1}{2} = 1\frac{1}{2}\,\text{m}^3$ aggregate.

Ratio 6　Changing in a ratio (4)

Example　Increase 16 in the ratio $11:6$.
An increase, so multiply by $\frac{11}{6}$.
$\dfrac{{}^8\cancel{16} \times 11}{\cancel{6}_3} = \dfrac{88}{3} = 29\frac{1}{3}$

Example　Decrease 35 in the ratio $6:11$.
A decrease, so multiply by $\frac{6}{11}$.
$35 \times \dfrac{6}{11} = \dfrac{210}{11} = 19\frac{1}{11}$

Example　6 men can paint a school in 46 days. How long should 8 men take?
The number of men has increased in the ratio $8:6$.
The time taken should *decrease* in the ratio $6:8$.
$\dfrac{{}^{23}\cancel{46} \times \cancel{6}^3}{\cancel{8}\cancel{4}_2} = \dfrac{69}{2} = 34\frac{1}{2}$ days

Calculators

Calculators 1   How a calculator carries out calculations (5)

All calculators have two kinds of keys:

● Digits and the decimal point 0 1 2 3 4 5 6 7 8 9 $\boxed{\cdot}$

● Function keys, e.g. $\boxed{+}$ $\boxed{-}$ $\boxed{\times}$ $\boxed{\div}$ $\boxed{x^2}$ $\boxed{\sqrt{}}$ $\boxed{\frac{1}{x}}$ $\boxed{+/-}$ $\boxed{\text{TAN}}$

Most calculators also have $\boxed{=}$ and memory, or store, keys, e.g. $\boxed{\text{MS}}$ $\boxed{\text{STO}}$ $\boxed{\text{Min}}$ $\boxed{x \to M}$ $\boxed{\text{MR}}$ $\boxed{\text{M+}}$

It is important to remember that all function keys operate on the number showing in the display at the moment that they are pressed (but see the note on 'BODMAS' below). This is why $\dfrac{12}{2 \times 3}$ will come to the wrong answer if you key in 12 $\boxed{\div}$ 2 $\boxed{\times}$ 3 $\boxed{=}$. The correct answer is 2, because 12 ÷ 6 = 2.

Key   12 $\boxed{\div}$ 2 $\boxed{\times}$ 3 $\boxed{=}$
Display   **12   12   2   6   3   18**

To get the right answer key in 12 $\boxed{\div}$ 2 $\boxed{\div}$ 3 $\boxed{=}$ or use brackets, 12 $\boxed{\div}$$\boxed{(}$ 2 $\boxed{\times}$ 3 $\boxed{)}$ $\boxed{=}$.

Notice that the calculator does not appear to do anything when the first function key is pressed, but in fact it puts the display into a hidden memory called the *y* register. When the next function key is pressed the calculator carries out the first operation, combining the number in the *y* register with the number on display (*x*). It then displays the answer (new *x*) and also stores the answer in the *y* register. ('BODMAS' calculators do not always follow this system; this is explained later.)

This is what happens when you work out $\dfrac{12}{2 \times 3}$:

Key   12 $\boxed{\div}$ 2 $\boxed{+}$ 3 $\boxed{-}$
Display (*x*)   **12   12   2   6   3** →**2**
Hidden (*y*)   0   12   12   6   6   0

Most scientific calculators follow the BODMAS (Brackets; Of; Divide; Multiply; Add; Subtract) rule, in as much as they save up additions and subtractions until any multiplications and divisions have been carried out. Mathematicians consider that the correct answer to 4 + 3 × 2 is 4 + 6 = 10, not 7 × 2 = 14.

Key in: 4 $\boxed{+}$ 3 $\boxed{\times}$. If your display is still 3, then your calculator is following the BODMAS rule; if it shows 7 it is not going to give the correct answer.

A BODMAS calculator stores up the pending operations in a 'stack', for which it has extra hidden memories. The example 4 + 3 × 2 only needs one extra memory (*z*):

Key   4 $\boxed{+}$ 3 $\boxed{\times}$ 2 $\boxed{=}$
Display (*x*)   **4   4   3   3   2**   →   6 → **10**
Stack (*y*)   0   4   4   3   3
Stack (*z*)   0   0   0   4+   4+   4+

To work out $(4 + 3) \times 2 = 14$:

Key	$[($	4	$[+]$	3	$[)]$	$[\times]$	2	$[=]$
Display (x)	**0**	**4**	**4**	**3**⎫	→**7**	**7**	**2**⎫	→**14**
Stack (y)	0	0	4	4⎭	0	7	7⎭	

Calculators 2   Constant function $\boxed{k}$ (5)

Most calculators will remember the last operation and keep repeating it. Some do this always, others need two presses of the function key or the use of $\boxed{k}$ key. Try the following. One of them, at least, will probably work out $2 \times 4 = 8$, $2 \times 6 = 12$ and $9 \times 2 \times 2 \times 2 = 72$

2 $\boxed{\times}$ or   2 $\boxed{\times}\boxed{\times}$ or   2 $\boxed{\times}\boxed{k}$ or $\boxed{\times}$ 2 $\boxed{=}$ or $\boxed{\times}$ 2 $\boxed{k}$
then   4 $\boxed{=}$ 6 $\boxed{=}$ 9 $\boxed{=}\boxed{=}\boxed{=}$

Now try the following. What is calculated?

2 $\boxed{\div}$ or   2 $\boxed{\div}\boxed{\div}$ or   2 $\boxed{\div}\boxed{k}$ or $\boxed{\div}$ 2 $\boxed{k}$ or $\boxed{\div}$ 2 $\boxed{=}$
then   4 $\boxed{=}$ 6 $\boxed{=}$ 9 $\boxed{=}\boxed{=}\boxed{=}$

The function is cancelled as soon as you press another function key.

Program your calculator for constant addition and subtraction.

Calculators 3   Memory keys (5)

There are two kinds of memory; one is called a 'store', the other is called an 'accumulator'.
Store keys are usually marked $\boxed{MS}$ or $\boxed{STO}$ or $\boxed{x \to M}$ or $\boxed{Min}$.
Accumulator keys are usually marked $\boxed{M+}$ or $\boxed{M+=}$ or $\boxed{M-}$ or $\boxed{M-=}$ or $\boxed{ACC}$ or $\boxed{SUM}$.
The memory content is recalled by a key usually marked $\boxed{RM}$ or $\boxed{MR}$ or $\boxed{RCL}$.

Examples

Key	6	$\boxed{MS}$	8	$\boxed{MR}$	9	$\boxed{MS}$	$\boxed{MR}$
Display	**6**	**6**	**8**	**6**	**9**	**9**	**9**
Store	0	6	6	6	6	9	9

Key	6	$\boxed{M+}$	8	$\boxed{M+}$	9	$\boxed{MR}$
Display	**6**	**6**	**8**	**8**	**9**	**14**
Store	0	6	6	14	14	14

You have to be careful to note whether or not the memory is empty when using the $\boxed{M+}$ key. Usually the display shows **M** to remind you that there is something in the memory. There is no problem with the $\boxed{MS}$ store key, because the memory is automatically emptied as soon as this key is pressed. There are various ways of cancelling memories. A common one is to store zero. Consult your handbook.

Calculators 4   Powers (5)

Powers like 2^3 can be worked out using the constant facility explained in point 2, but scientific calculators usually have a $\boxed{y^x}$ key[†], which is quicker if you have a large index, like 2^9.

Key	2	$\boxed{y^x}$	9	$\boxed{=}$
Display (x)	**2**	**2**	**9**⎫	→**512**
Store (y)	0	2	2⎭	0

†Labelled $\boxed{x^y}$ on Casio calculators.

Calculators 5   Roots (5)

Use the $\boxed{\sqrt{}}$ key for square roots (and the $\boxed{\sqrt[3]{}}$ key for cube roots if you have one). For further roots you need either a $\boxed{y^{\frac{1}{x}}}$ key (sometimes marked $\boxed{\sqrt[x]{y}}$) or a $\boxed{y^x}$ and $\boxed{\frac{1}{x}}$ key (although it is possible to use a 'trial and error' method that only uses basic functions). On Texas calculators use $\boxed{\text{INV}}\ \boxed{y^x}$
Remember that $8^{\frac{1}{2}}$ is an alternative way of writing $\sqrt{8}$, $8^{\frac{1}{3}} \equiv \sqrt[3]{8}$, and $8^{\frac{1}{4}} \equiv \sqrt[4]{8}$, so $y^{\frac{1}{x}} \equiv \sqrt[x]{y}$.

Example   Find $\sqrt[4]{2}$.

Key: $2\ \boxed{y^{\frac{1}{x}}}\ 4\ \boxed{=}$ **or** $2\ \boxed{y^x}4\ \boxed{\frac{1}{x}}\ \boxed{=}$ **or** $2\ \boxed{\text{INV}}\ \boxed{y^x}4\ \boxed{=}$

Calculators 6   Reciprocals (5)

The $\boxed{\frac{1}{x}}$ (or $\boxed{1/x}$) key is the **reciprocal** key. The reciprocal of x is the fraction $\dfrac{1}{x}$, usually changed to a decimal. The reciprocal of 4 is 0.25.

Reciprocals of fractions sometimes cause trouble to students, e.g. the reciprocal of $\frac{3}{4}$ is $\dfrac{1}{\frac{3}{4}}$.

Think of this as $1 \div \frac{3}{4} \rightarrow \dfrac{1 \times 4}{3} = 1.\dot{3}$. Your calculator would need the key sequence

$3\ \boxed{\div}\ 4\ \boxed{=}\ \boxed{\frac{1}{x}}$.

Remember that a function key operates on the number in the display. Function keys like $\boxed{x^2}$, $\boxed{\sqrt{}}$, $\boxed{\text{TAN}}$, etc. operate as soon as they are pressed. Others, like $\boxed{\times}$, $\boxed{+}$, $\boxed{y^x}$, etc., wait until another function key or $\boxed{=}$ is pressed before operating; this is because they combine the x and y registers.

To find the reciprocal of $\frac{3}{4}$ it would be wrong to key in $3\ \boxed{\div}\ 4\ \boxed{\frac{1}{x}}\ \boxed{=}$. Why?

Reciprocals are 'self-inverses'. Try $4\ \boxed{\frac{1}{x}}\boxed{\frac{1}{x}}\boxed{\frac{1}{x}}\boxed{\frac{1}{x}}$.

Calculators 7   Recurrence (5)

$a \div b$ always gives an exact answer or a recurring decimal unless a and/or b are irrational, but sometimes the calculator display is too short to show the recurrence.

The following method may be used to find the carry figure just before the display 'runs out' and hence continue the division until the recurrence becomes obvious.

Example   To find $5 \div 23$:

$5 \div 23 = 0.217\,391\,3$

Copy all but the last figure (which may be rounded up): $0.217\,391$

$0.217\,391 \times 23 = 4.999\,993$

The digit(s) you need to add to the end of this number to make the 5 we divided into tell you the carry figure. We need to add a 7, so now repeat the method starting with $7 \div 23$.

$7 \div 23 = 0.304\,347\,8$, so giving $0.217\,391\ \ 304\,347$
$0.304\,347 \times 23 = 6.999\,981$, giving carry figures 19

$19 \div 23 = 0.826\,086\,9$, so giving $0.217\,391\ \ 304\,347\ \ 826\,086$
$0.826\,086 \times 23 = 18.999\,978$, giving carry figures 22

$22 \div 23 = 0.956\,521\,7$, so giving $0.217\,391\ \ 304\,347\ \ 826\,086\ \ 956\,521$
$0.956\,521 \times 23 = 21.999\,983$, giving carry figure 17

$17 \div 23 = 0.739\,130\,4$, giving $0.217\,391\ \ 304\,347\ \ 826\,086\ \ 956\,521\ \ 739\,130$

This is recurring, so $5 \div 23 = 0.\dot{2}17\,391\ \ 304\,347\ \ 826\,086\ \ 956\,\dot{5}$

See also the notes: *Basic arithmetic 3* (page 202); *Percentages 6 and 8* (page 207); *Mensuration 1* (page 253); *Trigonometry 1, 5 and 6* (page 256); *Dispersion 1* (page 274).

Measure

Measure 1   The metric system (SI) (6)

Base units likely to be met in mathematics

Length: metre (m)
Mass (weight): kilogram (kg)
Time: second (s)

Prefixes in common use

mega (M) $= 10^6$ kilo (k) $= 10^3$ centi (c) $= 10^{-2} (\frac{1}{100})$
milli (m) $= 10^{-3}$ $(\frac{1}{1000})$ micro (μ) $= 10^{-6} (\frac{1}{1\,000\,000})$

Other units which may be used with SI

litre (best not abbreviated) = 1000 cm^3
tonne (best not abbreviated) = 1000 kg
hectare (best not abbreviated) = 10 000 m^2

Some metric prefixes are used with these, e.g. centilitre (cl) = $\frac{1}{100}$ litre = 10 cm^3, millilitre (ml) = $\frac{1}{1000}$ litre = 1 cm^3, megatonne.

Changing from one metric unit to another

Never insert zeros between figures. The figure in the units' column is rewritten in the correct column of the new unit.

Examples   (a) 108.9 mm → metres
                 The 8 in the units' column is 8 mm = $\frac{8}{1000}$ metre, so the 8 is rewritten in the thousandths' column, giving 108.9 mm → 0.1089 metres.

       (b) 3.06 cm → metres
           3.06 cm —————→ 0.0306 m
                3 cm = $\frac{3}{100}$ m

       (c) 0.003 06 m → km
           0.003 06 m —————→ 0.000 003 06 km
                0 m = $\frac{0}{1000}$ km

Measure 2   Imperial/metric equivalents (6)

It is useful to remember that:

1 inch = 2.54 cm
1 foot (12 inches) ≈ 30 cm
5 miles ≈ 8 km   3 metres ≈ 10 feet

1 pound (16 ounces) ≈ 454 g   1 kg ≈ $2\frac{1}{4}$ pounds

1 gallon (8 pints) ≈ $4\frac{1}{2}$ litres   1 litre ≈ $1\frac{2}{3}$ pints

Practical graphs

Practical graphs 1   Distance/time graphs (8)

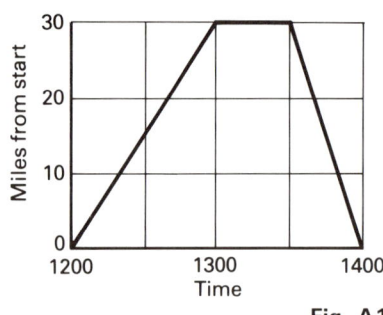

Fig. A1

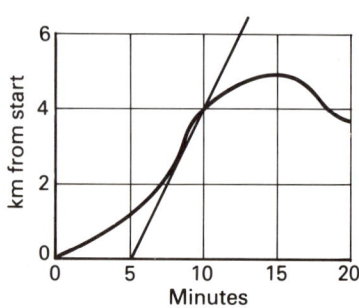

Fig. A2

A distance/time graph always has time plotted horizontally. The slopes of the lines, measured by vertical distance over horizontal distance, represent the average speed.

Figure A1 represents a journey of 60 miles in 2 hours. The average speeds are 30 m.p.h. for 1 hour, then a stop of $\frac{1}{2}$ hour, followed by 60 m.p.h. for $\frac{1}{2}$ hour.

In a real situation the journey line is a curve. The slope at any time is found by drawing a tangent to the curve. In Figure A2 the speed 8 minutes from the start was about $\frac{4}{5}$ or 0.8 km/min.

Practical graphs 2 (H)   Speed/time graphs (8)

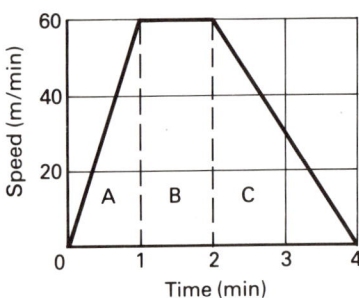

Fig. A3

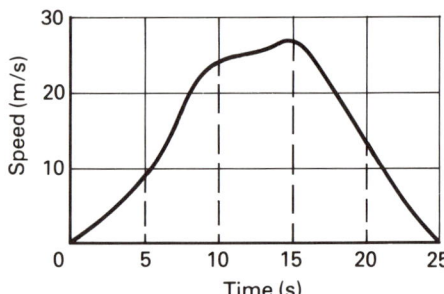

Fig. A4

In a speed/time graph, the slopes of the lines represent the average acceleration, while the area under the line represents the distance travelled.

In Figure A3, the object had an average acceleration of 60 metres per minute per minute (m/min^2) for 1 minute, then a constant speed of 60 m/min, followed by an average deceleration of 30 m/min^2. The distance travelled is the area A + B + C = 30 + 60 + 60 = 150 metres.

In Figure A4, we calculate the area by dividing the curve into a series of trapeziums whose areas are calculated and added to give an approximate distance travelled. These areas can be calculated individually, or you can use the trapezoidal formula:

Area $= \frac{1}{2}h(f + l + 2(m_1 + m_2 + m_3 + \ldots))$

where h is the width across the trapeziums, f and l are the heights of the first and last upright lines, and m_1, m_2, etc. are the heights of the intermediate lines. For Figure A4 this gives:

$2.5(0 + 0 + 2(8 + 24 + 27 + 16)) = 273$ metres.

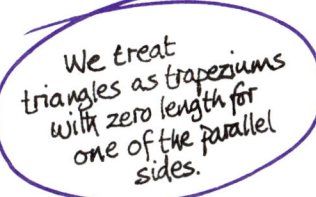

We treat triangles as trapeziums with zero length for one of the parallel sides.

Practical graphs 3   Conversion graphs (9)

When two quantities are in a constant ratio, a straight-line conversion graph may be used to convert one to the other. Figure A5 shows conversion from gallons to litres.

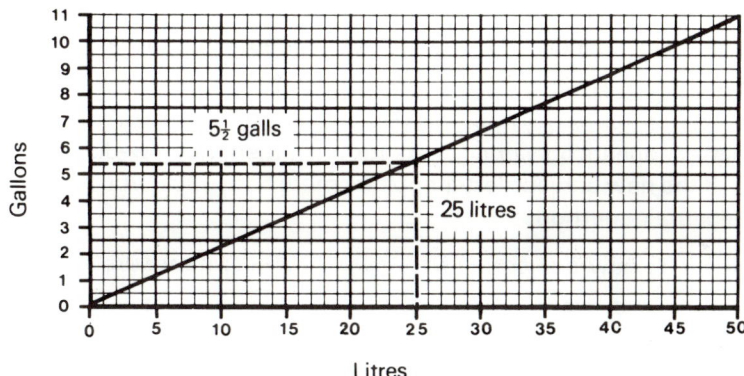

$5\frac{1}{2}$ galls

25 litres

Litres

Fig. A5

Basic algebra

Basic algebra 1   Basic algebraic notation (10)

When using letters to stand for numbers we do not need to write multiplication signs. So $3 \times a \times b$ is usually written as $3ab$. Note that $3 \times 4 \times 5$ can not be written as 345, and that a computer needs $3*4*5$ and $3*a*b$.

When using letters to stand for numbers the division sign is usually replaced by writing the expression as a fraction. $3 \div a$ is usually written as $\frac{3}{a}$. Note that $3 \div 4$ can also be written as $\frac{3}{4}$ or 3/4, and that a computer needs 3/4 for both $3 \div 4$ and $\frac{3}{4}$.

217

Basic algebra 2   Substitution (10)

Examples   If $a = -2$, $b = -3$ and $c = 4$, then:

$a + b = -2 + -3 \rightarrow -2 - 3 = -5$

> **Note** $-2 - 3$ can be thought of as 'down 2 then down another 3'. The two minuses do not make a plus here, as they are not multiplied.

$bc = -3 \times 4 = -12$

$ab = -2 \times -3 = 6$

$\dfrac{c}{a} = \dfrac{4}{-2} = -2$ **Note**   The same sign rules apply for division as for multiplication.

$\dfrac{a}{b} = \dfrac{-2}{-3} = \dfrac{2}{3}$

Basic algebra 3   Transposing formulae ('change of subject') (10)

$C = \pi d$ is the formula for the circumference of a circle; C is the **subject**.

When the formula is 'transposed', the subject is changed. The circumference formula can be transposed to make d the subject, giving $d = \dfrac{C}{\pi}$.

If the new subject-letter appears only once in the formula, then the flow-diagram method illustrated below may be used. If the new subject-letter appears more than once, you have to use the 'balance' or 'change sides' algebraic method.

Examples   (a) $u = s - t$; new subject s

$s \xrightarrow{-t} s - t$
$u + t \xleftarrow{+t} u$ so $s = u + t$

(b) $u = s - t$; new subject t

$t \xrightarrow{\text{taken from } s} s - t$
$s - u \xleftarrow{\text{taken from } s} u$ so $t = s - u$

Remember that 'taken from' does not change.

(c) $p = sr$; new subject s

$s \xrightarrow{\times r} sr$
$\dfrac{p}{r} \xleftarrow{\div r} p$ so $s = \dfrac{p}{r}$

(d) $u = sr - t$; new subject s

$s \xrightarrow{\times r} sr \xrightarrow{-t} sr - t$
$\dfrac{u + t}{r} \xleftarrow{\div r} u + t \xleftarrow{+t} u$ so $s = \dfrac{u + t}{r}$

(e) $t = 2\pi \sqrt{\dfrac{l}{g}}$; new subject g

$g \xrightarrow{\text{divided into } l} \dfrac{l}{g} \xrightarrow{\sqrt{\ }} \sqrt{\dfrac{l}{g}} \xrightarrow{\times 2\pi} 2\pi\sqrt{\dfrac{l}{g}}$

$\dfrac{l}{\left(\dfrac{t}{2\pi}\right)^2} \xleftarrow{\text{divided into } l} \left(\dfrac{t}{2\pi}\right)^2 \xleftarrow{(\ )^2} \dfrac{t}{2\pi} \xleftarrow{\div 2\pi} t$

so $g = \dfrac{l}{\left(\dfrac{t}{2\pi}\right)^2} = l \div \dfrac{t^2}{4\pi^2} = l \times \dfrac{4\pi^2}{t^2} = \dfrac{4\pi^2 l}{t^2}$

(f) $f = 2uf - v$; new subject f
The flow method cannot be used. Bring both f terms to the same side:
$f - 2uf = -v$
Take out the f as a common factor:
$f(1 - 2u) = -v$
Then $f = \dfrac{-v}{1 - 2u}$

If you multiply the top and bottom of the fraction by -1 you simplify the answer:

$f = \dfrac{v}{2u - 1}$.

Basic algebra 4   Brackets (11)

Any term written directly before a bracket multiplies each term in the bracket.

Examples $2(4 + a) \rightarrow 8 + 2a$ Working: $2 \times 4 = 8$; $2 \times a = 2a$

$-2(4 + a) \rightarrow -8 - 2a$ Working: $-2 \times 4 = -8$; $-2 \times a = -2a$

$-2(4 - a) \rightarrow -8 + 2a$ Working: $-2 \times 4 = -8$; $-2 \times -a = +2a$

$-(4 + a) \rightarrow -4 - a$ Working: $-(4 + a) \Rightarrow -1(4 + a)$; $-1 \times 4 = -4$; $-1 \times a = -a$

$a + 2(b - c) \rightarrow a + 2b - 2c$

$a - 2(b - c) \rightarrow a - 2b + 2c$

Read the → sign as 'becomes'.

⇒ is the sign for 'implies' or 'means'.

Note that a computer requires the * (multiply) sign between the term and the bracket; for example, $-2*(4 - a)$.

Basic algebra 5   Indices (11)

a^2 is shorthand for $a \times a$.
b^3 is shorthand for $b \times b \times b$.
$3a^2$ means $3 \times a^2$, so that if $a = 4$, $3a^2 \rightarrow 3 \times 16 = 48$.

$3a^2$ when $a = 4$ is not $12 \times 12 = 144$.

$3a^2b^3 \times 2ab^4 \rightarrow 6a^3b^7$. You may use the rule 'Add the indices when multiplying powers of the same letter', or think of the terms written out in full:

$3a^2b^3 \times 2ab^4 \rightarrow 3 \times a \times a \times b \times b \times b \times 2 \times a \times b \times b \times b \times b$

3×2 is 6; the three a's multiply together to give a^3; the seven b's multiply together to give b^7.

Example Simplify $4a^5b^3c \div 8a^2bc^3$.

$$\frac{4a^5b^3c}{8a^2bc^3} \rightarrow \frac{^1\cancel{4} \times \cancel{a} \times \cancel{a} \times a \times a \times a \times \cancel{b} \times b \times b \times \cancel{c}}{_2\cancel{8} \times \cancel{a} \times \cancel{a} \times \cancel{b} \times \cancel{c} \times c \times c}$$

Having cancelled as much as possible, this leaves $\dfrac{a^3b^2}{2c^2}$.

Or: Using the 'subtract indices' rule:

$$\frac{4a^5b^3c}{8a^2bc^3} \rightarrow \frac{1a^{5-2}b^{3-1}}{2c^{3-1}} \rightarrow \frac{a^3b^2}{2c^2}$$

Note: The c term is at the bottom in the answer because the bigger power of c (c^3) was at the bottom to start with.

Example $c^3(c^4 + 2) \rightarrow c^7 + 2c^3$

Basic algebra 6 (H)   Special indices (1)

The value of a^0

$a^x \div a^x$ must equal 1, but using the subtraction of indices rule, $a^x \div a^x \rightarrow a^{x-x} = a^0$.
Learn: $a^0 = 1$.

Fractional indices

$x^{\frac{1}{2}}$ is another way of writing $\sqrt{x}$ because $x^{\frac{1}{2}} \times x^{\frac{1}{2}} = x^{\frac{1}{2}+\frac{1}{2}} = x$.

Similarly $x^{\frac{1}{3}} \times x^{\frac{1}{3}} \times x^{\frac{1}{3}} = x$, so $x^{\frac{1}{3}}$ is the cube root of x, $\sqrt[3]{x}$.

Learn: $x^{\frac{1}{n}}$ is the same as $\sqrt[n]{x}$.

Example Simplify $8^{\frac{2}{3}}$.

As 8^2 means '8 squared', and $8^{\frac{1}{3}}$ means 'the cube root of 8', it follows that $8^{\frac{2}{3}}$ means the cube root of 8 squared, that is, $2^2 = 4$.

Always find the root first, then the power; this keeps the numbers smaller.

Combining indices

Example $(2a^2)^3 \rightarrow 2a^2 \times 2a^2 \times 2a^2 \rightarrow 8a^6$

Learn: $(x^m)^n \rightarrow x^{mn}$

Negative indices

By subtracting indices, $a^2 \div a^4 = a^{-2}$.

By cancelling, $a^2 \div a^4 \rightarrow \dfrac{a^2}{a^4} \rightarrow \dfrac{1}{a^2}$

Therefore $a^{-2} = \dfrac{1}{a^2}$.

Similarly $a^{-1} = \dfrac{1}{a}$ and $49^{-\frac{1}{2}} = \dfrac{1}{\sqrt{49}} = \frac{1}{7}$.

Learn: x^{-n} is the same as $\dfrac{1}{x^n}$.

Summary

$a^0 = 1$ for all values of a (except 0??!)

$a^{\frac{1}{n}} = \sqrt[n]{a}$

$a^{-n} = \dfrac{1}{a^n}$

Example Simplify $27^{-\frac{2}{3}}$.

The $-$ means 'one over', giving $\dfrac{1}{27^{\frac{2}{3}}}$.

The third means the cube root, giving $\dfrac{1}{3^2}$.

The two means squared, giving the final answer $\frac{1}{9}$.

Factors

Factors 1 Expansion of quadratic brackets (11)

Examples $(x + 2)(x + 3) \rightarrow x^2 + 3x + 2x + 6 \rightarrow x^2 + 5x + 6$

$(m - 3)(m + 2) \rightarrow m^2 + 2m - 3m - 6 \rightarrow m^2 - m - 6$

Note: The middle step should be done mentally when you have had some practice.

Example $(4 + x)(3 - x) \rightarrow 12 - 4x + 3x - x^2 \rightarrow 12 - x - x^2$

Note: Do not attempt to change the order of the given terms.

Factors 2   Common factors (11)

If all the terms in an expression have a common factor then the common factor may be 'taken out' and written in front of a bracket.

Examples
$3 + 6a \rightarrow 3(1 + 2a)$

$2a - ab \rightarrow a(2 - b)$

$3ax^2 + ax \rightarrow ax(3x + 1)$

Make sure that you take out the *highest* common factor.

Example   Factorise $3x - 3y + ax - ay$.

$3x - 3y + ax - ay \rightarrow 3(x - y) + a(x - y)$
Note that $(x - y)$ is itself now a common factor and may be 'taken out':
$3(x - y) + a(x - y) \rightarrow (x - y)(3 + a)$

Multiply $(x - y)(3 + a)$ to check the answer.

Example $\dfrac{6x + 6}{2xy + 2y} \rightarrow \dfrac{3(2x + 2)}{y(2x + 2)} \rightarrow \dfrac{3}{y}$

Factors 3   Difference of two squares (11)

If two terms in an expression are both squares and are connected by a minus, then they may be split into 'sum times difference'.

Examples
$a^2 - 16 \rightarrow (a + 4)(a - 4)$

$4x^2 - 25 \rightarrow (2x + 5)(2x - 5)$

$2a^2 - 18 \rightarrow 2(a^2 - 9) \rightarrow 2(a + 3)(a - 3)$

This can be useful in arithmetic.

Examples
$54^2 - 46^2 \rightarrow (54 + 46)(54 - 46) \rightarrow 100 \times 8 = 800$

Area of an annulus (e.g. a washer) is $\pi R^2 - \pi r^2$
$\pi R^2 - \pi r^2 \rightarrow \pi(R^2 - r^2) \rightarrow \pi(R + r)(R - r)$

Factors 4 (H)   Quadratics (11)

$(x + 4)(x + 3) \rightarrow x^2 + 3x + 4x + 12 \rightarrow x^2 + 7x + 12$

By reversing this we can factorise a quadratic expression.

By no means will all quadratic expressions factorise into two brackets, but an examiner will not ask you to factorise one that will not.

There is no 'golden rule' which will give you the correct answer first time, but the following will help:

If the last sign is $+$, then both brackets have the sign of the middle term. In checking with 'FOIL' (First, Outer, Inner, Last), O + I gives the middle term.

Examples $x^2 + 10x + 21 \rightarrow (x + 3)(x + 7)$
Check: $+7x + 3x \rightarrow 10x$
$\qquad +3 \times +7 \rightarrow +21$

$x^2 - 10x + 21 \rightarrow (x - 3)(x - 7)$
Check: $-7x - 3x \rightarrow -10x$
$\qquad -3 \times -7 \rightarrow +21$

If the last sign is $-$, then one bracket is $+$ and the other is $-$, and you must be very careful to get the correct sign with each number. For example $(2x + 1)(x - 2)$ gives $2x^2 - 3x - 2$, whilst $(2x - 1)(x + 2)$ gives $2x^2 + 3x - 2$.

Examples $x^2 + 4x - 12 \rightarrow (x + 6)(x - 2)$
Check: $-2x + 6x \rightarrow +4x$
$\qquad +6 \times -2 \rightarrow -12$

$x^2 - 4x - 12 \rightarrow (x - 6)(x + 2)$
Check: $+2x - 6x \rightarrow -4x$
$\qquad -6 \times +2 \rightarrow -12$

If there are a lot of factors to choose from, start with the pair closest together and work upwards; for example, for 12 try 3×4, then 2×6, then 1×12.

Factors 5   Summary (11)

This flow diagram summarises the rules for algebraic factorisation.

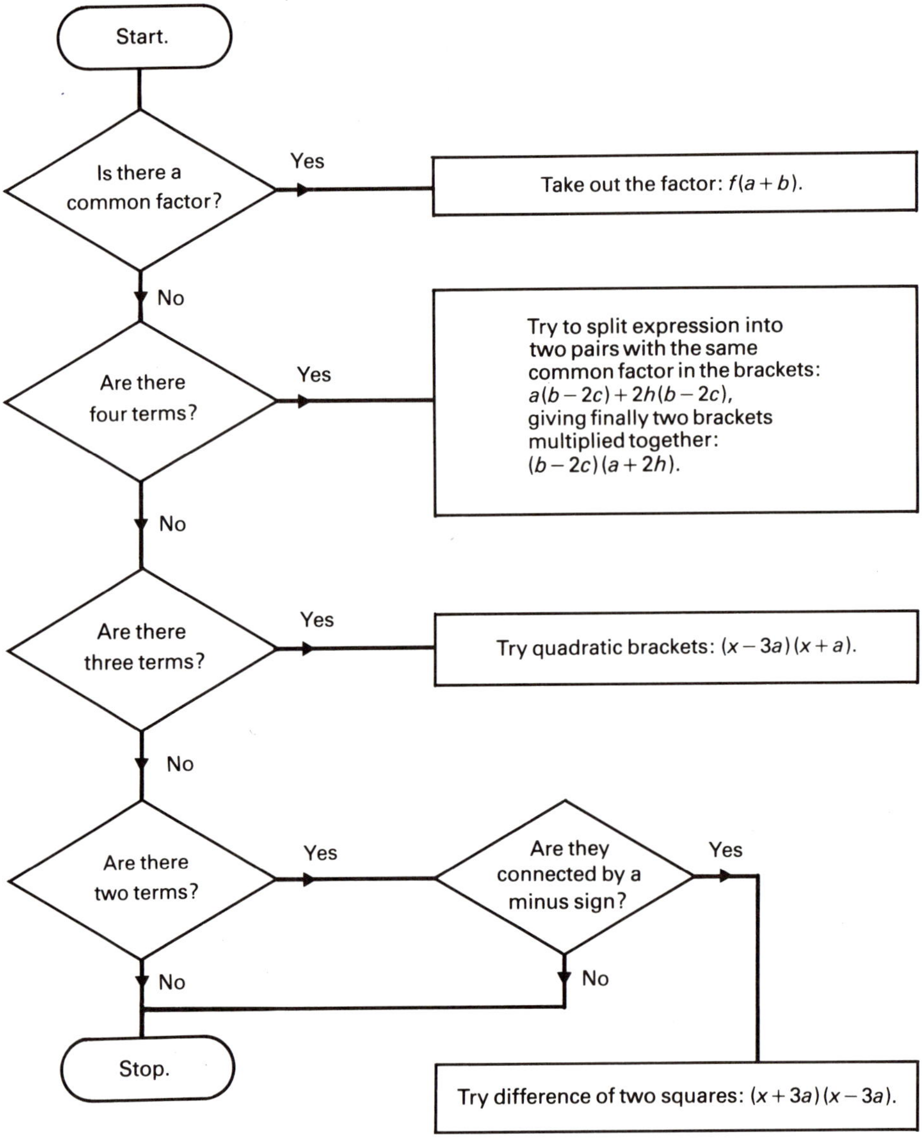

```
                    ┌──────────────┐
                    │    Start.    │
                    └──────────────┘
                            │
                            ▼
                      ╱ Is there a ╲        Yes      ┌──────────────────────────────────┐
                     ╱  common factor? ╲ ──────────▶ │  Take out the factor: f(a + b).   │
                      ╲              ╱               └──────────────────────────────────┘
                            │ No
                            ▼
                      ╱ Are there  ╲         Yes      ┌──────────────────────────────────┐
                     ╱ four terms?  ╲ ──────────────▶ │  Try to split expression into     │
                      ╲            ╱                  │  two pairs with the same          │
                            │ No                      │  common factor in the brackets:   │
                            │                         │  a(b − 2c) + 2h(b − 2c),          │
                            │                         │  giving finally two brackets      │
                            │                         │  multiplied together:             │
                            │                         │  (b − 2c)(a + 2h).                │
                            ▼                          └──────────────────────────────────┘
                      ╱ Are there  ╲         Yes      ┌──────────────────────────────────┐
                     ╱ three terms? ╲ ──────────────▶ │ Try quadratic brackets: (x − 3a)(x + a). │
                      ╲            ╱                  └──────────────────────────────────┘
                            │ No
                            ▼
                      ╱ Are there  ╲    Yes    ╱ Are they  ╲    Yes
                     ╱ two terms?   ╲ ──────▶ ╱ connected by a ╲ ──────▶
                      ╲            ╱           ╲ minus sign? ╱
                            │ No                    │ No
                            ▼                        ▼
                    ┌──────────────┐        ┌──────────────────────────────────┐
                    │    Stop.     │        │ Try difference of two squares: (x + 3a)(x − 3a). │
                    └──────────────┘        └──────────────────────────────────┘
```

Fig. A6

Is there a common factor?

Take out the factor: $f(a + b)$.

Are there four terms?

Try to split expression into two pairs with the same common factor in the brackets: $a(b - 2c) + 2h(b - 2c)$, giving finally two brackets multiplied together: $(b - 2c)(a + 2h)$.

Are there three terms?

Try quadratic brackets: $(x - 3a)(x + a)$.

Are there two terms?

Are they connected by a minus sign?

Try difference of two squares: $(x + 3a)(x - 3a)$.

Equations

Equations 1   One letter-term (12)

Equations with only one letter-term are nearly always solved (to find the value of the letter) most easily by the 'inspection' approach, not by using 'rules'.

Examples   If $b - 2 = 8$ then b must be $\underline{10}$ (as $10 - 2 = 8$).

If $8 + c = 6$ then c must be $\underline{-2}$ (as $8 + -2 = 6$).

If $7 - 2x = 8$ then $2x$ must be -1 (as $7 - -1 \rightarrow 7 + 1 = 8$)
so x must be $\underline{-\frac{1}{2}}$ (as $2 \times -\frac{1}{2} = -1$).

If $3e = 2$ then e must be $\frac{2}{3}$ (if 3 e's make 2, then 1 e must be a third of 2).

If $\dfrac{24}{1 - n} = 8$ then $1 - n$ must be 3 (as $24 \div 3 = 8$)
so n must be $\underline{-2}$ (as $1 - -2 \rightarrow 1 + 2 = 3$).

If $4(2a - 7) = 12$ then $2a - 7$ must be 3 (as $4 \times 3 = 12$)
so $2a$ must be 10 (as $10 - 7 = 3$)
so a must be $\underline{5}$ (as $2 \times 5 = 10$).

The above solutions can be written as follows:

$b - 2 = 8 \;\Rightarrow\; \underline{b = 10}$

$8 + c = 6 \;\Rightarrow\; \underline{c = -2}$

$7 - 2x = 8 \;\Rightarrow\; 2x = -1 \;\Rightarrow\; \underline{x = -\frac{1}{2}}$

$3e = 2 \;\Rightarrow\; \underline{e = \frac{2}{3}}$

$\dfrac{24}{1 - n} = 8 \;\Rightarrow\; 1 - n = 3 \;\Rightarrow\; \underline{n = -2}$

$4(2a - 7) = 12 \;\Rightarrow\; 2a - 7 = 3 \;\Rightarrow\; 2a = 10 \;\Rightarrow\; \underline{a = 5}$

Equations 2 Two letter-terms (12)

In solving equations like $3n - 2 = 5 + 2n$ we cannot use the inspection approach (see *Equations 1*, page 225), until one of the two letter-terms has been removed.

We remove one letter-term by adding to or subtracting from both sides of the equation a term which reduces it to zero, as in the following examples.

Examples (a) $3n - 2 = 5 + 2n$ $\xrightarrow{-2n \text{ on both sides}}$ $3n - 2 - 2n = 5 + 2n - 2n$ → $n - 2 = 5$

(b) $5 - 2w = 2 + 3w$ $\xrightarrow{+2w \text{ on both sides}}$ $5 - 2w + 2w = 2 + 3w + 2w$
→ $5 = 2 + 5w$

(c) $4 - 3x = 5 - 2x$ $\xrightarrow{+3x \text{ on both sides}}$ $4 - 3x + 3x = 5 - 2x + 3x$ → $4 = 5 + x$

(d) $4 - 3z = 5 + 2z$ $\xrightarrow{+3z \text{ on both sides}}$ $4 - 3z + 3z = 5 + 2z + 3z$ → $4 = 5 + 5z$

It does not really matter which letter-term you reduce to zero, but the remainder of the solution is usually easier if you leave a positive letter-term, as we did in the above examples. This can be remembered by the following rule, if you like rules!

Remove the term with the smaller coefficient.

The 'coefficient' is the number in front of it. Note that in example (c), -3 is smaller than -2.

Note: Because we perform the same operation on each side of the equation it remains true or 'in balance'. We can see this in number statements:

$3 + 2 = 5$ $\xrightarrow{-2 \text{ on both sides}}$ $3 + 2 - 2 = 5 - 2$ → $3 = 5 - 2$

$3 - 2 = 1$ $\xrightarrow{+2 \text{ on both sides}}$ $3 - 2 + 2 = 1 + 2$ → $3 = 3$

Examples Solve $6n + 7 = 4n + 13$.

$6n + 7 = 4n + 13$ $\xrightarrow{-4n}$ $2n + 7 = 13$ ⇒ $2n = 6$ ⇒ $\underline{n = 3}$

Solve $7 - 5x = 4x - 2$.

$7 - 5x = 4x - 2$ $\xrightarrow{+5x}$ $7 = 9x - 2$ ⇒ $9x = 9$ ⇒ $\underline{x = 1}$

Solve $3(x + 2) = 2(x - 1)$.

$3(x + 2) = 2(x - 1)$ → $3x + 6 = 2x - 2$ $\xrightarrow{-2x}$ $x + 6 = -2$ ⇒ $\underline{x = -8}$

If necessary, 'collect terms' on each side of the equation before beginning to solve it.

Example Solve $3(a - 2) - 1 = 4(a + 4) + 2a - 2$.

First multiply out the brackets and collect like terms:
$3(a - 2) - 1 = 4(a + 4) + 2a - 2$ → $3a - 6 - 1 = 4a + 16 + 2a - 2$
→ $3a - 7 = 6a + 14$.

Now solve the equation:

$3a - 7 = 6a + 14$ $\xrightarrow{-3a}$ $-7 = 3a + 14$ ⇒ $3a = -21$ ⇒ $\underline{a = -7}$

Equations 3 Simultaneous (12)

When there are two unknown letters to be found you need two equations. For instance: $x + y = 8$ is true for an infinite number of pairs of values for x and y. But if we also know that $x = y + 2$ then the only possible solution is $x = 5$ and $y = 3$.

Four methods are possible:

Method 1 **Draw intersecting graphs.**

Method 2 **Substitute for one letter its value in the other equation.** This is best used when one of the equations is in the form $x =$ or $y =$. For example, $y = x - 6$ and $3y - 2x = 8$.

Method 3 **Add or subtract the equations to eliminate one of the letter-terms,** having multiplied as necessary to make one letter-term the same absolute value in both. ('Absolute' means 'ignoring the sign'; ABS on a BASIC computer.)

Method 4 **Use matrices.** This is not covered in this book.

Method 1

To solve simultaneously $y = 2x - 1$ and $y - x = 1$.

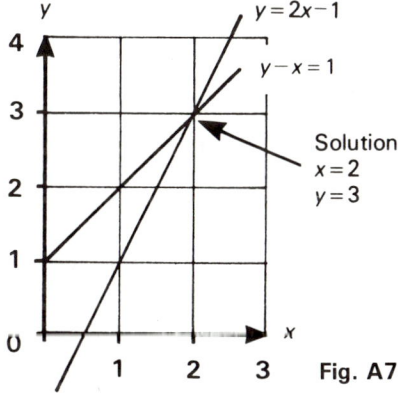

Fig. A7

Method 2

To solve $x = 2$ and $2x + 2y = 11$ simultaneously.

We can think of this as finding where the line $x = 2$ crosses the line $2x + 2y = 11$.
Substitute the value $x = 2$ into $2x + 2y = 11$, giving $4 + 2y = 11$.
Then $4 + 2y = 11 \rightarrow 2y = 7 \rightarrow y = 3\frac{1}{2}$.

Answer: $x = 2$, $y = 3\frac{1}{2}$ (The lines cross at $(2, 3\frac{1}{2})$.)

To solve $y = x - 6$ and $x + y = 4$ simultaneously.

Substitute the value $y = x - 6$ into $x + y = 4$, giving $x + (x - 6) = 4$.
Then $x + (x - 6) = 4 \rightarrow x + x - 6 = 4 \rightarrow 2x - 6 = 4 \rightarrow 2x = 10 \rightarrow x = 5$.
We know that $y = x - 6$, so if $x = 5$ then y must be -1.

Answer: $x = 5$, $y = -1$.

To solve $x = 2y - 1$ and $y + 2x = 2$ by substitution.

Substitute for x, giving $y + 2(2y - 1) = 2$.
$y + 2(2y - 1) = 2 \rightarrow y + 4y - 2 = 2 \rightarrow 5y - 2 = 2 \rightarrow 5y = 4 \rightarrow y = \frac{4}{5}$.
Now as we know that $x = 2y - 1$, then $x = \frac{8}{5} - 1 = \frac{3}{5}$.

Answer: $x = \frac{3}{5}$, $y = \frac{4}{5}$.

Method 3

To find the values of x and y that satisfy both $x + 3y = 9$ and $x - 2y = -1$.

$$\begin{array}{l} x + 3y = 9 \\ \underline{x - 2y = -1} \quad \text{SUB} \\ 5y = 10 \\ \underline{\underline{y = 2}} \end{array}$$

Note: By subtracting, the x terms have been eliminated. If the two given equations are true, then the result of the subtraction is true.

Substitute $y = 2$
into $x + 3y = 9$,
giving $x + 6 = 9$,
so $\underline{x = 3}$.

Compare: $2 + 3 = 5$
$\underline{1 + 2 = 3} \quad \text{SUB}$
$\underline{1 + 1 = 2}$

Check both equations are true
for $x = 3$, $y = 2$:
$3 + 6 = 9$; $\quad 3 - 4 = -1$.

When subtracting,
$+3y - -2y \rightarrow +3y + 2y = 5y$
and $9 - -1 \rightarrow 9 + 1 = 10$.

To solve $3x + 5y = 21$ and $7x - 2y = 8$ simultaneously.

$3x + 5y = 21 \xrightarrow{\times 2} 6x + 10y = 42$

$7x - 2y = 8 \xrightarrow{\times 5} \underline{35x - 10y = 40} \quad \text{ADD}$

$$\begin{array}{l} 41x = 82 \\ \underline{\underline{x = 2}} \end{array}$$

Do not forget to find the second letter.

Substitute $x = 2$ into $3x + 5y = 21 \rightarrow 6 + 5y = 21 \rightarrow 5y = 15 \rightarrow \underline{\underline{y = 3}}$.

Check this for yourself.

Equations 4   Inequalities (12)

Inequalities can be solved by treating them as equations (but see the note following the example).

Example   Find the range of values of x if $2x - 3 < 15 - 3x$.

$$2x - 3 < 15 - 3x \xrightarrow{\ +3x \text{ on both sides}\ } 5x - 3 < 15$$

If $5x - 3 = 15$, then $5x = 18$, giving $x = \dfrac{18}{5} = 3\tfrac{3}{5}$.

As $5x - 3 < 15$, then the answer is $x < 3\tfrac{3}{5}$.

Note   An inequality statement is reversed if you multiply or divide both sides by a negative amount. Remember that the letter itself may be negative in which case you have two possible solutions, only one of which will be correct.

For the equation $-3x = 6$:

$$-3x = 6 \xrightarrow{\ \text{multiply both sides by } -1\ } 3x = -6$$

Both equations are true when $x = -2$.

BUT for the inequality $-3x \geqslant 6$ the solution is NOT $x \geqslant -2$ (e.g. when $x = 4$, $-3x$ is -12, which is not more than 6).

You have to reverse the inequality sign.

so $-3x \geqslant 6 \xrightarrow{\ \text{divide both sides by } -3\ } x \leqslant -2$

or $-3x \geqslant 6 \xrightarrow{\ \text{multiply both sides by } -1\ } 3x \leqslant -6 \rightarrow x \leqslant -2$.

Inequalities arise in graphical work on regions; see *Algebraic graphs 3* (page 236).

The topic of linear programming involves inequalities and graphical regions; see *Linear programming* (page 280).

Equations 5 (H)   Quadratic (12)

If two factors multiply to give zero, one or both of them must be zero.

Example   If $x(x - 4) = 0$ then either $x = 0$ or $x - 4 = 0$.
Hence $x(x - 4) = 0$ when $x = 0$ and when $x = 4$.

Example   If $(x - 2)(x + 4) = 0$ then either $x - 2 = 0$ or $x + 4 = 0$.
Hence $(x - 2)(x + 4) = 0$ when $x = 2$ and when $x = -4$.

Example   If $(2x - 7)(4x + 2) = 0$ then either $2x - 7 = 0$ or $4x + 2 = 0$.
If $2x - 7 = 0$ then $2x = 7$, so $x = 3\tfrac{1}{2}$ is one solution.
If $4x + 2 = 0$ then $4x = -2$, so $x = -\tfrac{1}{2}$ is the other solution.

Example   Solve $x^2 - 3x + 2 = 0$.

$$x^2 - 3x + 2 = 0 \rightarrow (x - 2)(x - 1) = 0$$
$$\therefore x = 2 \text{ and } x = 1 \text{ are the solutions.}$$

Example Solve $x^2 - 4x = 0$.

Quadratics with no constant term are easy to factorise by taking out the common factor. Make a special point of remembering this.

$x^2 - 4x = 0 \rightarrow x(x - 4) = 0$
$\therefore x = 0$ and $x = 4$ are solutions.

The solutions of the equation $ax^2 + bx + c = 0$ are called its **roots**. These roots are usually referred to as α (alpha) and β (beta).

$$\alpha + \beta = -\frac{b}{a} \quad \text{and} \quad \alpha\beta = \frac{c}{a}$$

These two facts can be used to check solutions to a quadratic equation.

Example Find the roots of $2x^2 + 3x - 9 = 0$.

$2x^2 + 3x - 9 = 0 \rightarrow (2x - 3)(x + 3) = 0$

The roots are $x = 1\frac{1}{2}$ and $x = -3$.

Check by letting $\alpha = 1\frac{1}{2}$ and $\beta = -3$, and, knowing that $a = 2$, $b = 3$ and $c = -9$:

$\alpha + \beta = -\frac{b}{a} \rightarrow 1\frac{1}{2} - 3 = -\frac{3}{2}$ which is correct, and

$\alpha\beta = \frac{c}{a} \rightarrow 1\frac{1}{2} \times -3 = -\frac{9}{2}$ which is also correct.

Many quadratic equations will not factorise; the following two algebraic methods may then be used unless $b^2 < 4ac$ in which case there is no solution. For example, $x^2 + 1 = 0$ has no solution ($b^2 = 0$, $4ac = 4$).

Method 1 Completing the square

Example Solve $x^2 + 5x - 1 = 0$.

$$x^2 + 5x - 1 = 0 \rightarrow \left(x + \frac{5}{2}\right)^2 - \frac{25}{4} - 1 = 0$$

$$\left[\text{Note } \left(x + \frac{5}{2}\right)\left(x + \frac{5}{2}\right) = x^2 + 5x + \frac{25}{4}\right]$$

$$\rightarrow \left(x + \frac{5}{2}\right)^2 - \frac{29}{4} = 0$$

$$\rightarrow \left(x + \frac{5}{2}\right)^2 = \frac{29}{4} = 7.25$$

$$\rightarrow x + 2.5 = \sqrt{7.25}$$
$$\rightarrow x + 2.5 \simeq \pm 2.69 \text{ (plus or minus 2.69)}$$

So $x \simeq -2.5 + 2.69 = \underline{\underline{0.19}}$

or $x \simeq -2.5 - 2.69 = \underline{\underline{-5.19}}$

Example Solve $3x^2 - 2x - 2 = 0$.

$$3x^2 - 2x - 2 = 0 \xrightarrow{\text{divide all by 3}} x^2 - \tfrac{2}{3}x - \tfrac{2}{3} = 0$$

$\rightarrow \ (x - \tfrac{1}{3})^2 - \tfrac{1}{9} - \tfrac{2}{3} = 0$

$\rightarrow \ (x - \tfrac{1}{3})^2 - \tfrac{7}{9} = 0$

$\rightarrow \ (x - \tfrac{1}{3})^2 = \tfrac{7}{9}$

$\rightarrow \ x - 0.\dot{3} = \sqrt{0.\dot{7}}$

$\rightarrow \ x - 0.\dot{3} \simeq \pm 0.882$

Either $x \simeq 0.\dot{3} + 0.882 \simeq \underline{1.22}$

or $x \simeq 0.\dot{3} - 0.882 \simeq \underline{-0.55}$

We divide by 3 to reduce the coefficient of x^2 to unity.

Method 2   The quadratic equation formula

The formula is obtained by applying the completion of the square method to the general quadratic equation $ax^2 + bx + c = 0$. Your teacher will illustrate this.

$$x = \frac{-b \pm \sqrt{b^2 - 4ac}}{2a}$$

Learn: 'x equals minus b plus or minus the square root of b squared minus four ac, all divided by two a.'

Example Solve $x^2 + 5x - 1 = 0$.

$a = 1 \qquad b = 5 \qquad c = -1$

$$x = \frac{-5 \pm \sqrt{25 + 4}}{2} = \frac{-5 \pm \sqrt{29}}{2}$$

$$x \simeq \frac{-5 \pm 5.385}{2}$$

Solutions are $x \simeq 0.19$ and $x \simeq -5.19$

Example Solve $3x^2 - 2x - 2 = 0$.

$a = 3 \qquad b = -2 \qquad c = -2$

$$x = \frac{+2 \pm \sqrt{4 + 24}}{6} = \frac{2 \pm \sqrt{28}}{6}$$

$$x \simeq \frac{2 \pm 5.29}{6}$$

Solutions are $x \simeq 1.22$ and $x \simeq -0.55$

The quadratic equation formula shows that unless $b^2 > 4ac$ there can be no solutions, for the square root of a negative number does not exist.

Example $x^2 + x + 1 = 0$ cannot be solved as $b^2 = 1$ and $4ac = 4$.

Equations 6 (H) Fractional (12)

Example Solve $\dfrac{x^2}{2} - x - \dfrac{3}{4} = 0$.

First remove the fractions by multiplying every term by 4, which is the common denominator.
As $4 \times \dfrac{x^2}{2} \rightarrow 2x^2$, and $4 \times \dfrac{3}{4} \rightarrow 3$, then the equation becomes:

$2x^2 - 4x - 3 = 0$.
This can now be solved as in *Equations 5*.

Example Solve $\dfrac{x}{x-1} - \dfrac{x+2}{x} = \dfrac{1}{2x}$

Multiply every term by $2x(x-1)$ and cancel the denominators.

$$2x^2 - 2(x - 1)(x + 2) = x - 1$$
$$2x^2 - 2x^2 - 2x + 4 = x - 1$$
$$-3x = -5$$
$$\underline{\underline{x = \tfrac{5}{3}}}$$

Equations 7 (H) Problems (12)

Many problem questions are best answered by letting the required answer be a letter (x is traditional), then forming an equation which can be solved to find x.

Example The length of a rectangle is 5 cm more than its breadth. If its area is 24 cm^2 find its dimensions.

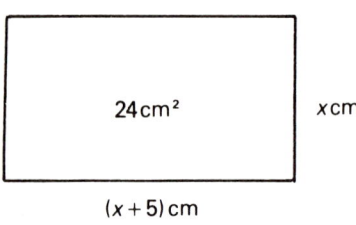

Let its breadth be x cm.
Then its length is $x + 5$ cm.
Area of rectangle = length × breadth
so $x(x + 5) = 24 \rightarrow x^2 + 5x = 24$
$\rightarrow x^2 + 5x - 24 = 0$
$\rightarrow (x - 3)(x + 8) = 0$
$\therefore x = 3$ or $x = -8$.
x cannot be -8, so the only solution is $x = 3$, making the rectangle of breadth 3 cm and length 8 cm.

Fig. A8

Equations 8 (H)   Iteration (12)

Iteration means repetition. It describes a method of solving equations by finding a formula, the value of which 'homes in' on a solution. At GCSE you will usually be given the iteration formula to use, as finding one that works is not easy.

Example   Find a solution to $x^3 + 3x^2 - 2x = 4$.

$$x^3 + 3x^2 - 2x = 4 \rightarrow x^3 = 4 - 3x^2 + 2x$$

$$\rightarrow x = \frac{4 - 3x^2 + 2x}{x^2}$$

The formula is usually written like this:

$$x_{n+1} = \frac{4 - 3x_n^2 + 2x_n}{x_n^2}$$

Choose a starting value for x, say 1.

When $x_n = 1$, $x_{n+1} = \dfrac{4 - 3 + 2}{1} = 3$

When $x_n = 3$, $x_{n+1} = \dfrac{4 - 27 + 6}{9} = -1.\dot{8}$

When $x_n = 1.\dot{8}$, $x_{n+1} \approx -2.938$

When $x_n = -2.938$, $x_{n+1} \approx -3.217$

When $x_n = -3.217$, $x_{n+1} \approx -3.235$

When $x_n = -3.235$, $x_{n+1} \approx -3.236$

When $x_n = -3.236$, $x_{n+1} \approx -3.236$

An approximate solution is $x \approx -3.236$.

To find the other two solutions you have to find other ways to rewrite the original equation, to give new iteration formulae. Do you like a challenge?!

The iteration method is ideally suited to a BASIC computer program. Here is the program for the given example:

```
10  INPUT X
20  X = (4 - 3*X↑2 + 2*X)/X↑2
30  PRINT X
40  GOTO 20
```

X↑2 means X raised to the power 2.

Algebraic graphs

Algebraic graphs 1   Co-ordinates (13)

In Figure A9:

Point A has co-ordinates (2, 0).
Point B has co-ordinates (0, 1).
Point C has co-ordinates (-1, -1).

Line DE is the x-axis. Its equation is $y = 0$.
Line FG is the y-axis. Its equation is $x = 0$.
The x-axis crosses the y-axis at the origin.

Line HI has the equation $y = x$. Each point on $y = x$ has its x co-ordinate the same as its y co-ordinate; e.g. (3, 3); (-2, -2); (1.5, 1.5).

Line JK has the equation $y = -x$. Each point on $y = -x$ has its y co-ordinate equal to its x co-ordinate times -1, e.g. (3, -3); (0, 0); (-2, 2).

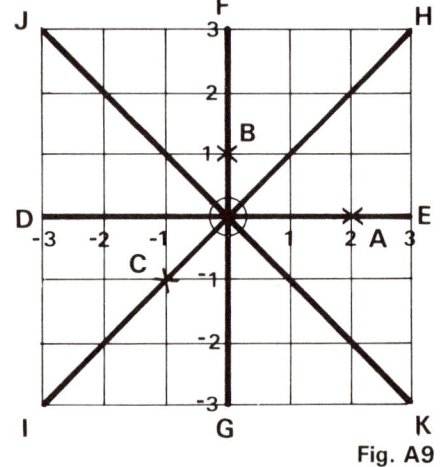

Fig. A9

Algebraic graphs 2   Linear (straight line) (13)

All straight lines can be expressed in equation form as $y = mx + c$, though the three terms may be moved around; for example, $y = 2x$, $y = 3$, $y = 2x + 3$, $x + y = 2$, $2y + 3x = 4$ and $x = 2 - 7y$ are all equations of straight-line graphs.

Linear graphs are drawn by one of the following methods.

Plotting method for $y + 2x = 3$

Choose three values for x (including zero) and find y for each. For example:

If $x = 0$, then $y + 0 = 3$ → $y = 3$. Plot (0, 3).
If $x = 2$, then $y + 4 = 3$ → $y = -1$. Plot (2, -1).
If $x = -1$, then $y - 2 = 3$ → $y = 5$. Plot (-1, 5).

See Figure A10.

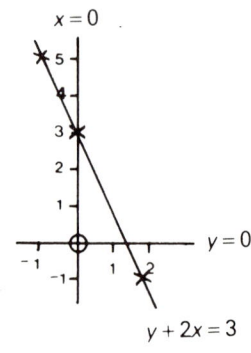

$y + 2x = 3$

Fig. A10

Slope/crossing ($y = mx + c$) method for $y + 2x = 3$

When the equation is expressed in the form $y = mx + c$, then c gives the crossing point on the y-axis $(0, c)$ and m gives the slope (see Figures A11 (a) and (b)).

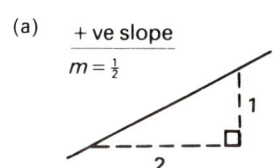

(a) + ve slope
$m = \frac{1}{2}$

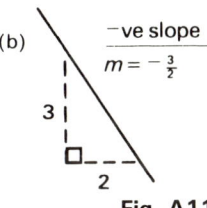

(b) −ve slope
$m = -\frac{3}{2}$

Fig. A11

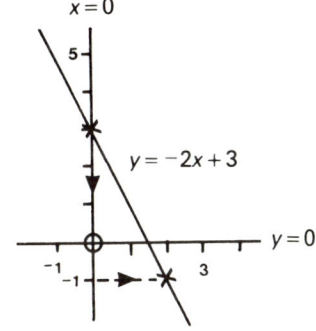

$x = 0$

$y = -2x + 3$

$y = 0$

Fig. A12

To use this method for $y + 2x = 3$ we first rearrange the terms to change it to $y = -2x + 3$. We now know that $m = -2$ and $c = 3$. Start from 3 on the y-axis, then go down and across to give a slope of -2. See Figure A12.

The slope/crossing method may also be used to find the equation of a drawn line, and an extension of the method finds the equation of the line through two points.

Example Find the equation of the line through $(-2, 4)$ and $(3, -1)$.

Sketch the position of the points (Figure A13).

The slope is $-\dfrac{5}{5} = -1$.

$y = mx + c \rightarrow y = -x + c$
When $x = -2$, $y = 4$, so $4 = 2 + c \Rightarrow c = 2$.
The equation of the line is $y = -x + 2$.

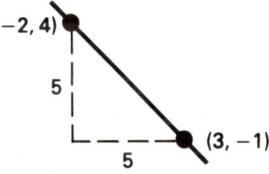

$(-2, 4)$

$(3, -1)$

Fig. A13

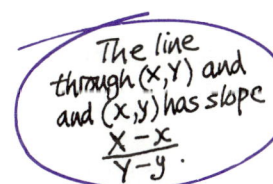

The line through (X,Y) and and (x,y) has slope $\dfrac{X-x}{Y-y}$.

See also *Equations 3* Method 1 (page 227) for the use of linear graphs to solve simultaneous equations.

Algebraic graphs 3   Regions (13)

Figure A14 shows the line $y = -x + 2$.

For every point on the line, the y co-ordinate is equal to $-x + 2$, where x is the x co-ordinate of the point.

Above the line, y is more than $-x + 2$.

Below the line, y is less than $-x + 2$.

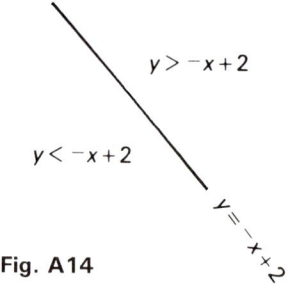

$y > {}^-x + 2$

$y < {}^-x + 2$

$y = {}^-x + 2$

Fig. A14

Figure A15 shows the region

$\{(x, y):\ x < 3;\quad -x + 2 < y \leqslant x\}.$

to the left ↗ above ↗ below or on ↑
of $x = 3$ $y = -x + 2$ $y = x$

Dotted lines are used when the region does not include the line.

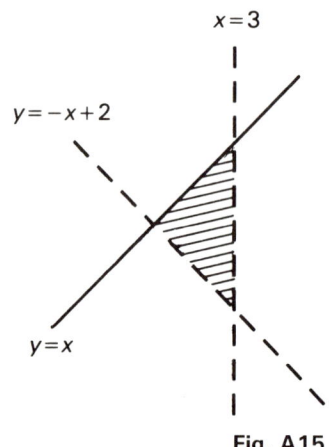

$x = 3$

$y = -x + 2$

$y = x$

Fig. A15

Algebraic graphs 4   Parabolas (13)

Equations like $y = x^2$, $y = x^2 + 3$ and $y = 2x^2$ give **parabolas** when drawn as graphs.

Sometimes the equation is given in function notation, e.g. $y = f(x)$ where $f(x):\ x \rightarrow x^2$. This has the same meaning as $y = x^2$.

To draw the graphs, plot values of y for a series of values of x. (You are usually told which values of x to use.) It is best to work out the values in a table.

Example   Draw the parabola $y = 2x^2 - 3$.

	x	-2	-1	0	1	2
Values for x:						
Working out:	$2x^2$	8	2	0	2	8
	-3	-3	-3	-3	-3	-3
Values for y:	y	5	-1	-3	-1	5

Values for x: (row x)
Working out: $\left\{\begin{array}{l}2x^2 \\ -3\end{array}\right.$ $\left.\begin{array}{l}\\ \end{array}\right\}$ Add these two rows to find y.
Values for y: (row y)

The points $(-2, 5)$, $(-1, -1)$, etc. are now plotted and joined with a smooth continuous curve. See Figure A16.

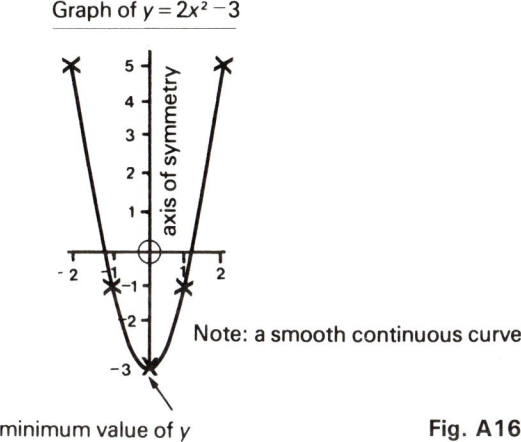

Graph of $y = 2x^2 - 3$

axis of symmetry

Note: a smooth continuous curve

minimum value of y

Fig. A16

Algebraic graphs 5 (H)   Using parabolas to solve quadratic equations (15)

A quadratic equation is one involving x^2. A very simple quadratic equation is $x^2 = 1$. A quadratic equation may have two solutions. Both $x = 1$ and $x = -1$ make $x^2 = 1$ true. It may also have no solutions, like $x^2 + 1 = 0$.

Example   Figure A17 shows the graph of $y = x^2 - 3x + 1$.

We want to solve the equation $x^2 - 3x + 1 = 0$ (or $x^2 - 3x = -1$).

The parabola's equation, $y = x^2 - 3x + 1$, becomes the equation we have to solve, $x^2 - 3x + 1 = 0$, when $y = 0$. Therefore we find the solutions where the parabola crosses the line $y = 0$ (the x-axis). They are $x = 2.6$ and $x = 0.38$ approximately. (You often do not have exact answers to a quadratic equation.)

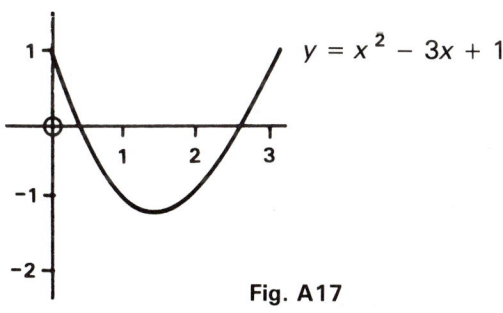

$y = x^2 - 3x + 1$

Fig. A17

Where two graphs cross their y-values are equal.

Example   Where $y = x + 1$ crosses $y = x^2 + 2x - 1$ the equation solved is
$x + 1 = x^2 + 2x - 1 \rightarrow x^2 + x - 2 = 0$.

To solve an equation from a given parabola, change the given equation to the drawn one, then draw a straight line to cross it.

Example Solve $x^2 - 3x + 2 = 0$ using the graph of $y = x^2 - 3x + 1$ (Figure A17).

First we change the left-hand side of the given equation to make it the same as the graph equation; that is, we have to change $x^2 - 3x + 2$ into $x^2 - 3x + 1$. We do this by subtracting 1:

$$x^2 - 3x + 2 = 0 \xrightarrow{\text{subtract 1 from both sides}} x^2 - 3x + 1 = -1.$$

Now the solutions may be read where $y = -1$.

Answer: $x = 1$ and $x = 2$.

Check that $x^2 - 3x + 2 = 0$ when $x = 1$ and when $x = 2$.

Example Solve $x^2 + 3x - 4 = 0$ from $y = x^2 + x - 3$.

$$x^2 + 3x - 4 = 0 \xrightarrow{-2x + 1} x^2 + x - 3 = -2x + 1$$

Draw $y = -2x + 1$ and read values of x where the graphs cross.

Algebraic graphs 6 (H)   Tangents (slope of curve) (13)

The slope of the tangent to a curve at a point A gives the slope of the curve at A.

In Figure A18 the slope of the curve at A is $-\dfrac{u}{b}$.

Note that u and b are the distances given by the axes' scales, not necessarily the distances measured in squares.

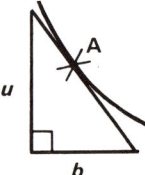

Fig. A18

Tangents to curves can be drawn 'by eye', but for a parabola the method given below enables an exact slope to be found.

In Figure A19, ST is the axis of symmetry of the parabola, PM is the tangent to the parabola at point P. Note that TM = MF. This fact means that you can draw accurate tangents to parabolas and calculate their slopes.

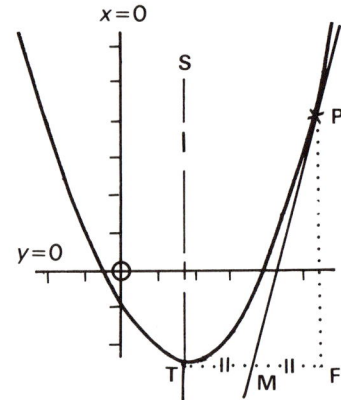

Fig. A19

Algebraic graphs 7 (H)   Area under a curve (8, 13)

The area under a curve may be found approximately by dividing it into a series of trapeziums. See *Practical graphs 2* (page 216).

Algebraic graphs 8 (H)   Graph sketching (13)

You need to learn the shape of the following graph families:

$y = ax + b$ A straight line

$y = ax^2 + bx + c$ A parabola (see *Algebraic graphs 4*)

$y = ax^3 + bx^2 + cx + d$ A cubic (see Figure A20)

$y = \dfrac{a}{x^n}$ A reciprocal curve (see Figure A21)

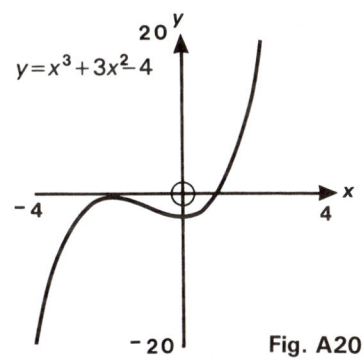

Fig. A20

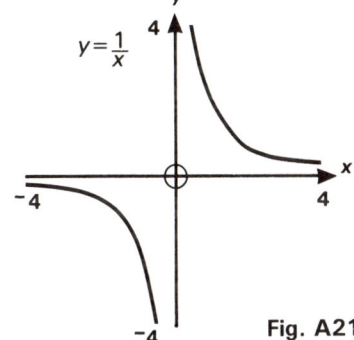

Fig. A21

To sketch a graph, work out values of (x, y) until you can see roughly where the curve fits on the grid, then remembering the general shape draw in the curve.

Comparing graphs

Given $y = f(x)$, like $y = -x^2 + x$,

$y = f(x) + a$, like $y = -x^2 + x + 3$, moves the curve vertically a units; up if a is +ve.

$y = f(x + b)$, like $y = -(x + 2)^2 + (x + 2)$, moves the curve horizontally b units; left if b is +ve.

$y = f(cx)$, like $y = -(3x)^2 + 3x$ alters the width and 'way up' of the curve; narrower if $c > 1$; wider if $0 < c < 1$; upside down if c is negative.

There is no better way to learn about this than to investigate using a computer graph-drawing program or a graphics calculator.

Sequences

Sequences 1   *n*th terms (14)

Sequences are number patterns whose terms follow a rule. The rule is given either in words or by stating the nth term, i.e. the term that is n from the start.

If the nth term is $3n - 2$, then the sequence is made by letting $n = 1$, $n = 2$, $n = 3$, etc. giving
1, 4, 7, 10, ...

Sequences 2   Standard sequences (14)

Powers   Squares 1, 4, 9, 16, ..., n^2

$$\sum_{1}^{n} \text{(squares)} = \tfrac{1}{6}n(n + 1)(2n + 1)$$

Cubes 1, 8, 27, 64, ..., n^3

$$\sum_{1}^{n} \text{(cubes)} = [\tfrac{1}{2}n(n + 1)]^2$$

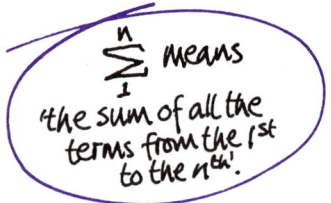

$\sum_{1}^{n}$ means the sum of all the terms from the 1st to the nth.

Fibonacci   1, 1, 2, 3, 5, 8, ... or any sequence where the next term is made by adding the previous two. (The nth term is too complex for GCSE work.)

Triangular   1, 3, 6, 10, 15, ..., $\tfrac{1}{2}n(n + 1)$

$$\sum_{1}^{n} \text{(triangular numbers)} = \tfrac{1}{6}n(n + 1)(n + 2)$$

Arithmetic   Made by adding the same number to the previous term. When a is the first term and d is the number added, the sequence is:

$a, a + d, a + 2d, a + 3d, ..., a + (n - 1)d$

$$\sum_{1}^{n} \text{(arithmetic sequence)} = n[a + \tfrac{1}{2}d(n - 1)]$$

Geometric   Made by multiplying the previous term by the same number. When a is the first term and r is the number multiplied by, the sequence is:

$a, ar, ar^2, ar^3, ..., ar^{n-1}$

$$\sum_{1}^{n} \text{(geometric sequence)} = \frac{a(r^n - 1)}{r - 1}$$

Sequences 3   nth terms of the form $an^x + bn^{x-1} + cn^{x-2} + ...$ (14)

These are best found by seeking a **common difference** (CD). See Chapter 14 page 70.

If a CD appears at the first line, try for an nth term of the form $an + b$, where $a = $ CD and $a + b = $ 1st term.

If the CD appears at the second line, try for an nth term of the form $an^2 + bn + c$, where $2a = $ CD, $3a + b = $ first difference in line one, and 1st term is $a + b + c$.

If the CD appears at the third line, try for an nth term of the form $an^3 + bn^2 + cn + d$, where $6a = $ CD, $12a + 12b = $ first difference in line two, $7a + 3b + c = $ first difference in line one, and $a + b + c + d = $ 1st term of the sequence.

The reasons for all this will be seen if you work question 11 in Chapter 14.

Variation

Variation 1   Direct and inverse proportion (15)

Two quantities are in **direct proportion** when a multiplicative increase in one leads to the same increase in the other. For example, if one quantity becomes four times bigger, then the other becomes four times bigger as well.

Two quantities are in **inverse proportion** when a multiplicative increase in one leads to the inverse multiplicative decrease in the other. For example, if one quantity becomes four times bigger, the other becomes a quarter of itself.

Example of direct proportion
The number of turns made by a car wheel is directly proportional to the number of miles travelled.

Example of inverse proportion
The time taken for a journey is inversely proportional to the average speed for the journey.

Variation 2   Direct (15)

When a quantity y is directly proportional to another quantity x (written $y \propto x$) then $y = kx$, where k is a constant. We also say that y 'varies directly as' x.

A simple example is the number of 10p's you have (n) and their value (p pence). If n doubles then p doubles; if n trebles then p trebles, and so on. Therefore p is directly proportional to n. We can write $p \propto n$ and $p = kn$. In this case we know that $k = 10$.

Slightly more difficult to understand is that the area of a circle varies directly as the square of the radius. We can write $A \propto r^2$ and $A = kr^2$. In this case we know that $k = \pi$. (Note that the area varies directly as the *square* of the radius, not directly as the radius. If you double the radius the area increase *four* times.)

Many laws of science involve direct variation. For example:
Hooke's law for a spring: extension $\propto$ load
Ohm's law for a conductor: voltage $\propto$ current
Snell's law of refraction: $\sin i \propto \sin r$.

Variation 3   Inverse (15)

$$\text{Time} = \frac{\text{distance}}{\text{speed}} \rightarrow t = \frac{d}{s}$$

If the distance travelled is constant, then $t \propto \dfrac{1}{s}$; t is 'inversely proportional' to s. If you double the speed, then you halve the time taken.

If the area (A) of a rectangle is constant, then the length (l) is inversely proportional to the width (w):

$$A = lw \rightarrow l = \frac{A}{w} \rightarrow l \propto \frac{1}{w} \text{ when } A \text{ is constant.}$$

Say the area is 36 cm^2. If the length is 9 cm then the width is 4 cm. If the width doubles to 8 cm, then the length must halve, to 4.5 cm.

Variation 4 Joint (15)

Often one variable depends on several others.

The volume of a cylinder depends on both its base radius and its height, as $V = \pi r^2 h$. We say that V 'varies jointly as the square of the radius and the height'. (The word 'jointly' is often missed out.)

The volume (V) of a given mass of gas varies directly as the temperature (T) and inversely as the pressure (P). Note that the 'and' does not imply 'add' here.

$$V \propto \frac{T}{P} \;\rightarrow\; V = \frac{kT}{P}$$

If the pressure increases then the volume decreases. If the temperature increases then the volume also increases.

Many other laws of science involve joint variation. For example:
in electrolysis: mass $\propto$ (current $\times$ time)
in thermal and electrical conductivity: conductivity $\propto$ (radius)^2/time

Example p varies as t squared and inversely as the cube root of w.
$p = 1$ when $t = 1$ and $w = 8$.
(a) Find p when $t = 2$ and $w = 1$.
(b) State how w varies with p and t.

Variation statement: $p \propto \dfrac{t^2}{\sqrt[3]{w}}$

Equation with constant k: $p = \dfrac{kt^2}{\sqrt[3]{w}}$

(a) Substitution to find k: $1 = \dfrac{k \times 1^2}{\sqrt[3]{8}} \;\rightarrow\; 1 = \dfrac{k}{2} \;\rightarrow\; k = 2$

Equation with k known: $p = \dfrac{2t^2}{\sqrt[3]{w}}$

Substitution for t and w: $p = \dfrac{2 \times 2^2}{\sqrt[3]{1}} = 8$

(b) $p = \dfrac{kt^2}{\sqrt[3]{w}}$

Change subject to w:

$w \xrightarrow{\;\sqrt[3]{}\;} \sqrt[3]{w} \xrightarrow[\text{divide into } kt^2]{} \dfrac{kt^2}{\sqrt[3]{w}}$

$\left(\dfrac{kt^2}{p}\right)^3 \xleftarrow{\;(\;)^3\;} \dfrac{kt^2}{p} \xleftarrow[\text{divide into } kt^2]{} p$

So $w = \left(\dfrac{kt^2}{p}\right)^3 = \dfrac{k^3 t^6}{p^3} \;\rightarrow\; w \propto \dfrac{t^6}{p^3}$

Reference notes: Shape

Basic geometry

Basic geometry 1   Straight-line and parallel-line angles (16)

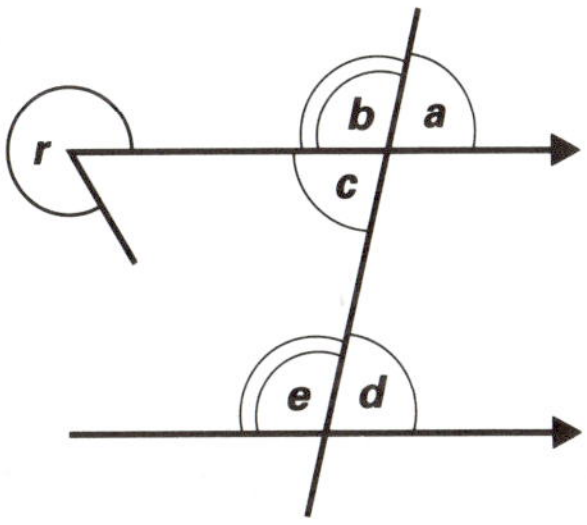

In Figure S1: angle *a* is **acute**,
      angle *b* is **obtuse**,
      angle *r* is **reflex**.

$a + b = 180°$ **(adjacent angles on a straight line)**

  $a = c$ **(vertically opposite angles)**

  $a = d$ **(corresponding angles)**

  $c = d$ **(alternate angles)**

$c + e = 180°$ **(allied angles)** [Also known as interior angles]

Hints: 'Adjacent' means 'next-door'.
    'Corresponding' means 'in the same position'.
    'Alternate' means 'on opposite sides' ('Z angles').
    'Allied' means 'joined together'.

Angles that add up to 180° are said to be **supplementary**.

Fig. S1

Basic geometry 2   Special quadrilaterals (16)

Name	Sides	Angles	Diagonals
Trapezium	1 pair parallel	—	—
Isosceles trapezium	1 pair parallel 1 pair non-parallel equal	2 equal pairs	equal
Kite	2 adjacent equal pairs	1 equal pair	one bisected, cross at 90°
Parallelogram	2 pairs equal and parallel	opposites equal	bisect each other
Rhombus	4 equal	opposites equal	bisect at 90°
Rectangle	as parallelogram	all 90°	equal
Square	as rhombus	all 90°	equal, cross at 90°

Basic geometry 3   Polygons (16)

polygon: many-sided
pentagon: 5 sides   octagon: 8 sides
hexagon: 6 sides   nonagon: 9 sides
heptagon: 7 sides   decagon: 10 sides

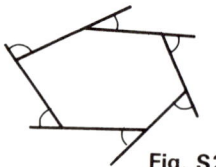

Fig. S2

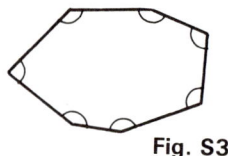

Fig. S3

The exterior angles of all polygons total 360° (Figure S2).

The interior angles of an *n*-sided polygon total $(n - 2) \times 180°$ (Figure S3).

Plane symmetry

Plane symmetry 1   Line (16)

A figure has line symmetry if it can be divided into two parts by a mirror line.

An H has two lines of symmetry.

Plane symmetry 2   Rotational (16)

The order of rotational symmetry is the number of times a figure fits into a tracing of itself (or 'maps onto itself') in one revolution (one full turn).

All figures have at least order 1 rotational symmetry, but often this is not counted, so an M can be said to have no rotational symmetry. An H has rotational symmetry of order 2.

Plane symmetry 3   Point (16)

The simplest way to find out if a figure has point symmetry is to see if it looks different when rotated through 180°. An H has point symmetry, but an M has not.

Movement

Movement 1   Bearings (17)

Cardinal bearings

These use N, S, W, E. Exact directions are measured from north and south towards east and west, for example N 35° E.

The sixteen-point compass defines halfway between N and W as NW, and halfway between NW and W as WNW (west of north-west), etc.

Three-figure bearings

These are measured clockwise from north.

Back bearings   Add 180°; if answer more than 360°, subtract 360°.

Example   Town A is on a bearing of 235° from town B. Find the bearing of town B from town A
235 + 180 = 415 $\xrightarrow{-360}$ 055°.

Movement 2   Earth navigation (TRB)

Longitude   A circle drawn round the Earth that passes through both the poles. The longitude through Greenwich, London, is 0°. All positions round the Earth are measured east and west of this longitude, which is also called the **Greenwich meridian**.

Great circles   A great circle is the largest circle that can be drawn on the surface of a sphere. The equator is a great circle, so are all lines of longitude through the north and south poles. On the equator or any line of longitude (meridian), 1 minute of arc is 1 nautical mile long.

Latitudes   Circles drawn parallel to the equator. On latitude θ°N or θ°S, 1 minute of arc is $\cos\theta$ nautical miles long.

Knot   A speed of one nautical mile per hour.

Movement 3   Loci (17)

A **locus** (plural 'loci') is the path traced by a moving point.

The locus of a point moving so that it is a constant distance from:
(a)   a line – is a **parallel line** (Figure S4);
(b)   a fixed point – is a **circle**;
(c)   two fixed points – is the **perpendicular bisector of the line joining the points** (Figure S5);
(d)   the arms of an angle – is the **bisector of the angle** (Figure S6).

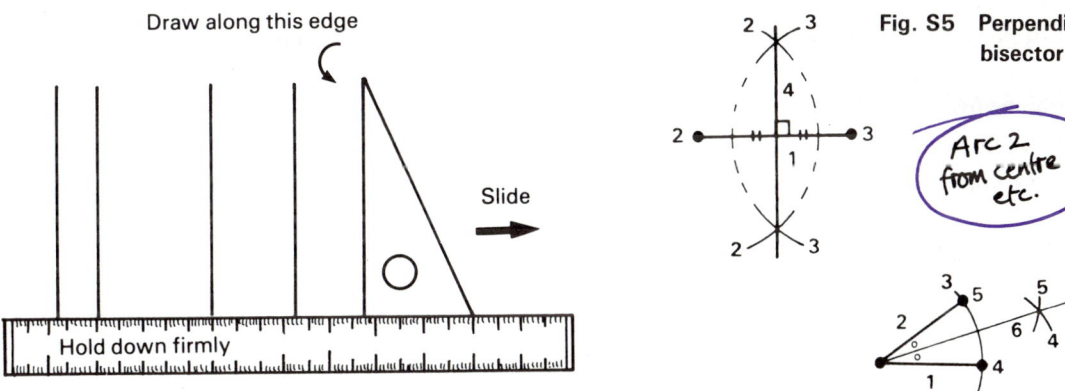

Fig. S4   **Using a set-square and a straight edge to draw parallel lines**

Fig. S5   **Perpendicular bisector**

Arc 2 from centre 2, etc.

Fig. S6   **Angle bisector**

Movement 4   Construction methods (17)

Figure S6 shows the way to bisect an angle.

Figure S7 shows the compass construction for a 60° angle.

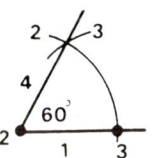

Fig. S7   **Angle of 60°**

These two constructions can be combined to draw many other angles, e.g. 30° by bisecting 60°.

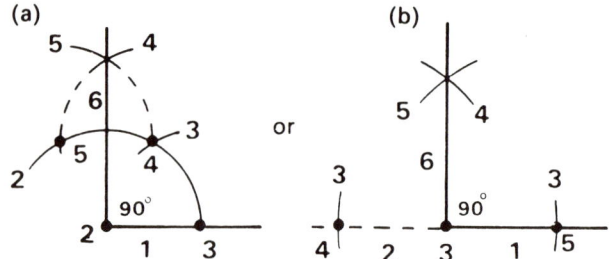

(a) or **(b)**

Fig. S8   Right angle (two methods)

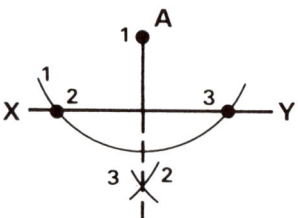

Fig. S9   Dropping a perpendicular

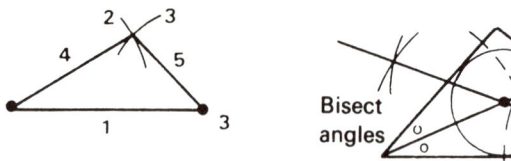

Fig. S10   △ given 3 sides

Bisect angles

Fig. S11   Incircle of △

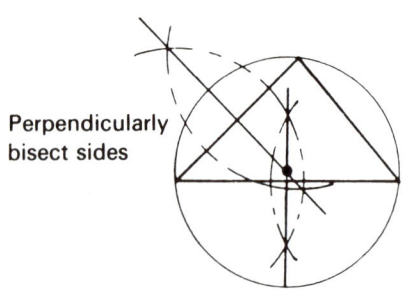

Perpendicularly bisect sides

Fig. S12   Circumcircle of △

Movement 5   Route matrix to describe a network (17)

Figure S13 shows a two-way network. Q is a **node** of order 3 (3 **arcs** leave it). Order-2 nodes, like P, only exist if specially marked. Figure S13 has 4 nodes (N), 6 arcs (A) and 4 **regions** (R). Note that the loop counts as one arc and the regions include the outside of the network.

For all networks $N + R = A + 2$ [Euler's (pronounced 'Oiler') Theorem].

A **route matrix** describes a network by stating how many ways there are of passing from one node directly to another. This is the route matrix for Figure S13:

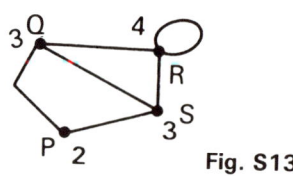

Fig. S13

		to			
		P	Q	R	S
from	P	0	1	0	1
	Q	1	0	1	1
	R	0	1	2	1
	S	1	1	1	0

A two-way route matrix is always symmetrical about the leading (＼) diagonal. Note that the loop at R is shown as a 2, clockwise and anticlockwise.

Figure S14 shows a one-way network and its matrix. Note that a one-way matrix is not symmetrical. When drawing a one-way network from a matrix be careful that you do not draw a new arc if you can put a second arrow on an existing one. Figure S15 illustrates this.

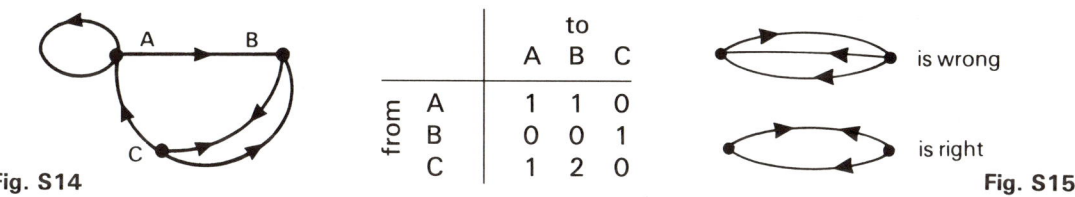

	to		
from	A	B	C
A	1	1	0
B	0	0	1
C	1	2	0

Fig. S14

is wrong

is right

Fig. S15

A network is **traversable** (can be drawn with one continuous line) if it has two odd nodes or no odd nodes. An odd node must be either a start or a finish. An even node may be both the start and the finish, or an intermediate point.

Similarity

Similarity 1   Similar and congruent shapes (18)

Two shapes are *similar* when one is an exact enlargement of the other.
Two shapes are *congruent* when they are exactly the same shape and size.

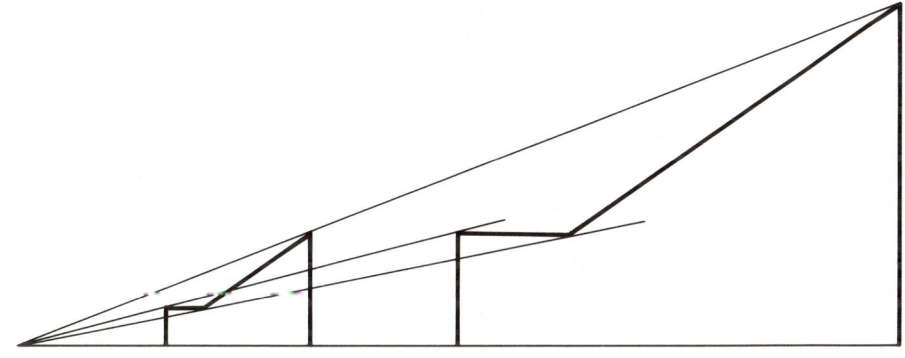

Fig. S16   Similar shapes by ray enlargement; scale factor 3

When two shapes are similar, their corresponding (in the same position) sides are in the same ratio, and their angles are the same sizes. For a triangle, and only for a triangle, it is sufficient to know *one* of these two facts to know that the shapes are similar. In Figure S17 the two shapes have angles of the same sizes, but they are not similar. In Figure S18 the two shapes have their sides in the same ratio (2:1) but they are not similar.

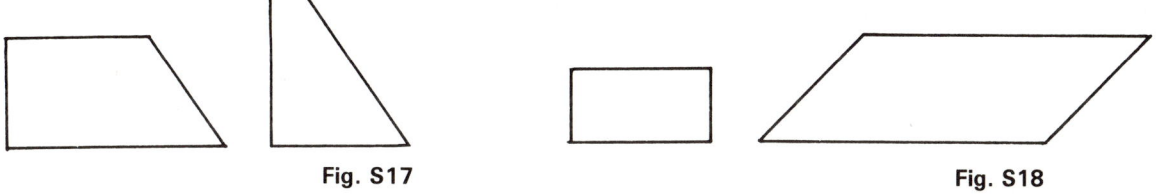

Fig. S17

Fig. S18

Similarity 2   Calculating sides (18)

The two triangles in Figure S19 are similar.

Looking at the positions of the sides we can see that
4 cm → 6 cm, *a* cm → 5 cm, and 2 cm → *b* cm.
Therefore their sides are in the ratio 4 : 6 = 2.3.

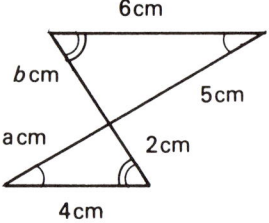

The scale factor of the enlargement is $\frac{2}{3}$ or $\frac{3}{2}$.

a is smaller than 5 so $a = \frac{2}{3} \times 5 = 3\frac{1}{3}$ cm.

b is larger than 2, so $b = \frac{3}{2} \times 2 = 3$ cm.

Fig. S19

Similarity 3 (H)   Areas (18)

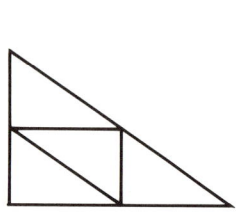 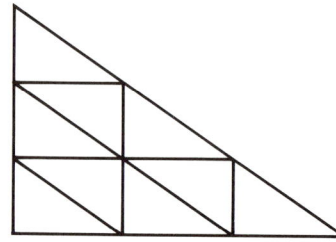

Fig. S20

In Figure S20 the ratio of the sides of the similar triangles is 2 : 3. However, the areas of the triangles are in the ratio $2^2 : 3^2 = 4 : 9$. That is, the larger triangle has sides only half as big again as the smaller, but its area is $2\frac{1}{4}$ times as big.

Remember: Areas' ratio = Lengths' ratio squared

Similarity 4 (H)   Volumes (18)

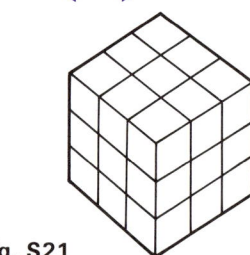  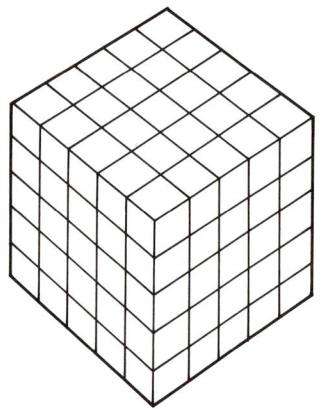

Fig. S21

In Figure S21 the ratio of the sides of the similar cuboids is 3 : 5. However, the volumes of the cuboids are in the ratio $3^3 : 5^3 = 27 : 125$.

The larger is $\frac{125}{27} = 4\frac{17}{27}$ times as large in volume, but only $\frac{5}{3} = 1\frac{2}{3}$ times as large in lengths.

Volume of smaller $= \frac{27}{125} \times$ volume of larger.

Remember: Volumes' ratio = Lengths' ratio cubed

248

Summary
Ratio of lengths = $x:y$
Ratio of areas = $x^2:y^2$
Ratio of volumes = $x^3:y^3$

Similarity 5 (H)   Congruent triangles (18)

Congruent figures are exactly the same shape and size

For all shapes except triangles you must know that all their angles are the same size and that all their sides are the same length and in the same relative position to the angles (we say they 'correspond').

For triangles there are four 'cases of congruency' where you only need to know three equalities, not the six you would expect.

3 sides (Figure S22)

Fig. S22

2 sides, included angle (Figure S23)
(Included means 'between the sides'.)

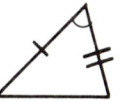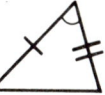

Fig. S23

Right angle, hypotenuse, side (Figure S24)
(The hypotenuse is the side opposite the right angle.)

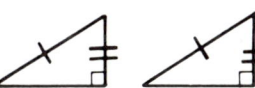

Fig. S24

2 angles, corresponding side (Figure S25)
(Corresponding means in the same position: opposite the same angle.)

Fig. S25

Example   In Figure S26, △s $\begin{matrix} ABC \\ CDA \end{matrix}$ are congruent (3 sides).

   Note   AC is the third side. It is **common** to both triangles, therefore equal in both. We mark a common side with a wavy line.

From $\begin{matrix} ABC \\ CDA \end{matrix}$: covering up $\begin{matrix} A \\ C \end{matrix}$ gives BC = DA

covering up $\begin{matrix} B \\ D \end{matrix}$ gives AC = CA

covering up $\begin{matrix} C \\ A \end{matrix}$ gives AB = CD.

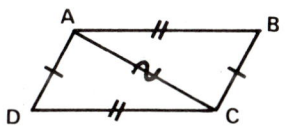

Fig. S26

Solid geometry

Solid geometry 1   Symmetry (19)

Planes of symmetry (reflection symmetry)

In the cuboid shown in Figure S27, BDHF is a plane of symmetry (see Figure S28(a)), and IKSQ is another (see Figure S28(b)). ABCD is a square.

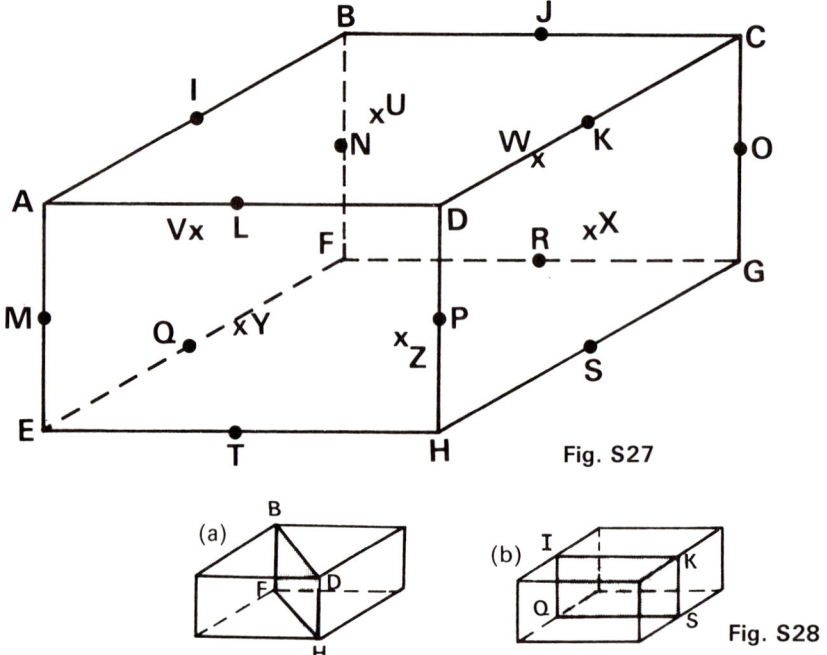

Fig. S27

(a)   (b)

Fig. S28

A plane of symmetry acts as a mirror.

Why is ADGF in Figure S27 not a plane of symmetry?

Which are the two diagonal planes of symmetry, and the three planes of symmetry parallel to the faces, for the cuboid in Figure S27?

Axes of symmetry (rotational symmetry)

In Figure S27, UZ is an axis of symmetry (see Figure S29). The cuboid in Figure S27 has 5 axes of symmetry. Name each of them.

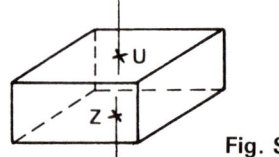

Fig. S29

Symmetry number

The symmetry number of a solid is the number of ways that it can be put into a mould of itself. The cuboid in Figure S27 has symmetry number 8.

Solid geometry 2   Representation in 2-D (19)

Oblique projection

Figure S30 shows an equi-triangular prism in oblique projection. The front face is seen 'square-on'. The edges truly at right-angles to this face go away at 45° and are drawn half the scale of the front face. A view like this is impossible in 3-D.

Isometric projection

Figure S31 shows the same prism in isometric projection on an isometric grid. The lines go away at 30° from the nearest point and are all drawn to the same scale. Apart from not allowing for perspective (parallel lines seeming to converge to a point in the distance) this is close to a true 3-D view.

Orthographic projection

The prism is viewed from three directions – from the front (the elevation), the side (the side elevation), and from above (the plan). All three views must 'line up'.

Note   In all projections upright lines **must** be upright.

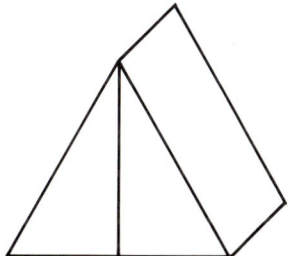

Fig. S30   Oblique

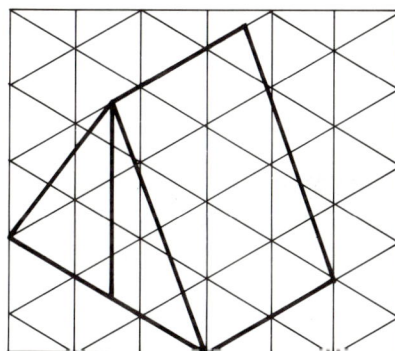

Make sure you have the grid the correct way up.

Fig. S31   Isometric

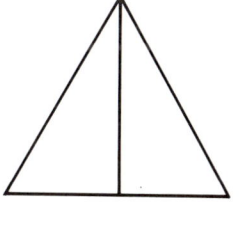

Fig. S32   Orthographic

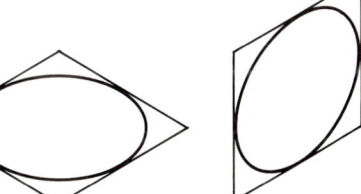

Fig. S33   Circles in isometric

251

Solid geometry 3   Nets (19)

A net is the 2-D shape which, when cut out, can be folded up to make a 'hollow' solid. If you add tabs, they are needed on every other edge. Figure S34 is a net for the prism in *Solid geometry 2*.

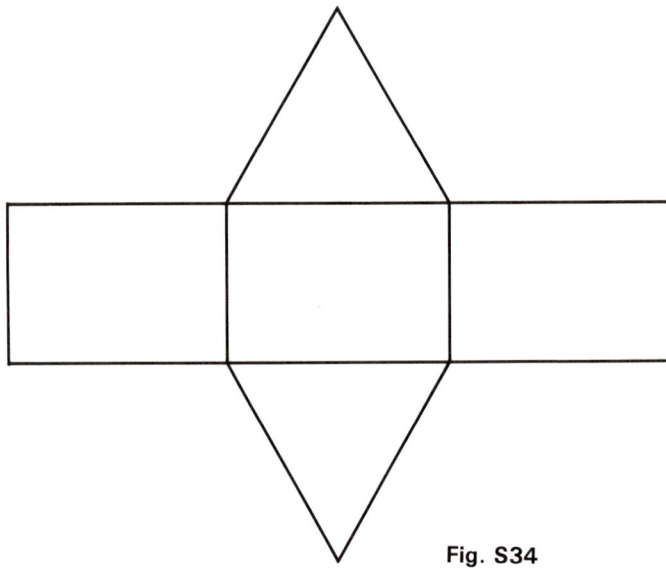

Fig. S34

Solid geometry 4   3-D co-ordinates (19)

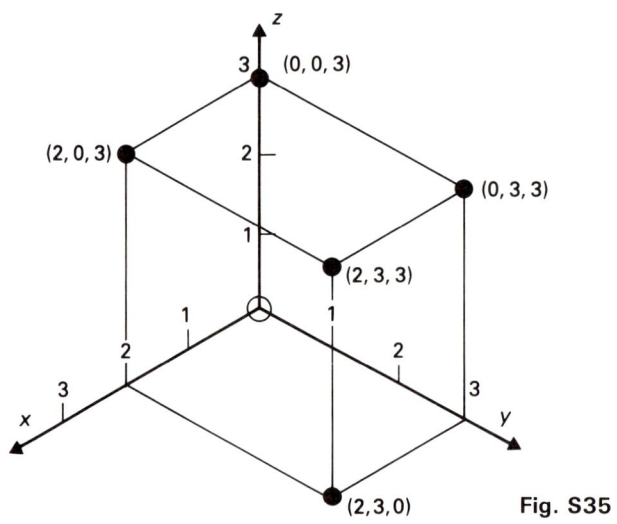

Fig. S35

Mensuration

Mensuration 1 Pythagoras' theorem (20)

In all right-angled triangles, the square on the hypotenuse is equal to the sum of the squares on the other two sides.

In Figure S36, $h^2 = a^2 + b^2$

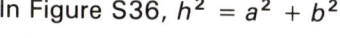

Fig. S36

The $\boxed{x^2}$ key is useful for finding the squares.

Examples   In Figure S37,
$$h^2 = 5.6^2 + 8.2^2$$
$$h^2 = 98.60$$
$$h = \sqrt{98.60} \simeq 9.9 \text{ cm}$$
Key: $5.6 \boxed{x^2} \boxed{+} 8.2 \boxed{x^2} \boxed{=} \boxed{\surd}$

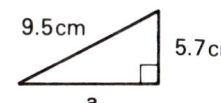

Do not forget to press the $\boxed{\surd}$ key.

Fig. S37

In Figure S38,
$$9.5^2 = 5.7^2 + a^2$$
$$a^2 = 9.5^2 - 5.7^2$$
$$a^2 = 57.76$$
$$a^2 = \sqrt{57.76} \simeq 7.6 \text{ cm}$$
Key: $9.5 \boxed{x^2} \boxed{-} 5.7 \boxed{x^2} \boxed{=} \boxed{\surd}$

Fig. S38

Mensuration 2 Areas (21)

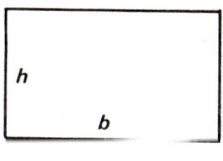

Rectangle

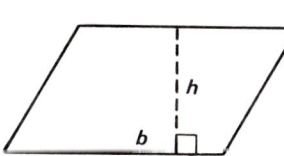

Parallelogram

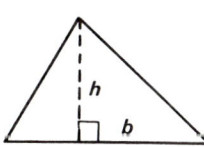

Triangle

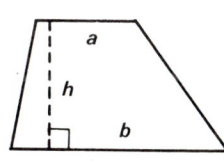

Trapezium

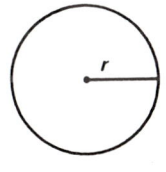

Circle   Fig. S39

Rectangle/parallelogram   Base times height: $A = bh$

Triangle   Half base times height: $A = \frac{1}{2}bh$

Trapezium   Half the sum of the parallel sides times the distance between them: $A = \frac{1}{2}(a + b)h$

Circle   Pi times the square of the radius: $A = \pi r^2$

π is defined as circumference divided by diameter. Hence circumference of a circle $= \pi d$.
π is irrational; it cannot be written as an exact common or decimal fraction, not even a recurring one.

Mensuration 3 (H)   Arc and sector of a circle (21)

Length of arc

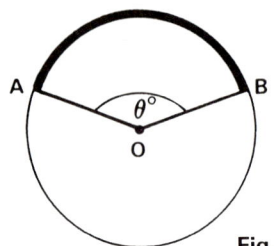

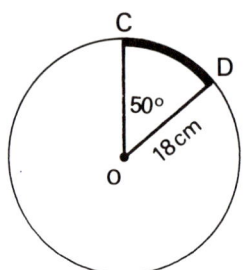

Fig. S40

Fig. S41

To find the length of an arc we first find what fraction of the circumference it is by considering the angle it subtends at the centre.

In Figure S40:

$$\text{arc AB} = \frac{\theta}{360} \times \text{circumference} = \frac{\theta}{360} \times \pi \times \text{diameter}$$

θ is a Greek letter, pronounced 'theta'.

Example   In Figure S41, $\theta = 50$ and the diameter is 36 cm.

$$\text{Arc CD} = \frac{50}{360} \times \pi \times 36 \;\rightarrow\; \frac{5\cancel{0}}{\cancel{360}_1} \times \frac{22}{7} \times \cancel{36}^1 = \frac{110}{7} \;\rightarrow\; 15\tfrac{5}{7} \text{ cm}$$

Area of sector

In Figure S42, P is the major sector and Q is the minor sector.

$$\text{Area of sector Q} = \frac{\theta}{360} \times \text{area of circle} = \frac{\theta}{360} \times \pi \times r^2.$$

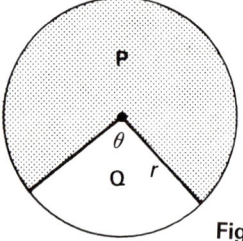

Fig. S42

Mensuration 4 (H)   Surface areas of solids (21)

Cylinder   Curved surface area is circumference of base circle times height of cylinder: πdh.
Total surface area is curved surface area plus the areas of the top and bottom circles: $\pi dh + 2\pi r^2$.

Cone   Curved surface area is *pi* times the radius of the base times the slant height: πrl.
Total surface area is curved surface area plus the area of the base circle: $\pi rl + \pi r^2$.

Sphere   Surface area is 4 times *pi* times the square of the radius: $4\pi r^2$.

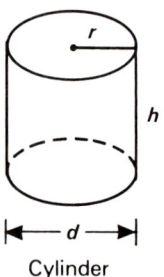

Cylinder

Cone

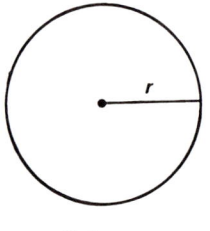

Sphere

Fig. S43

Area formulae must involve two lengths multiplied together.

Mensuration 5 Volumes (21)

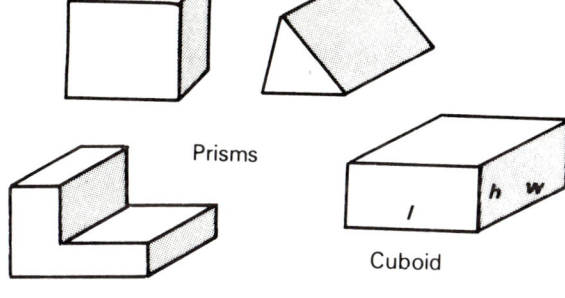

Prisms

Cuboid

Cylinder

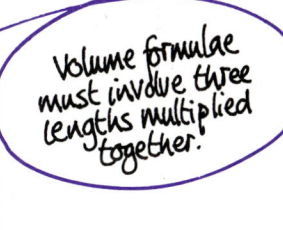

Volume formulae must involve three lengths multiplied together.

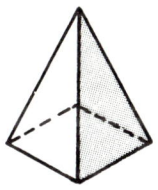

Pyramids

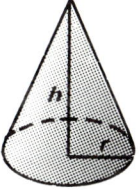

Cone

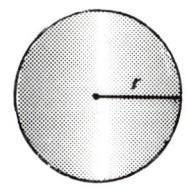

Sphere   **Fig. S44**

Name	Definition	Volume
Prism	A solid with a constant cross-section; that is, it has the same shape all through it. Examples: breeze block, kitchen roll, Toblerone box, unsharpened pencil, wedge of cheese.	Area of cross-section times length (or height): $V = Al$ or Ah
Cuboid	A prism with rectangular cross-sections, like a brick	Length times width times height: $V = lwh$
Cylinder	A prism whose constant cross-section is circle.	Area of circular cross-section times height: $V = \pi r^2 h$
Pyramid	May have any shape base, with sloping sides coming to a point. A pyramid on a square base is usually called an Egyptian pyramid. A pyramid with four triangular faces is called a tetrahedron (if the triangles are equilateral it is a regular tetrahedron).	One third of the base area times height: $V = \frac{1}{3}Ah$
Cone	A pyramid with a circular base.	One third of the circular base area times height: $V = \frac{1}{3}\pi r^2 h$
Sphere	A ball.	Four-thirds times *pi* times the cube of the radius: $V = \frac{4}{3}\pi r^3$

255

Trigonometry

Trigonometry 1 Solution of right-angled triangles (22)

$$o = a \times \tan \theta \quad\quad o = h \times \sin \theta \quad\quad a = h \times \cos \theta$$

$$\frac{o}{a} = \tan \theta \quad\quad \frac{o}{h} = \sin \theta \quad\quad \frac{a}{h} = \cos \theta$$

one ancient teacher of history swore at his class!

Finding sides

$$x = 5 \times \tan 54°$$
$$\simeq 6.88$$

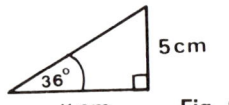

Fig. S45

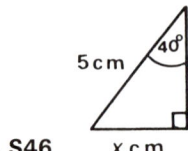

Fig. S46

$$x = 5 \times \sin 40°$$
$$\simeq 3.21$$

$$x = 5 \times \cos 25°$$
or $\quad x = 5 \times \sin 65°$
$$\simeq 4.53$$

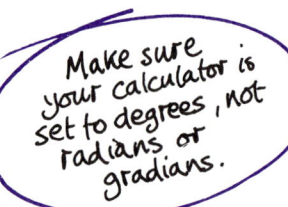

Fig. S47

Sample key sequence: 5 ⊠ 54 TAN ⊟

Make sure your calculator is set to degrees, not radians or gradians.

Finding angles

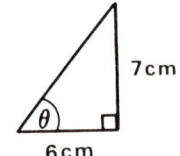

Fig. S48

$$\frac{7}{6} = \tan \theta$$
$$\theta \simeq 49.4°$$

$$\frac{7}{9} = \sin \theta$$
$$\theta = 51.1°$$

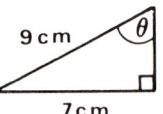

Fig. S49

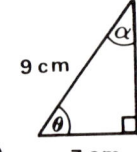

Fig. S50

$$\frac{7}{9} = \cos \theta$$
or $\frac{7}{9} = \sin \alpha \rightarrow \alpha \simeq 51.1°$
$$\theta \simeq 38.9°$$

Sample key sequence: 7 ⊟ 6 ⊟ ARCTAN   (Note: ARCTAN may be INV TAN or TAN⁻¹ .)

256

Trigonometry 2   Minutes

Fractions of a degree are now usually given as decimals. In the past, minutes were used where 60 minutes = 1 degree (60′ = 1°). They are still vital in navigation (see *Movement* 2, page 245).

Many scientific calculators have a degree/minute/second key marked $\boxed{°\,'\,''\to}$ or $\boxed{\text{DMS}}$ to convert between the two systems of measuring angle. Consult your handbook for its use.

Trigonometry 3 (H)   Applications (22)

Depression = looking down

Elevation = looking up

Angle of elevation = angle of depression

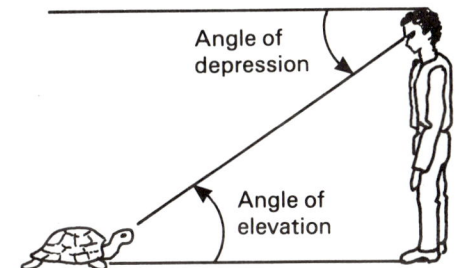

Fig. S51

The **altitude of the sun** is its angle of elevation.

At noon the sun is at its highest point; its angle of elevation is at a maximum.

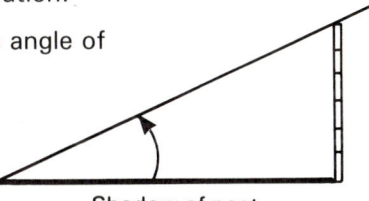

Shadow of post   Fig. S52

Figures S53 (a) and (b) show the **vertical angle** of an isosceles triangle and of a cone. Half this angle is called the **semi-vertical angle**.

(a) 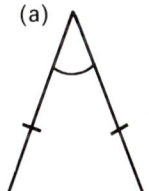   (b)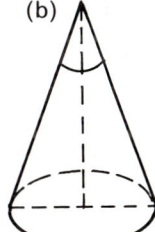

Fig. S53

Inclination is usually the angle made with the horizontal, but it can be with a vertical, or any other, plane.

Slope is usually the tangent of the angle made with the horizontal.

Gradient is mathematically the same as slope, but for surveying (roads, railways, etc.) it is taken as the sine of the angle. For small values of inclination, θ, which will usually be the case for road and railway slopes, $\sin\theta \simeq \tan\theta$, so the difference is unimportant.

Fig. S54 Gradient = $\sin\theta = \frac{1}{100}$ = 1%   **Fig. S55** Gradient = $\tan\theta = \frac{2}{5}$

Northings and Eastings

Ship B is 1 nautical mile (n.m.) north and 5 n.m. east of ship A.

One nautical mile is $\frac{1}{3600}$ of the length of the equator, so 1 n.m. subtends 1′ at the centre of the Earth.

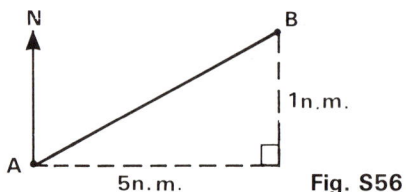

Fig. S56

Trigonometry 4 (H)   Angles beyond 90° (22)

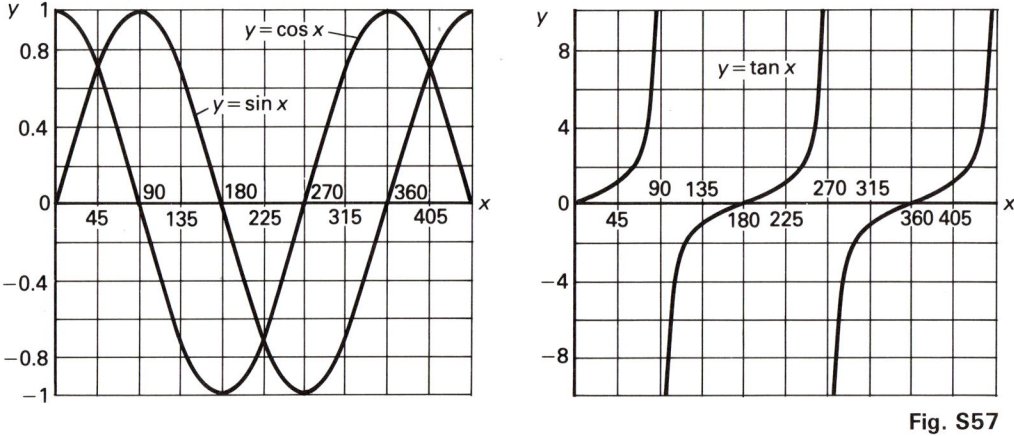

Fig. S57

Figure S57 shows how the values of $\sin x$, $\cos x$ and $\tan x$ change as x increases. For each the cycle repeats every 360°. Note how we deal with $\tan 90°$ being infinity.

For any angle θ, the trigonometric functions of $(180 - \theta)°$, $(180 + \theta)°$ and $(360 - \theta)°$ have the same absolute (ignoring $\pm$) value, but the sign changes as shown in Figure S58, where A means 'all positive', S means 'sine positive' etc.

Example

sin 30° = 0.5
sin 150° = 0.5
sin 210° = −0.5
sin 330° = −0.5

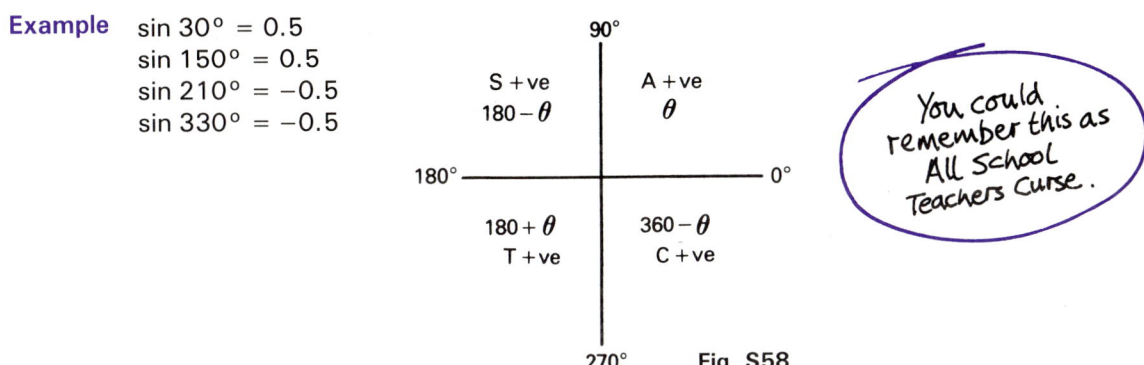

Fig. S58

Trigonometry 5 (H)   Sine rule for a non-right-angled triangle (22)

When a triangle is not right-angled, the basic trig. functions cease to be true. However it is possible to obtain a rule, called the **sine rule**, that can be used when the triangle is not right-angled. You must, however, know one side and the angle opposite it, together with another side or angle.

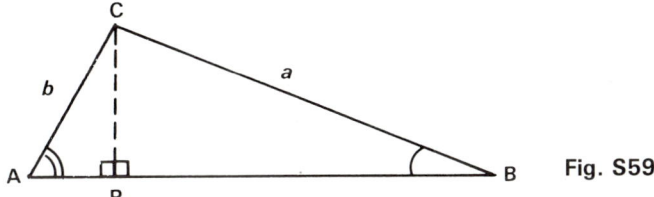

Fig. S59

In Figure S59, △ABC is not right-angled, but △APC and △CPB are. From these triangles:

$$CP = a \times \sin B \quad \text{and also} \quad CP = b \times \sin A.$$

Therefore $a \times \sin B = b \times \sin A$. This is the **sine rule**, usually written:

$$\frac{a}{\sin A} = \frac{b}{\sin B}$$

to find a side

or

$$\frac{\sin A}{a} = \frac{\sin B}{b}$$

to find an angle

Do not use the sine rule to find the largest angle, which might be obtuse.

Always start the rule with the side, or the sine of the angle, that you are trying to find.

You can only use the sine rule if you know a side and the angle opposite to it.

Example   In Figure S60:

$$\frac{x}{\sin 27°} = \frac{12\,cm}{\sin 48°}$$

$$x = \frac{12\,cm}{\sin 48°} \times \sin 27°$$

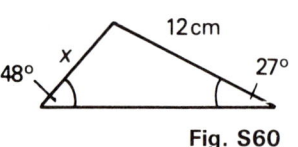

Fig. S60

Key: 12 $\boxed{÷}$ 48 $\boxed{SIN}$ $\boxed{×}$ 27 $\boxed{SIN}$ $\boxed{=}$

Answer: $x = 7.33\,cm$ correct to 3 s.f.

Example   In Figure S61:

$$\frac{\sin\theta}{3.6\,cm} = \frac{\sin 37°}{3.9\,cm}$$

$$\sin\theta = \frac{\sin 37°}{3.9\,cm} \times 3.6\,cm$$

Fig. S61

Key: 37 $\boxed{SIN}$ $\boxed{÷}$ 3.9 $\boxed{×}$ 3.6 $\boxed{=}$ $\boxed{ARCSIN}$

Answer: $\theta = 33.8°$ to the nearest $0.1°$.

Trigonometry 6 (H)   Cosine rule for a non-right-angled triangle (22)

The cosine rule may be used when there is no opposite side/angle pair (required by the sine rule, see *Trigonometry 5*). It is the general case of Pythagoras' theorem, and starts in the same way.

In Figure S62:

$$a^2 = b^2 + c^2 - 2bc\cos A$$

The cosine rule may also be used to find an angle given the three sides.

In Figure S62:

$$\cos A = \frac{b^2 + c^2 - a^2}{2bc}$$

Fig. S62

This may be learnt, or worked out from the first version.

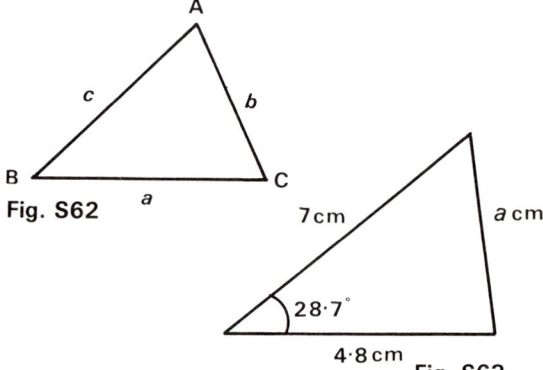

Examples   In Figure S63:

$$a^2 = 7^2 + 4.8^2 - 2 \times 7 \times 4.8 \times \cos 28.7°$$

Key: 7 $\boxed{x^2}$ $\boxed{+}$ 4.8 $\boxed{x^2}$ $\boxed{-}$ 2 $\boxed{×}$ 7 $\boxed{×}$ 4.8 $\boxed{×}$ 28.7 $\boxed{COS}$ $\boxed{=}$ $\boxed{\sqrt{\ }}$

In Figure S64:

$$\cos\theta = \frac{6^2 + 5.4^2 - 8.9^2}{2 \times 6 \times 5.4}$$

Don't forget the $\boxed{\sqrt{\ }}$ and the $\boxed{INV}\boxed{COS}$!

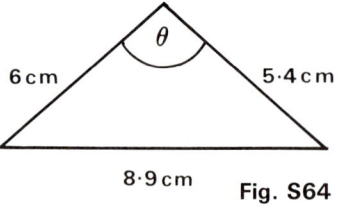

Fig. S64

$\boxed{COS^{-1}}$

Key: 6 $\boxed{x^2}$ $\boxed{+}$ 5.4 $\boxed{x^2}$ $\boxed{-}$ 8.9 $\boxed{x^2}$ $\boxed{÷}$ 2 $\boxed{÷}$ 6 $\boxed{÷}$ 5.4 $\boxed{=}$ $\boxed{INV}$ $\boxed{COS}$

$\boxed{ARCCOS}$

Trigonometry 7 (H)   Which trigonometry rules to use (22)

Right angle?
Use tan/sin/cos.

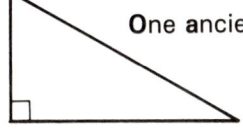

One ancient teacher of history swore at his class.

Angle and opposite side?
Use sine rule.

Fig. S65

$$\frac{a}{\sin A} = \frac{b}{\sin B} \text{ to find side } a$$

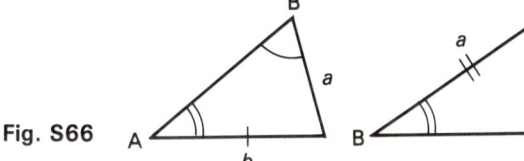

Fig. S66

$$\frac{\sin A}{a} = \frac{\sin B}{b} \text{ to find angle A}$$

You must know a side/angle pair, e.g. side b and angle B.

Note: Do not use the sine rule to find the biggest angle in a triangle; it might be obtuse, but your calculator will only give you an acute angle. Find a smaller angle first, then use the angle sum of the triangle.

Two sides and included angle?
Use cosine rule.

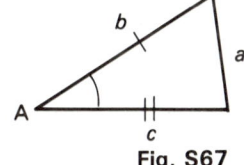

Fig. S67

$$a^2 = b^2 + c^2 - 2bc \cos A \text{ to find side } a$$

Three sides but no angles?
Use cosine rule.

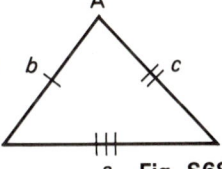

Fig. S68

$$A = \arccos \frac{b^2 + c^2 - a^2}{2bc} \text{ to find angle A}$$

Trigonometry 8 (H)   Area of a triangle (23)

The area of $\triangle ABC$ is $\frac{1}{2}ab \sin C$.

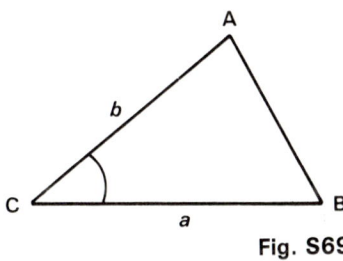

Fig. S69

Figure S70 shows the proof of this formula.
Area $\triangle ABC = \frac{1}{2}ah$
$h = b \sin C$
$\therefore$ Area $\triangle ABC = \frac{1}{2}ab \sin C$

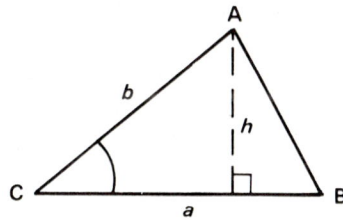

Fig. S70

Vectors

Vectors 1   Definition (23)

A **vector** is a line with both length and direction. It is described by a column matrix, showing how far the end point of the vector is from the start point, measured horizontally and vertically. Positive and negative directions are the same as for graph axes. Figures S71 and S72 show two examples.

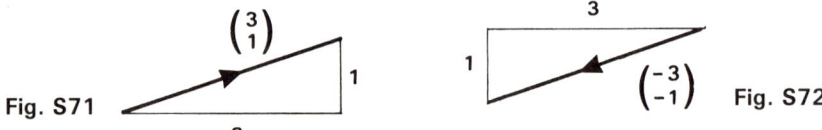

Fig. S71   Fig. S72

Vectors 2   Position vectors (23)

Position vectors always start at the origin, so the position vector $\begin{pmatrix} -4 \\ 2 \end{pmatrix}$ starts at (0, 0) and ends at (−4, 2).

Vectors 3   Shift vectors (23)

Shift vectors describe translations (slidings). In Figure S73, the hatched square is translated to the shaded square by the vector $\begin{pmatrix} 3 \\ 1 \end{pmatrix}$. Each corner of the square moves 3 units to the right and 1 unit upwards.

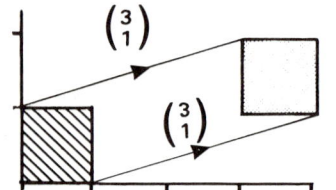

Fig. S73

Shift vectors can start at any point, and the same vector can be at several places on the same grid.

Vectors 4   Definition by a matrix; by end letters; by a̰ (23)

Vectors may be described as a matrix, $\begin{pmatrix} 3 \\ 1 \end{pmatrix}$, or by their end letters, $\overrightarrow{AB}$, or by a single letter distinguished with a wavy line underneath (or sometimes, in print, by being thicker, e.g. **a**).

The matrix notation may be summarised as:

$$\begin{pmatrix} +\,\text{Right} & -\,\text{Left} \\ +\,\text{Up} & -\,\text{Down} \end{pmatrix}$$

Example $\begin{pmatrix} -4 \\ 2 \end{pmatrix}$ moves 4 units to the left and 2 units up.

Vectors 5 Parallel vectors (23)

The same letter (or letters) can be used for parallel vectors, but different letters *must* be used for non-parallel vectors.

Figures S74 to S77 show some examples of parallel vectors.

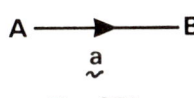

A——▶——B

a

Fig. S74

C——◀——D

$-a$

Fig. S75

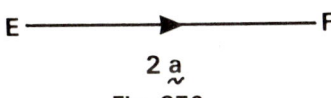

E————————▶————————F

$2\,a$

Fig. S76

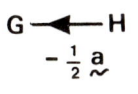

G——◀——H

$-\frac{1}{2}\,a$

Fig. S77

Vectors 6 (H)   Resultant (23)

A **resultant vector** is the single vector that gives the same translation as all the others added together.

In Figure S78, $\overrightarrow{OA}$ is the resultant of a and b and c.

The resultant vector may be calculated by adding the matrices of the given set of vectors.

In Figure S78, $\overrightarrow{OA} = a + b + c$

$$= \begin{pmatrix} 4 \\ 0 \end{pmatrix} + \begin{pmatrix} -2 \\ -1 \end{pmatrix} + \begin{pmatrix} 0 \\ -3 \end{pmatrix} = \begin{pmatrix} 2 \\ -4 \end{pmatrix}.$$

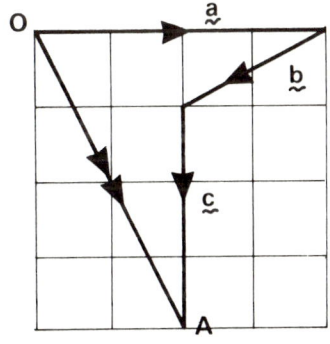

Fig. S78

Vectors 7 (H)   Magnitude (23)

The **magnitude** of a vector is its length. It is often written as $|\overrightarrow{OA}|$, and it is called the **modulus** of $\overrightarrow{OA}$.

In Figure S79:

$(|\overrightarrow{OA}|)^2 = 2^2 + 4^2$ (Pythagoras' theorem).

Hence $|\overrightarrow{OA}| = \sqrt{20} = 4.47$ to 3 s.f.

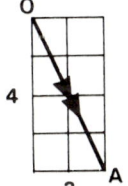

Fig. S79

Vectors 8 (H)   The golden rule of vectors (23)

Most vector proofs depend on the rule that parallel vectors must be expressed in terms of the same vector letters.

Example   If $\overrightarrow{AC} = (1 - k)p + \frac{1}{3}kq$
and $\overrightarrow{AD} = p + q$
where AC//AD,
then $1 - k = \frac{1}{3}k \Rightarrow k = \frac{3}{4}$
so that $\overrightarrow{AC} = \frac{1}{4}(p + q) = \frac{1}{4}\overrightarrow{AD}$.

Example   If $\overrightarrow{EF} = 3(m + n)$ and $\overrightarrow{GH} = 5(m + n)$
then EF//GH and EF:GH = 3:5.

Transformations

Transformations 1   Rotation (24)

To describe a rotation state the centre of rotation and the angle turned (positive rotations are anticlockwise).

Example   In Figure S80, triangle ABC has rotated through 270° about O.

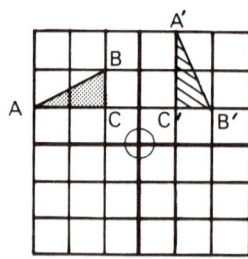

Fig. S80

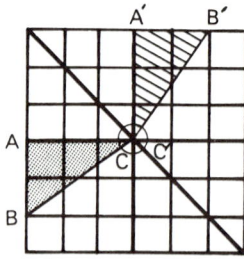

Fig. S81

Transformations 2   Reflection (24)

To describe a reflection state the equation, or the name, of the mirror line.

Example   In Figure S81, triangle ABC has been reflected in the line $y = -x$.

Transformations 3   Enlargement (24)

To describe an enlargement state the scale factor by which the figure has been enlarged and the centre of the enlargement.

Figures S82 to S85 illustrate four kinds of enlargement. In each diagram △ABC is mapped onto △A′B′C′.

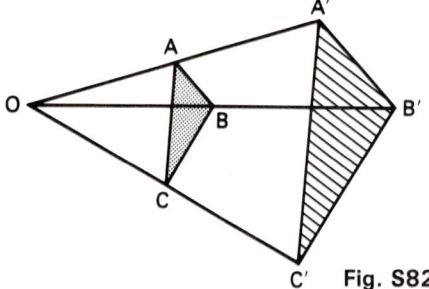

Fig. S82

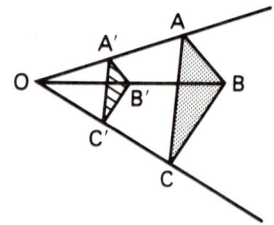

Fig. S83

Figure S82 is an enlargement of scale factor 2 from centre O.

Figure S83 is an enlargement of scale factor $\frac{1}{2}$ from centre O.
(Note that we refer to the transformation as an enlargement even when the image is smaller than the object.)

Figure S84 is an enlargement of scale factor −1 from centre O.
(Note that in a negative enlargement the image is on the opposite side of the centre and upside down.)

Figure S85 is an enlargement of scale factor −2 from centre O.

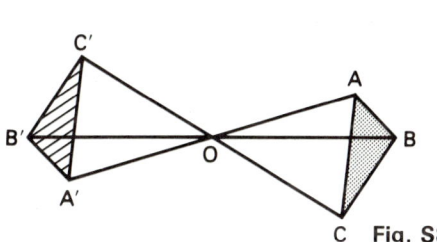

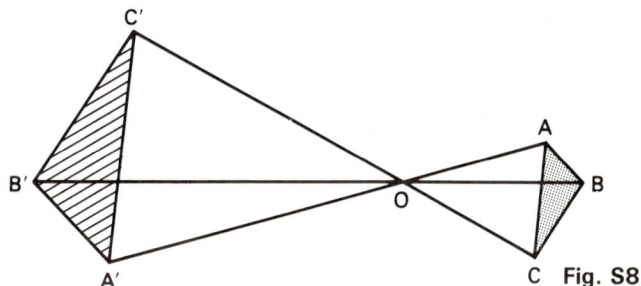

Fig. S84

Fig. S85

Transformations 4 Translation (24)

A sliding movement, able to be described by a vector. See *Vectors 3* (page 262).

Circle geometry (H)

Circle geometry 1 (H) Perpendicular bisector of a chord (30)

The perpendicular bisector of a chord is a diameter. See Figure S86.

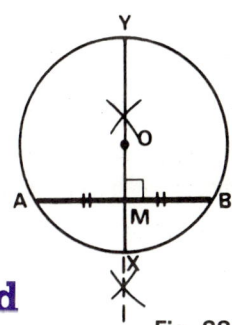

Fig. S86

Circle geometry 2 (H) Angles at centre and circumference (30)

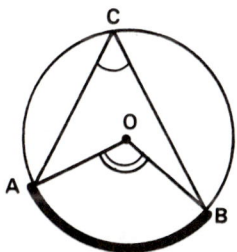

Fig. S87

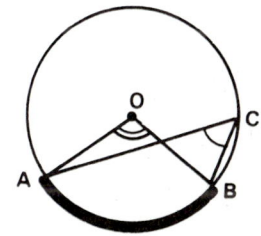

Fig. S88

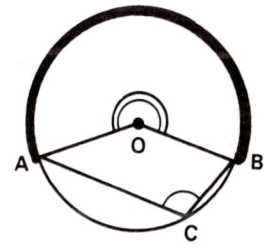

Fig. S89

In each of Figures S87, S88 and S89, the arc AB subtends ∠AOB at the centre O and subtends ∠ACB at the circumference.

∠AOB = 2∠ACB (Angles at centre and circumference)

Note that in Figure S89 it is the major (longer) arc which subtends the angles, and ∠AOB is reflex in this case.

265

(a) The pair of angles must start and finish with the same letters, e.g. ∠POQ = 2∠PXR could not be correct.

(b) Both angles must 'open out' the same way. This is especially important when the centre angle is reflex, as in Figure S89. This is illustrated in Figure S90.

 and 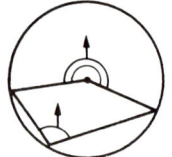 but NOT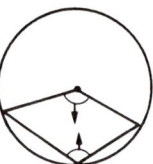

A trap often set by examiners!

Fig. S90

Circle geometry 3 (H)   Angle in a semicircle (30)

The angle in a semicircle is a right angle. See Figure S91.

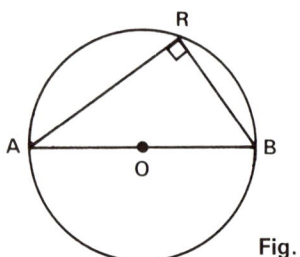

Fig. S91

Circle geometry 4 (H)   Angles in the same segment (30)

A chord cuts a circle into two segments. In Figure S92 angles AXB and AYB are **angles in the same segment** (the segment cut off by chord AB). These angles are equal: ∠AXB = ∠AYB.

If a chord was drawn from X to Y, then angles XAY and XBY would also be 'equal angles in the same segment', though of course it is a different segment to the first one.

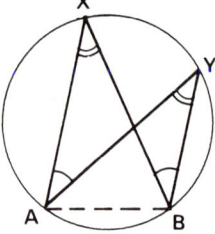

Fig. S92

Angles in the same segment must start and finish with the same letter, and all the letters must be points on the circumference.

Circle geometry 5 (H)   Cyclic quadrilaterals (30)

A cyclic quadrilateral has its four corners on the circumference of a circle.

The opposite angles of a cyclic quadrilateral add up to 180° (they are 'supplementary').

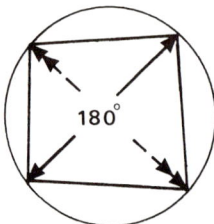

Fig. S93

Why are the quadrilaterals in Fig S90 not cyclic quadrilaterals?

Circle geometry 6 (H)   Tangents (30)

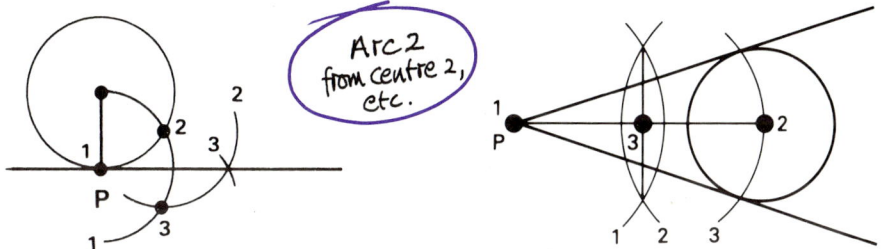

Fig. S94   Tangent to a circle   **Fig. S95   Two tangents from a point**

A tangent to a circle makes an angle of 90° with the radius to the point of contact (see Figure S94).

The two tangents from a point to a circle are equal (see Figure S95).

The angle between a tangent and a chord is equal to the angle the chord subtends at the circumference of the alternate segment (Figure S96).

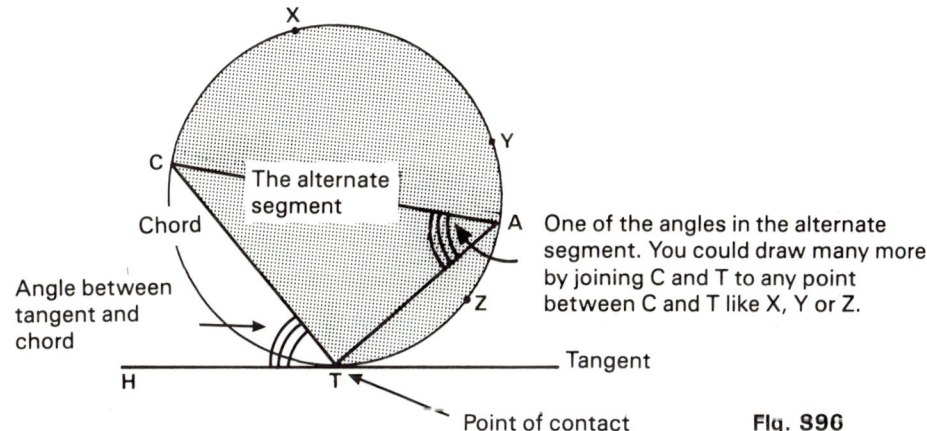

The alternate segment

Chord

Angle between tangent and chord

One of the angles in the alternate segment. You could draw many more by joining C and T to any point between C and T like X, Y or Z.

Tangent

Point of contact   **Fig. S96**

Matrices (H)

Matrices 1 (H)   Multiplication of matrices (31)

Example
$$\begin{pmatrix} 2 & 3 \\ 2 & 5 \end{pmatrix}\begin{pmatrix} 4 & 2 \\ 1 & 5 \end{pmatrix} = \begin{pmatrix} 11 & 19 \\ 13 & 29 \end{pmatrix}$$

Step 1 $(2 \quad 3)\begin{pmatrix} 4 \\ 1 \end{pmatrix} \rightarrow 8 + 3 = 11$

Step 2 $(2 \quad 3)\begin{pmatrix} 2 \\ 5 \end{pmatrix} \rightarrow 4 + 15 = 19$

Step 3 $(2 \quad 5)\begin{pmatrix} 4 \\ 1 \end{pmatrix} \rightarrow 8 + 5 = 13$

Step 4 $(2 \quad 5)\begin{pmatrix} 2 \\ 5 \end{pmatrix} \rightarrow 4 + 25 = 29$

Be especially careful when negative (minus) numbers are involved. Check the following example carefully:

$$\begin{pmatrix} 2 & -1 \\ -3 & 1 \end{pmatrix}\begin{pmatrix} -1 & -2 \\ 1 & 4 \end{pmatrix} = \begin{pmatrix} -3 & -8 \\ 4 & 10 \end{pmatrix}$$

Remember that the most matrices give different answers when multiplied together, depending on which one is at the front (we say multiplication of matrices is 'not commutative').

For example,
$$\begin{pmatrix} 1 & 0 \\ 1 & 2 \end{pmatrix}\begin{pmatrix} 2 & 1 \\ 0 & 0 \end{pmatrix} = \begin{pmatrix} 2 & 1 \\ 2 & 1 \end{pmatrix}$$

but
$$\begin{pmatrix} 2 & 1 \\ 0 & 0 \end{pmatrix}\begin{pmatrix} 1 & 0 \\ 1 & 2 \end{pmatrix} = \begin{pmatrix} 3 & 2 \\ 0 & 0 \end{pmatrix}$$

Matrices 2 (H)   Matrix transformations (31)

The following three methods can be used to define the position of points A, B and C on the grid in Figure S97.

Co-ordinates

A   (1, 1)

B   (1, 2)

C   (3, 1)

Position vectors

$\overrightarrow{OA} = \begin{pmatrix} 1 \\ 1 \end{pmatrix}$

$\overrightarrow{OB} = \begin{pmatrix} 1 \\ 2 \end{pmatrix}$

$\overrightarrow{OC} = \begin{pmatrix} 3 \\ 1 \end{pmatrix}$

Matrix

$$\begin{matrix} & A & B & C \\ x & \\ y & \end{matrix}\begin{pmatrix} 1 & 1 & 3 \\ 1 & 2 & 1 \end{pmatrix}$$

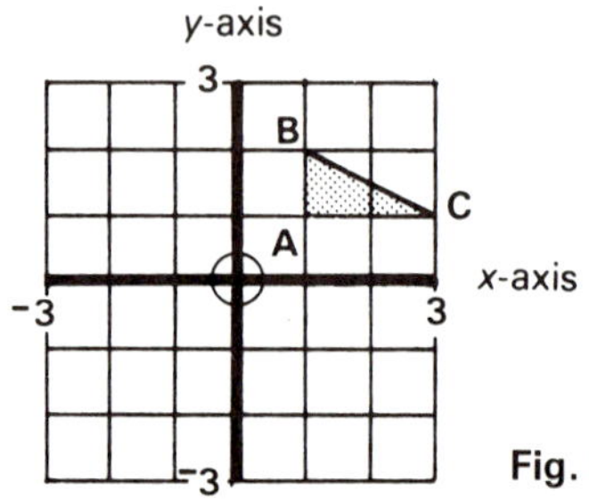

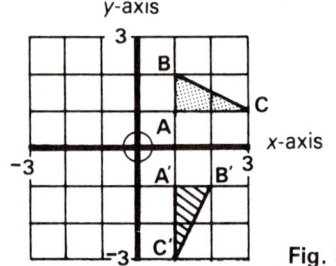

Fig. S97

Using the matrix method to define the triangle, the co-ordinates of the triangle after a rotation of 270° about (0, 0) can be found by multiplying by the matrix $\begin{pmatrix} 0 & 1 \\ -1 & 0 \end{pmatrix}$:

$$\begin{pmatrix} 0 & 1 \\ -1 & 0 \end{pmatrix}\begin{matrix} A & B & C \\ \begin{pmatrix} 1 & 1 & 3 \\ 1 & 2 & 1 \end{pmatrix} \end{matrix} = \begin{matrix} A' & B' & C' \\ \begin{pmatrix} 1 & 2 & 1 \\ -1 & -1 & -3 \end{pmatrix} \end{matrix}$$

Fig. S98

Reference notes: Data handling

Charts

Charts 1   Pictogram (25)

Figure D1 is a pictogram. Each complete symbol represents 10 cars. The whole picture represents 25 cars.

 **Fig. D1**

Charts 2   Line graph (25)

Figure D2 is a line graph. The points plotted with a cross are from correct data, but points on the lines joining the crosses may have no meaning, as in this line graph. It is then best to use a dashed line.

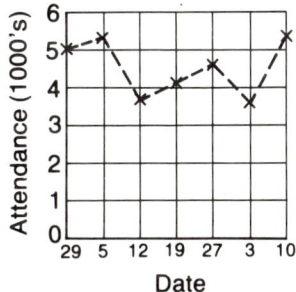

Fig. D2

Charts 3   Bar chart (25)

Figure D3 is a bar chart, or block graph. It illustrates the data in the table, which shows the choices of 100 pupils in a survey at Upside School.

Life aim	Frequency
An interesting job	18
Happiness	20
Start a family	15
Plenty of money	25
Caring for people	10
Excitement	12

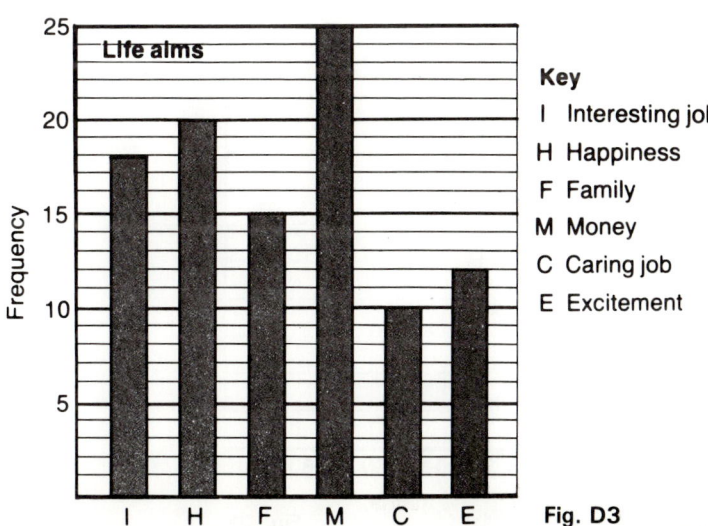

Key
I   Interesting job
H   Happiness
F   Family
M   Money
C   Caring job
E   Excitement

Fig. D3

Charts 4   Proportionate bar chart (25)

Figure D4 shows the above bar chart data as a proportionate bar chart.

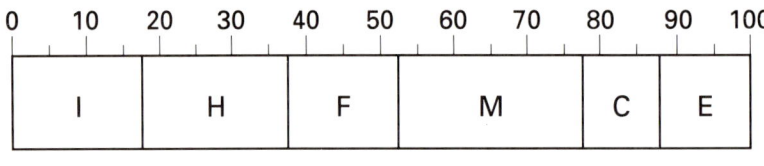

Fig. D4

Key

I	Interesting job
H	Happiness
F	Family
M	Money
C	Caring job
E	Excitement

Charts 5   Pie chart (25)

Figure D5 shows the same data as a pie chart. As there were 100 pupils in the survey, 1 pupil is represented by 360° ÷ 100 = 3.6°, so the 18 pupil sector has an angle of 18 × 3.6° = 64.8° at its centre.

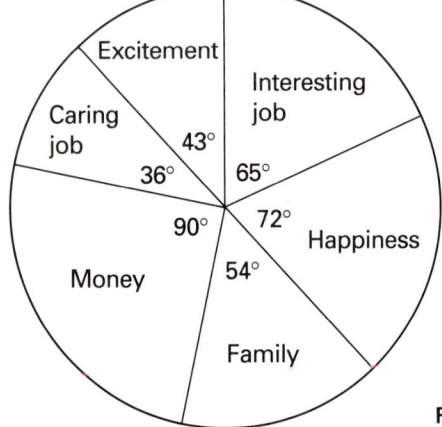

Fig. D5

Charts 6   Scatter graph (25)

A scatter graph compares two sets of data. From it we can see if the data is linked or 'correlated'. The scatter graph in Figure D6 shows a correlation between ability in maths and ability in music.

The random points in the scatter graph in Figure D7 indicates that there is no correlation between height and ability to score runs in cricket.

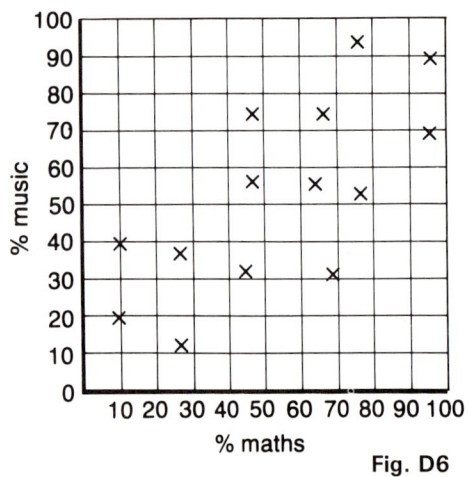

Fig. D6

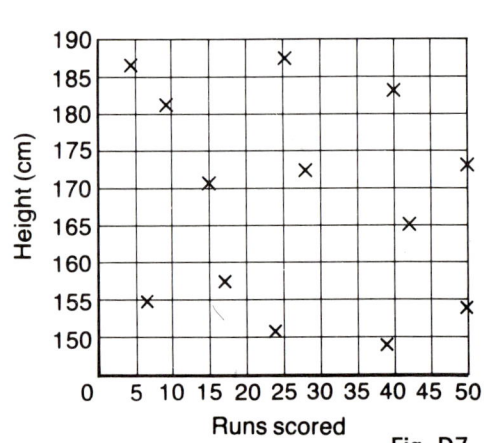

Fig. D7

Charts 7 Growth and decay (25)

These are graphs illustrating the way a measure increases or decreases. Figure D8 shows some examples. At any point on the curve the slope of a tangent to the curve gives the rate of growth or decay; see Figure D9.

Catching a cold

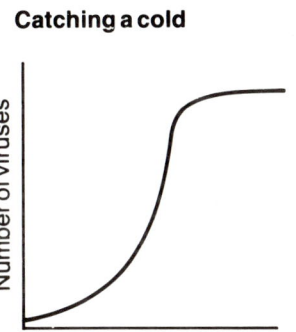

Paying off a loan

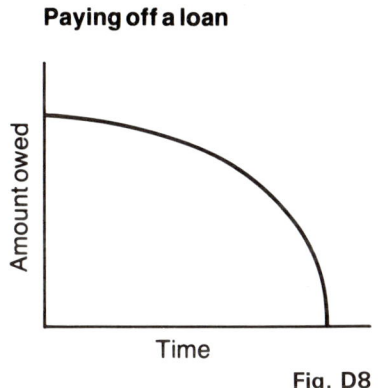

Fig. D8

Leaving the traffic-lights
Speed at A = 15 m/s ≃ 54 km/h
Speed at B = 35 m/s ≃ 126 km/h

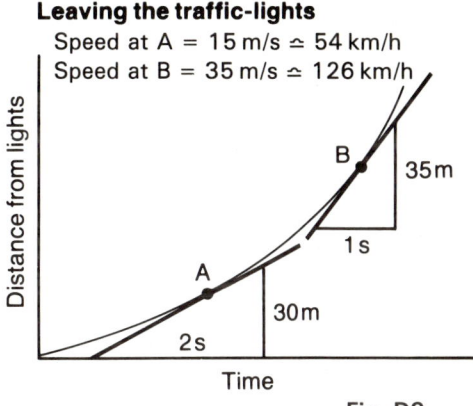

Fig. D9

Charts 8 (H) Histogram (25)

These appear at first to be bar charts, and are often confused with them in other subjects, and even by some mathematicians! In a histogram the **area** of the bar represents the frequency, not the height. They are only used when the data is continuous and are intended to avoid giving a wrong impression when the bars are of unequal width. However, most people still look at the heights and reach the wrong conclusion! Figures D10 and D11 show a bar chart and a histogram representing the same data.

Age	Number of people
0–5	6
5–20	6
20–60	8
60–80	6

'0–5' means 'up to but not including 5'.

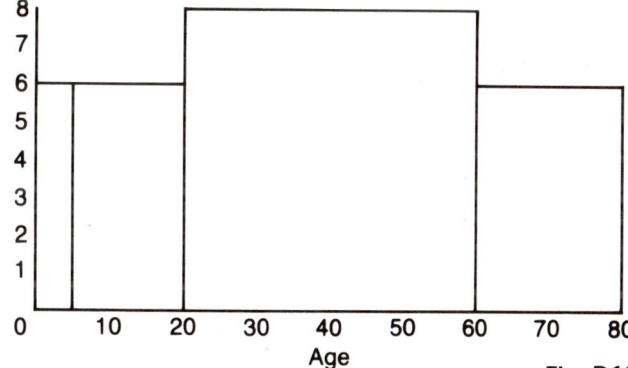

Fig. D10

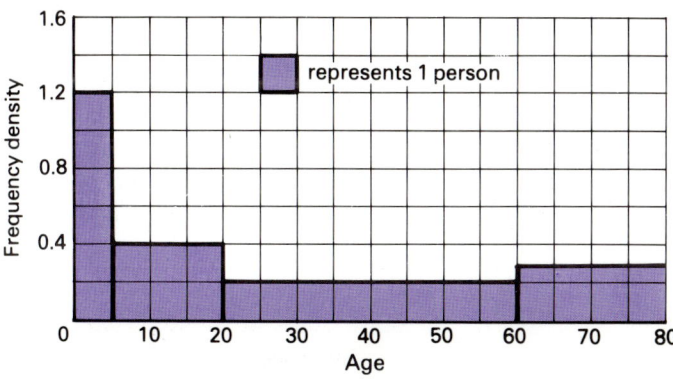

Fig. D11

Frequency = frequency density × class interval
For the 5–20 class, the class interval is 15.
Frequency = 0.4 × 15 = 6.

271

Charts 9 Cumulative frequency curve (25)

Marks	Frequency
1 to 5	1
6 to 10	2
11 to 15	4
16 to 20	7
21 to 25	9
26 to 30	5
31 to 35	2
36 to 40	1

Mark	Cumulative frequency
5 or less	1
10 or less	1 + 2 = 3
15 or less	3 + 4 = 7
20 or less	7 + 7 = 14
25 or less	14 + 9 = 23
30 or less	23 + 5 = 28
35 or less	28 + 2 = 30
40 or less	30 + 1 = 31

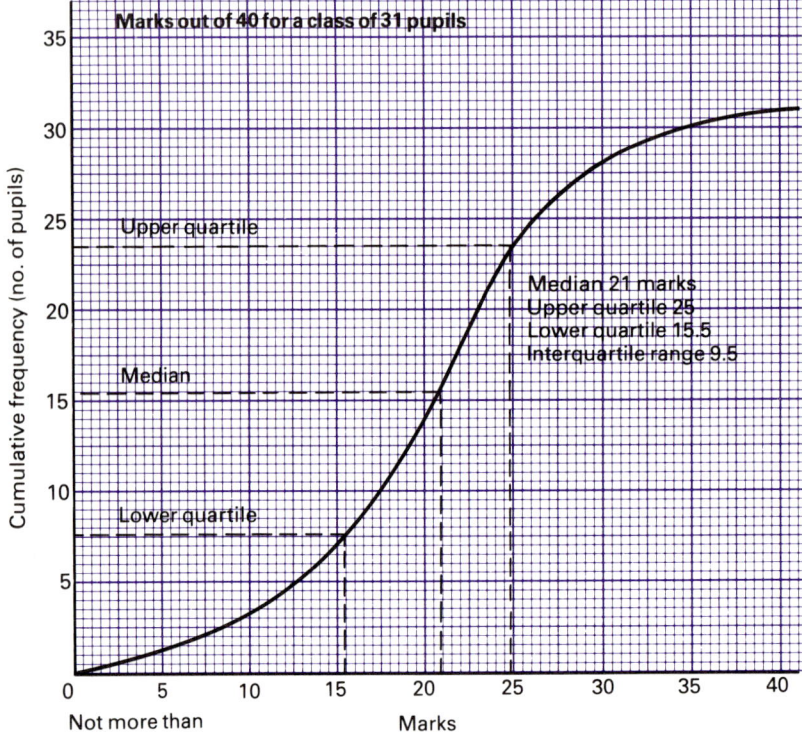

Fig. D12

Statisticians usually use 'percentiles', which are more meaningful than quartiles. As the name indicates, percentiles divide the total frequency into 100 equal divisions. However, the National Curriculum mentions only quartiles.

Averages

Averages 1   The three statistical averages (26)

Mean

The total score distributed equally to each item. The mean of 2, 7 and 9 is the total (18) divided by 3, giving 6.

Mode

The item that occurs most frequently. There may be several modes, or no mode.

Median

The middle score of an odd number of items. For an even number of items it is usually taken as the mean of the two middle scores, although it could be anywhere between them.

Averages 2   Range (26)

The range is the difference between the highest and lowest scores in a set of data. It indicates how spread out the data is, but does not allow for exceptional scores.

Averages 3   Frequency tables (26)

These are used to simplify the arithmetic with a large amount of data. (Processing such large amounts takes a long time, so examples in books usually apply the method to a small amount of data!)

Example Find the mean of £16, £16, £17, £17, £17, £18, £18, £18, £18, £19, £19, £19, £20.

Score (x)	Frequency (f)	Total ($f \times x$)
£16	2	£32
£17	3	£51
£18	4	£72
£19	3	£57
£20	1	£20
TOTALS	13	£232

Mean $= \dfrac{£232}{13} \simeq £18$

Mode $= £18$ (highest frequency)

Median $= £18$ (score of the 7th item, which is clearly in the £18 frequency: $2 + 3 = 5$; $2 + 3 + 4 = 9$).

Range $= £4$ ($£20 - £16$).

Averages 4   Mean from grouped data; modal and median class (26)

To calculate the mean from grouped data you assume that each frequency scores the mid-value (middle) of the class. This mid-value is the mean of the upper and lower limits of the class. For example, the middle of 0–9 is $\dfrac{0 + 9}{2}$ = 4.5, and the middle of 31–35 is $\dfrac{31 + 35}{2}$ = 33.

Example

Class	Frequency (f)	Middle (x)	Total (fx)
30–39	23	34.5	793.5
40–49	55	44.5	2447.5
50–59	32	55	1760
	110		5001

Mean $\simeq \dfrac{5001}{110} \simeq 45$

The class which has the highest frequency is called the **modal class**. As with the mode, there can be more than one modal class. In the last example the modal class is 40–49.

The **median class** is the one in which the middle item of the data lies. In the example this is at 55/56, so class 40–49.

Dispersion

Dispersion 1 (H)   Standard deviation (26)

There are various ways of expressing the standard deviation formula. Check if your examination board gives you one on its formula sheet and learn to use that one if it does. One way is:

$$\sigma = \sqrt{\frac{\sum (x - \bar{x})^2}{n}}$$

Steps in using this formula:

(a) Find the mean ($\bar{x}$).

(b) Find the positive differences between the mean and each score ($x - \bar{x}$).

(c) Square the differences, then add them ($\sum$ is the sign for 'the sum of').

(d) Divide by the number of items (n), then square root to give the standard deviation.

If the data is grouped follow the same process, using the middle of a class value where there is a class interval, but in step (c) multiply each square of the differences by the frequency for that class before adding them.

We can write this as

$$\sigma = \sqrt{\frac{\sum f(x - \bar{x})^2}{n}}$$

Using a statistics function calculator

A statistics calculator in SD mode will calculate σ instantly once you have typed in all the data, which is much easier than the above!

Many calculators will allow you to type in grouped data as $x \times f$; with others you have to enter the x value f times. Consult your handbook (if you can find it and understand the English in it!).

Collecting data

Collecting data 1   Sampling (27)

You cannot ask every voter who they will vote for when trying to predict the outcome of a general election.

Sampling attempts to build up a picture of the whole of a set of data by selecting a few of them. The selection should not be biased. To try to ensure the selected data is not biased we can either select at random or select representatives of different groups within the data, a method called **stratified sampling**.

For example, in a survey of your whole school you could pick all the pupils whose surname is two letters longer than their first name. This should be random, though maybe the girls' names are longer than the boys' on average, so you end up picking more girls than boys. You could select a stratified sample by picking one boy and one girl from every form in the school, selecting names by their order in the register, or by using balls numbered from 1 to 40 picked from a box.

Collecting data 2   Surveys (27)

A survey gathers the required information. It might be simply a case of observing and recording (e.g. men who hold open doors for women) or you may need to ask people questions, using a questionnaire.

Collecting data 3   Questionnaires (27)

A questionnaire is difficult to design. People might think the question means something different to what you intended, or the question might be 'leading' so that they feel they should answer the way you seem to want. If you ask the questions in front of another person, the person may be embarrassed to give an honest answer.

Ideally, questions should have one simple answer, such as 'Yes', 'No', or perhaps a place or a number.

If one-word answers are not possible then questionnaires can have multi-choice answers. The person chooses the answer that fits their view best, although it is difficult to cover every possible viewpoint. Or you could ask them to choose 'Agree/Disagree/No strong feeling' in response to a statement.

Probability theory

Probability theory 1   Definitions (28)

- A **trial** is any process which, when repeated, generates a set of results.

  Examples   Picking a playing card from a pack and recording its colour.
  Asking a man if he is short-sighted.
  Tossing a coin and recording how it lands (heads or tails).

- An **outcome** is the result of a trial.

  Examples   (referring back to the above trials)
  The picked card is black.
  The man is short-sighted.
  The coin falls heads.

- An **event** is one or more of the possible outcomes of a trial or sequence of trials.

  Examples   A playing card being red or an ace.
  A person being a man and being short-sighted.
  A tossed coin falling heads twice in two tosses.

The probability line

> The probability of an event is a fraction between 0 (impossible) and 1 (certain).

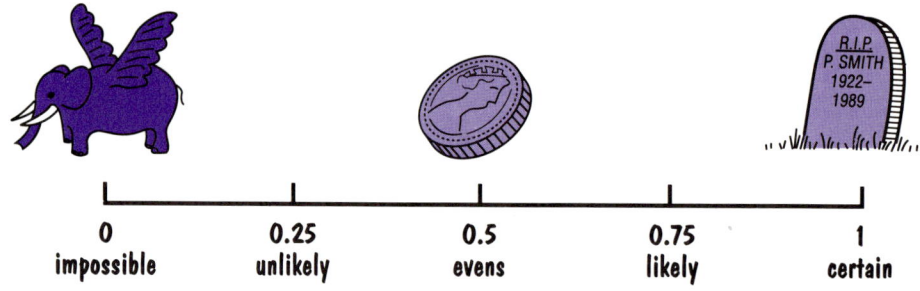

| 0 | 0.25 | 0.5 | 0.75 | 1 |
| impossible | unlikely | evens | likely | certain |

Fig. D13

Probability theory 2   Probability of single events

A successful trial is one that gives you the outcome you want.

> The probability of an event is:
> $$\frac{\text{the number of successful outcomes possible}}{\text{the total number of possible outcomes}}$$

For this to be true, all the outcomes must be equally likely. For instance, although a loaded die can still fall in six ways, the chance of a 6 is not $\frac{1}{6}$.

For any one trial the probability of success + the probability of failure is 1, as it is certain to be one or the other.

Example   When throwing a fair die, $P(\text{six}) = \frac{1}{6}$ so $P(\text{not six}) = \frac{5}{6}$.

Probability calculations

Probability calculations 1   Mutually exclusive events (29)

Events are **mutually exclusive** when they cannot happen together in one trial.

Example   The throw of a die can result in six events, i.e. 1, 2, 3, 4, 5, 6. Only one of these can happen in one throw, so they are mutually exclusive events.

> If two events are mutually exclusive, we add their probability fractions to find the probability that one or the other happens.

For a die, $P(1)$ is $\frac{1}{6}$ and $P(\text{odd})$ is $\frac{1}{2}$.

So $P(1 \text{ or odd})$ is $\frac{1}{6} + \frac{1}{2} \rightarrow \frac{2}{3}$.

Remember $P(A \text{ or } B) = P(A) + P(B)$

> For any trial, the total of the probabilities of all the possible mutually exclusive events is 1.

Probability calculations 2   Independent events (29)

A bag contains three red and two green balls. In a trial one ball is picked from the bag.

Then: P(the ball is red)   = 0.6
P(the ball is green) = 0.4

Now consider a trial in which a ball is picked then put back in the bag before another ball is picked.

There are four possible events which could be the outcome of this trial: 'first ball red, second ball green', 'first ball green, second ball red', 'both balls red', 'both balls green'.

These four events are mutually exclusive, for only one of them can be the outcome of the trial.

But each event in itself consists of two other events, 'picking a red' or 'picking a green'.

Both these events can happen in the one trial, and neither happening alters the probability of the other happening. They are called **independent** events.

> For independent events we multiply their probabilities to find the probability of the combined event.

So P(first ball red and second ball green) = $0.6 \times 0.4 = 0.24$

This can be clearly shown on a tree diagram; see Figure D14.

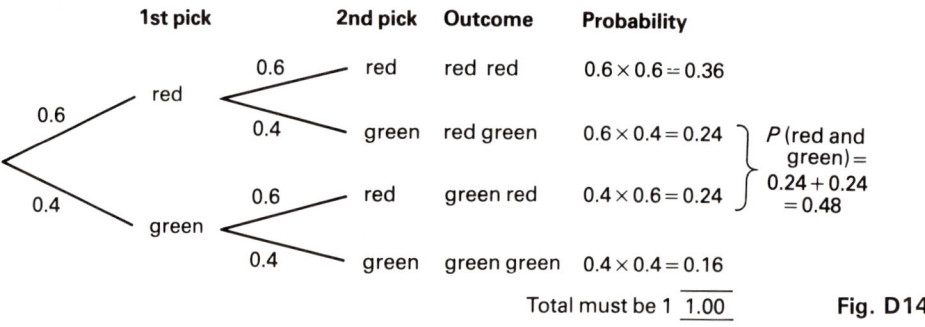

Fig. D14

The order the individual events occur has no effect on their combined probability, so if you know
P(red then green) = 0.24
you also know
P(green then red) = 0.24.

This means if $P(A$ then $B)$ is x, $P(A$ and $B)$ is $2x$.

Four events can happen in $4 \times 3 \times 2 = 24$ ways,
so if $P(A$ then B then C then $D) = x$,
$P(A$ and B and C and $D) = 24x$.

Probability calculations 3 (H)   Not-independent events

In our trial above we put the first picked ball back before we picked the second. Now consider picking two balls without replacing the first one. The probability that the second ball is green depends on the colour of the first ball.

P(first ball is red) and P(second ball is green) are called **'not-independent'** events.

(In English grammar the opposite of 'independent' is 'dependent', but the National Curriculum and most statisticians prefer 'not-independent'.)

If the first ball is red, then P(second ball is green) is $\frac{2}{4} = 0.5$.

If the first ball is green, then P(second ball is green) is $\frac{1}{4} = 0.25$.

If the first ball is red, then P(second ball is red) is $\frac{2}{4} = 0.5$.

If the first ball is green, then P(second ball is red) is $\frac{3}{4} = 0.75$.

We still multiply the probabilities, but must use the correct probability for the second event.

To find P(one ball is red and one ball is green) we have to consider both of the mutually exclusive events (first red, second green) and (first green, second red).

P(first red, second green) $= 0.6 \times 0.5 = 0.3$
P(first green, second red) $= 0.4 \times 0.75 = 0.3$

Note that again the order of the individual events does not affect their combined probability.

One or the other of the two events give us a successful outcome, so:

P(red and green) $= 0.3 + 0.3 = 0.6$

Figure D15 shows how helpful a tree diagram is in understanding conditional probabilities like these.

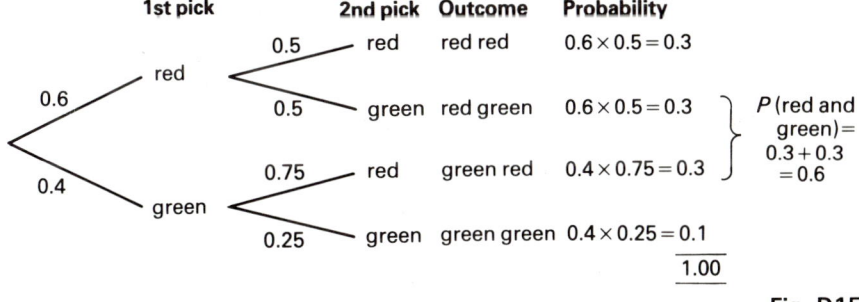

Note Each new branch fork must total 1

Fig. D15

Linear programming (H)

Linear programming 1 (H)   Graph regions (32)

When a complex situation, perhaps involving millions of pounds, arises in business, the managers employ mathematicians to suggest the best solution: possibly that which does the job as quickly and cheaply as possible within the constraints imposed by the availability of resources such as material, finance, machines and labour. This is called 'linear programming', because the mathematician will attempt to connect the variables by linear equations, that is, equations which give straight lines when plotted. A computer is usually used in real life, but at GCSE you will only meet problems with two variables and these can be solved by drawing graphs.

Steps in answering a linear programming question

1 Decide on the two variables – often the examiner will tell you.

2 Draw up inequalities based on the conditions given.

3 Draw a graph based on the inequalities of stage 2.

4 Shade the regions which are *not* allowable.

5 Consider the points in the unshaded region to find the best solution – it is often at or near the corner furthest from the origin.

Example

Figure D16 has been drawn to represent the following inequalities.

$y \geqslant 2$ (Shade under $y = 2$, where $y < 2$.)

$x > 1$ (A dotted line, as $x = 1$ is not allowed. Shade to the left, where $x < 1$.)

$y \leqslant \frac{1}{2}x + 4$ (Shade above $y = \frac{1}{2}x + 4$, where $y > \frac{1}{2}x + 4$.)

$x + y < 8$ (Shade above $x + y = 8$, where $x + y > 8$.)

Important note   More than $\Rightarrow$ above or to the right.
Less than $\Rightarrow$ below or to the left.

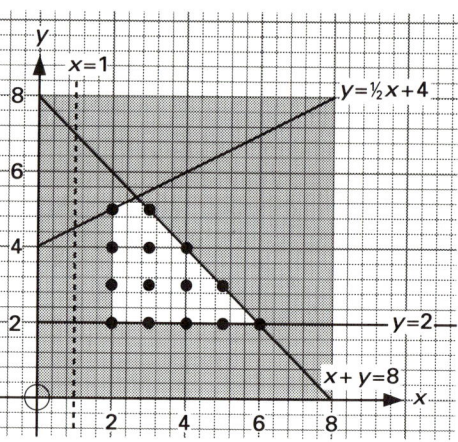

Fig. D16

If x and y have to be integers then only fourteen points are possible in this example. These are marked with dots, e.g. (4, 3). Remember that points are not allowed on $x = 1$ as $x > 1$ **not** $x \geqslant 1$.

Maximising and minimising

(a) If x has to be a maximum (as big as possible) then $x = 6$, $y = 2$ is the best answer.

(b) If $x + y$ has to be a minimum (as small as possible) then $x = 2$, $y = 2$ is the best answer.

(c) If $y - x$ has to be maximum then $x = 2$, $y = 5$ is the best answer.

Note Parts (b) and (c) can be answered by considering all possible points. Alternatively you can consider a set of parallel lines that share the condition given and choose the last possible point that is on one of these lines before the line 'misses' the region altogether.

For (b) you need $x + y = a$ (where a is the answer), so you consider lines of the family $x + y = a$, or $y = -x + a$. These are shown in Figure D17.

For (c) you need $y - x = a$, so consider the lines of the family $y = x + a$ (Figure D18).

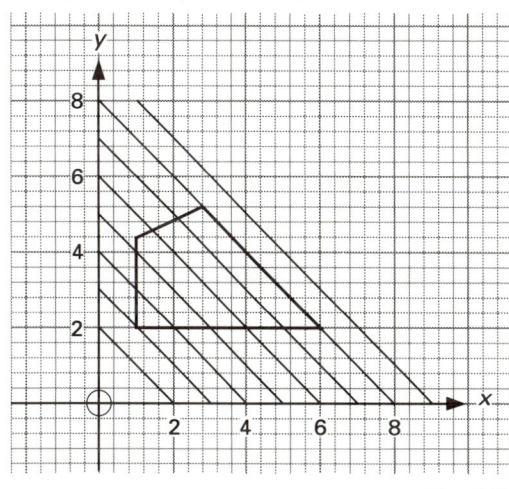

Fig. D17

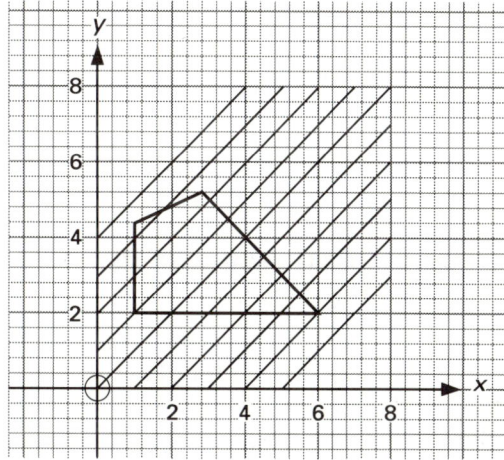

Fig. D18

Critical-path analysis (H)

Critical-path analysis 1 (H)   Strategy (33)

Imagine you had to arrange the building of a new motorway. You would have to break the job down into smaller jobs, then decide how long each would take, how many people you would need, and when each job would start and finish. Obviously some jobs have to be done before others, while some could be going on at the same time.

Even for much simpler jobs, drawing a clear correct diagram is not easy! Be prepared to have lots of tries. The following strategy will help:

First Decide which jobs must follow each other and which can be done simultaneously. It saves time to label the jobs A, B, C etc., though of course they do not have to be done in alphabetical order.

Then Use the longest chain of consecutive jobs as your starting line, then fit the other jobs onto the diagram to show the earliest point at which they can start and the latest point at which they can finish.

D

Critical-path analysis 2 (H)   Planning networks (33)

Figure D19 shows a planning network to do a job made up of the smaller jobs A, B, C and D.

A must be done before B.

C must be done before D.

However, A and B can be done at the same time as C and D.

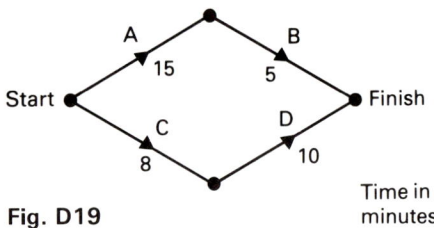

Fig. D19

Two jobs can be done at the same time when there are two people working, or there is a machine (like a cooker) doing some jobs, or the job may be a waiting time (like waiting for paint to dry).

Critical-path analysis 3 (H)   Critical paths (33)

The critical path is the one that gives the fastest possible time for the complete job. In Figure D19 it is A then B, giving 20 minutes.

Critical-path analysis 4 (H)   Starting times (33)

In Figure D19, jobs C and D take only 18 minutes, so job C need not start until 2 minutes after A. It is helpful to count back from the time the job has to be completed by and call this zero hour. Then the latest start jobs for the four jobs are:

B: zero minus 5
D: zero minus 10
C: zero minus 18
A: zero minus 20

These can be shown with flags on the lines near their start, as shown in Figure D20.

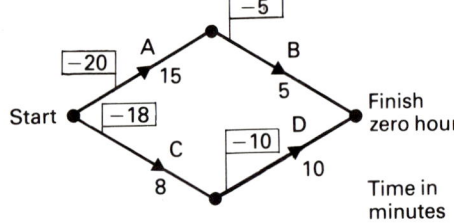

Fig. D20

Information for aural tests

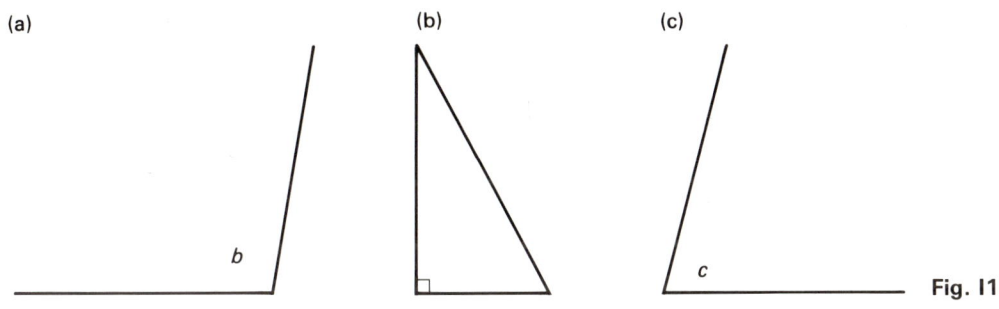

(a)   (b)   (c)

b   *c*

Fig. 11

EXHAUST BARGAINS

Model	Prices From	Model	Prices From
METRO to 1985 - 1.0, 1.3	£35.15	CAVALIER, fwd, ex FP	£37.00
MAESTRO 1.3 ex FP	£33.50	ASTRA MK 1 1.3, box & tail	£17.50
MONTEGO 1.6 ex FP 1984-1987	£40.55	ASTRA MK 11 1.3, box & tail	£22.95
FIESTA 950, 1100 to 1983	£21.40	NOVA 1.0/1.2, box & tail	£19.85
FIESTA 950, 1100, 1983-86, ex FP	£26.45	VW GOLF, box & tail to 1984	£19.85
ESCORT 1.3/1.6 1980-85, ex FP	£33.50	VW POLO, box & tail to 1984	£21.05
SIERRA 1.3/1.6 to 1985, ex FP	£35.85	NISSAN MICRA, box & tail	£25.30
ALPINE, ex FP	£28.80	SUNNY/CHERRY, 1982-1986 reg box & tail	£24.45
HORIZON, ex FP	£28.80	SUNNY 86, on box & tail	£33.50
		NISSAN BLUEBIRD, box & tail	£41.00

Fig. 12

Radio 1

(MW 1053/1089kHz, 285/275m; FM 97.6-99.8)
7.00am Gary Davies. **10.00** Dave Lee Travis. **1.00** Adrian Juste. **2.00** One Step Beyond Madness. **3.00** Johnnie Walker. **6.00** The Saturday Rock Show. **8.30** Songlines. **9.00** Andy Kershaw. **11.00** *(FM only after 12m't)* John Peel. **2.00** *(FM)* Lynn Parsons. **4.00-7.00am** *(FM)* Neale James.

Radio 2

(FM 88-90.2)
6.00am Barbara Sturgeon. **8.05** Brian Matthew: Sounds of the Sixties. **10.00** Anne Robinson. **12.00** Mark Wynter. **1.00** Howerd's Way. **2.00** Sounds of the Fifties. **3.00** Steve Race. **4.00** Judith Chalmer's Hi Days and Holidays: Margate. **5.00** Cinema 2. **5.30** The Movie Quiz. **6.00** If Wet, Under the Pier. **7.00** You Can't Have One Without the Other. **7.30** Natalie Cole in Concert. **8.55** Easy Does it. **10.00** Martin Kelner. **1.00** Charles Nove. **4.00-7.00am** Barbara Sturgeon.

Fig. 13

ACCOMODATION

SINGLE JOURNEY PRICES IN £'s
All cabins have air conditioning with en-suite shower and toilet

	LOWER DECK CABINS		DECK	DAY	NIGHT
	4-berth inside	per berth	J/K	5.00	8.00
	UPPER DECK CABINS				
	4-berth inside	per berth	C/D/E	7.00	12.00
	4-berth outside	per berth	B/C/D/E	8.00	13.00
	2-berth inside - lower twin berths	per berth	C/D/E	9.50	16.00
	2-berth outside - bunk beds	per berth	E	9.50	16.00
	2-berth outside - lower twin berths	per berth	B/C/D	11.50	18.00
	2-berth outside - adapted for physically disabled	per cabin	E	19.00	32.00
	Single outside - lower single berth and sofa	per cabin	B/C/D/E	24.00	40.00

Fig. 14

283

London→Reading→Westbury→Taunton→Exeter→Torbay→Plymouth

Mondays to Fridays

All trains shown are InterCity 125. Notes above columns: FO = Fridays only; FX = Fridays excepted (FO over one afternoon train; FO FO FX over the last three evening trains).

Station		1	2	3	4	5	6	7	8	9	10	11	12	13	14	15	16	17	18	19	20
Paddington	d	23.59	—	07 25	08 45	09 05	09 40	10 05	10 25	11 40	12 40	13 40	14 05	14 45	14 47	15 45	16 45	17 45	17 47	18 57	19 00
Reading E	d	00 40	—	07 53	09 09	09 30	10 14	10 29	10 58	—	13 04	14 14	14 30	15 09	15 17	16 09	17 09	18 10	18 19	19 21	19 24
Newbury	d	—	—	—	09 24	—	—	—	11 18	—	13 23	—	—	15 24	15 34	—	—	—	18 41	—	—
Pewsey	d	—	—	—	—	—	—	—	11 41	—	—	—	—	—	—	—	17 39	—	19 03	—	19 54
Westbury	d	—	06 30	08 04	10 00	—	—	—	12 01	—	13 59	—	—	16 00	16 13	—	18 00	—	19 24	—	20 13
Castle Cary	d	—	—	09 23	—	—	—	—	12 17	—	14 11	—	—	—	—	—	18 17	—	—	—	20 32
Taunton	a	03 06	08 32	09 44	10 40	11 22	—	—	12 45	—	14 39	—	16 23	16 40	16 56	17 31	18 40	19 32	20 08	20 43	20 55
Tiverton Junction	a	—	08 50	—	—	—	—	—	—	—	—	—	—	—	17 13	—	18 56	—	—	—	—
Exeter St David's	a	03 48	09 06	10 12	11 08	11 50	12 21	12 44	13 20	13 40	15 07	16 26	16 51	17 08	17 30	17 59	19 20	20 01	20 41	21 11	21 23
Barnstaple	a	05 07	—	—	—	12 28	—	—	—	14 36	—	16 27	—	—	18 35	—	—	20 49	—	22 38	22 38
Exmouth	a	06 55	—	10 52	11 46	—	13 00	14 10	15 25	16 00	—	17 41	18 11	—	18 41	19 49	20 41	—	—	22 14	22 14
Dawlish	a	—	09 22	11 10	—	12 04	—	—	13 38	14 49	15 30	16 58	—	17 22	17 57	18 29	—	20 32	—	—	22 15
Teignmouth	a	—	09 28	11 15	—	12 09	—	—	13 44	14 54	15 35	17 03	—	17 27	18 02	18 35	—	20 37	—	—	22 15
Newton Abbot	a	04 15	09 35	10 34	11 30	12 16	—	13 53	14 43	15 29	16 52	17 19	17 34	18 08	18 20	—	19 35	20 23	21 06	21 33	21 46
Torquay	a	05 56	10 19	—	11 46	12 48	—	14 04	15 15	16 14	17 25	—	18 13	—	18 55	19 54	20 58	—	—	22 02	22 02
Paignton	a	06 01	10 27	—	11 53	12 56	—	14 12	15 22	16 20	17 32	—	18 20	—	19 02	20 00	21 05	—	—	22 07	22 07
Totnes	a	—	—	—	12 30	—	13 16	—	—	15 56	—	17 48	18 24	—	19 49	—	21 21	22 16	—	—	22 00
Plymouth	a	05 05	10 17	11 13	12 09	12 58	13 26	13 44	—	14 40	16 08	17 37	17 58	18 16	18 55	18 59	20 17	21 01	21 52	22 12	22 28

Fig. I5

NOVEMBER 1993

S		7	14	21	28
M	1	8	15	22	29
T	2	9	16	23	30
W	3	10	17	24	
T	4	11	18	25	
F	5	12	19	26	
S	6	13	20	27	

Fig. I6

STANDARD VEHICLE & PASSENGER TARIFFS

SINGLE JOURNEY PRICES IN £'s	DAY	NIGHT	SINGLE JOURNEY PRICES IN £'s		DAY	NIGHT
Cars	23.00	38.50	Adults		38.50	38.50
Motorcycles/Scooters	9.50	15.50	Children – 6-14 years		19.00	19.00
Mopeds/Bicycles	3.50	6.50	Infants – under 6 years		FREE	FREE
Coaches (empty coaches excluded)	78.00	105.00	Students/Senior Citizens		29.00	38.50
Caravans/Trailers/ Mobile Homes/Minibuses – per metre	21.50	21.50	Party Tariff group of 10 or more	Adults	31.00	35.00
				Children – 6-14 years	15.50	17.50
				Students/Senior Citizens	23.50	26.00

Fig. I8

BBC 2

6.35 Open University. 12.00 The Other Olympians. 12.00 The Other Olympians. 12.30 Sunday Grandstand. 12.35 Motor Racing. 3.00 The British formula 3 Championship. 3.35 Golf. 4.30 Bowls.
6.00 Golf
6.30 One Man and His Dog
7.15 The Living Planet
8.10 A Visit from Mrs Protheroe
8.50 Night Time Architecture
9.10 The Hungarian Grand Prix
9.50 Moviedrome
9.55 Walker Film (1988)
11.25 International Golf
1.35 Close

Fig. I10

1991 (J) VAUXHALL ASTRA 1.7 DIESEL L 5 DOOR ESTATE, china blue, 5 speed, sunroof, radio cassette **£8,950**
1991 (J) RENAULT CLIO 1400 RT HATCHBACK, 5-speed, willow metallic, sunroof ... **£8,595**
1990 (G) VOLKSWAGEN POLO 1300 CL 3-DOOR, midnight blue **£5,195**
1990 (G) NISSAN BLUEBIRD 1600LX 5 DOOR, 5 speed, power steering, blue metallic **£5,995**
1989 (F) PEUGEOT 205 GTi 1.9, cherry red, alloy wheels **£6,195**
1988 (E) VAUXHALL BELMONT 1300 4-DOOR JUBILEE, red metallic, radio **£4,395**
1988 (E) VAUXHALL CARLTON 2.0 GLi ESTATE, red metallic, sunroof, 5 speed, power steering **£5,750**
1987 (E) AUSTIN METRO 1300 L 3-DOOR, blue, radio.................... **£3,150**
1985 (C) CAVALIER 1600 L 5-DOOR, 5-speed, white **£2,595**
1985 (C) VAUXHALL SENATOR 2.5 INJ SALOON, white, sunroof, alloy wheels, power steering **£4,495**
1984 (A) AUSTIN METRO 1300 L 3-DOOR, red, tan trim, radio **£1,995**
1984 (A) PORSCHE 924 LUX COUPE, Guards red, 5-speed, electric sunroof, radio/cassette **£5,995**

Fig. I11

READY RECKONER FOR MONTHLY REPAYMENTS

RATE		6%			8%			10%			12%		
PERIOD		2 yrs	3 yrs	4 yrs	2 yrs	3 yrs	4 yrs	2 yrs	3 yrs	4 yrs	2 yrs	3 yrs	4 yrs
SUM BORROWED	£50	£2.33	£1.64	£1.29	£2.42	£1.73	£1.38	£2.50	£1.81	£1.46	£2.59	£1.89	£1.55
	£100	£4.67	£3.28	£2.58	£4.84	£3.45	£2.75	£5.00	£3.62	£2.92	£5.17	£3.78	£3.09
	£200	£9.33	£6.56	£5.17	£9.67	£6.89	£5.50	£10.00	£7.23	£5.84	£10.34	£7.56	£6.17
	£500	£23.33	£16.39	£12.92	£24.17	£17.23	£13.75	£25.00	£18.06	£14.59	£25.84	£18.89	£15.42
	£600	£28.00	£19.67	£15.50	£29.00	£20.67	£16.50	£30.00	£21.67	£17.50	£31.00	£22.67	£18.50
	£800	£37.33	£26.22	£20.67	£38.67	£27.56	£22.00	£40.00	£28.89	£23.34	£41.34	£30.23	£24.67
	£1000	£46.67	£32.78	£25.83	£48.34	£34.45	£27.50	£50.00	£36.12	£29.17	£51.67	£37.78	£30.84
	£2000	£93.34	£65.56	£51.66	£96.68	£68.90	£55.00	£100.00	£72.24	£58.34	£103.34	£75.56	£61.68
	£3000	£140.01	£98.34	£77.49	£145.02	£103.35	£82.50	£150.00	£108.36	£87.51	£155.01	£113.34	£92.52
	£4000	£186.68	£131.12	£103.32	£193.36	£137.80	£110.00	£200.00	£144.48	£116.68	£206.68	£151.12	£123.36
	£5000	£233.35	£163.90	£129.15	£241.70	£172.25	£137.50	£250.00	£180.60	£145.85	£258.35	£188.90	£154.20

Fig. I12

BLINDS
AUGUST SALE

0% OFF VERTICALS*
0% OFF VENETIANS*
5% OFF ROLLERS*
0% OFF PLEATEDS*

FREE
MEASURING & FITTING
FAST
LOCAL SERVICE

Fig. I13

		FRANCE							GERMANY				HOLLAND				
		Bayuex	Bordeaux	Caen	Cannes	Marseille	Paris	Rouen	Cologne	Frankfurt	Hanover	Hamburg	Amsterdam	Eindhoven	Rotterdam	The Hague	Utrecht
HOOK OF HOLLAND	KM	644	1063	616	1350	1225	481	491	298	493	443	884	77	143	41	20	87
	MILES	400	661	383	839	761	299	305	185	306	275	549	48	89	25	12	54
CALAIS	KM	365	874	337	1193	1067	292	212	415	604	647	995	371	296	312	338	335
	MILES	272	543	209	741	663	181	132	258	375	402	618	231	184	194	210	208
BOULOGNE	KM	332	881	303	1200	1074	299	178	449	638	681	961	405	330	346	372	369
	MILES	206	547	188	746	667	186	111	299	396	423	597	252	205	215	231	229
DIEPPE	KM	198	767	170	1078	952	173	62	496	685	779	928	517	406	458	484	481
	MILES	123	477	106	670	592	108	39	308	426	484	577	321	252	285	301	299

Fig. I14

I

MILEAGE FROM PORTS	Boulogne	Calais	Cherbourg	Dieppe	Dunkerque	Hoek van Holland	Oostende
Amsterdam	245	224	489	300	209	48	172
Barcelona	833	854	804	786	859	938	869
Basel	475	478	572	473	473	449	439
Biarritz	619	640	508	527	645	752	656
Bordeaux	504	525	400	411	527	637	540
Bruxelles	146	127	362	195	112	107	71
Esbjerg	664	643	897	771	828	508	589
Firenze	878	867	950	834	852	890	841
Frankfurt	387	367	585	426	356	292	320
Genève	476	484	545	446	489	546	509
Hannover	446	425	647	500	410	297	368
København	886	664	918	781	649	528	611
Köln	268	253	488	312	239	187	202
Lisboa	1269	1285	1167	1179	1290	1417	1308
Luxembourg	252	256	433	280	250	231	211
Lyon	454	469	508	409	457	536	472
Madrid	942	963	843	855	964	1075	984
Marseille	653	664	703	604	669	735	678
Milano	666	679	751	645	668	682	649
München	598	607	725	594	594	529	563
Napoli	1190	1194	1251	1143	1179	1182	1157
Nice	732	747	801	702	749	804	764
Paris	151	172	221	122	182	284	186
Roma	1054	1059	1115	1007	1044	1046	1021
St-Malo	298	319	121	220	332	487	363
Salzburg	684	693	814	680	678	615	649
Strasbourg	386	393	518	372	388	381	348
Trieste	892	894	999	900	878	864	860
Venezia	837	836	929	809	831	818	824
Wien	864	858	992	860	843	769	797

Fig. I15

Distances—miles and kilometres

miles	km	miles	km
0.62	1	1	1.61
1.86	3	3	4.83
3.11	5	5	8.05
4.34	7	7	11.27
5.59	9	9	14.48
6.20	10	10	16.10
12.40	20	20	32.20
31.00	50	50	80.50
43.00	70	70	112.70
62.00	100	100	161.00

For a quick conversion, km to miles, divide the km distance bt 8, then multiply the result by 5.

Sample Distances

Ostend to:

Berlin	601 miles
Cologne	212 miles
Frankfurt	326 miles
Freiburg	499 miles
Hamberg	425 miles
Munich	586 miles
Trier	241 miles

Boulogne to:

Cologne	276 miles
Frankfurt	421 miles
Munich	639 miles

Hook of Holland to:

Cologne	205 miles
Frankfurt	319 miles
Munich	561 miles

Pounds into kilograms

lb	kg	lb	kg
1	0.45	20	9.07
2	0.91	30	13.61
3	1.36	40	18.14
5	2.27	60	27.22
10	4.54	112	50.80

Kilograms into pounds

kg	lb	kg	lb
1	2.20	20	44.09
2	4.41	30	66.14
3	6.61	40	88.19
5	11.02	60	132.28
7	15.43	80	176.37
9	19.84	100	220.46
10	22.05	250	551.15

Petrol and Oil—gallons and litres

litres	gallons	litres	gallons
4.55	1	1	0.22
13.65	3	3	0.66
22.75	5	5	1.10
31.82	7	7	1.54
40.92	9	9	1.98
45.46	10	10	2.20
90.92	20	20	4.40
227.30	50	50	11.01
454.60	100	100	22.00

Temperatures

°F	°C	°F	°C
212	100	59	15
104	40	50	10
102	38.9	41	5
101	38.3	32	0
100	37.8	28	−2
98.4	37	23	−5
97	36.1	18	−8
86	30	12	−11
80	26.7	5	−15
77	25	0	−18
68	20	−4	−20
64	17.8		

Continental and British Clothing Sizes

SHOES

Children's

U.K.	1	2	3	4	5	6	7	8	9	10
Cont.	17	18	19	20	22	23	24	25	27	28

U.K.	11	12	13
Cont.	29	30	31

Woman's

U.K.	1	2	3	4	5	6	7	8
Cont.	33	34	35	36	37	38	39	40

Men's

U.K.	1	2	3	4	5	6	7	8	9	10
Cont.	35	36	37	38	39	40	41	42	43	44

U.K.	11	12	13
Cont.	45	46	48

SHIRTS AND COLLARS

U.K.	14	$14\frac{1}{2}$	15	$15\frac{1}{2}$	16	$16\frac{1}{2}$	17
Cont.	36	37	38	41	41	42	43

Fig. I16

SUITS AND OVERCOATS (Men's)

U.K.	36	38	40	42	44	46
Cont.	46	48	50	52	54	56

DRESSES AND SUITS (Women's)

U.K.	8	10	12	14	16	18
Cont.	34	36	38	40	42	44

Hats (Men's)

U.K.	$6\frac{1}{2}$	$6\frac{5}{8}$	$6\frac{3}{4}$	$6\frac{7}{8}$	7	$7\frac{1}{8}$	$7\frac{1}{4}$	$7\frac{3}{8}$	$7\frac{1}{2}$
Cont.	53	54	55	56	57	58	59	60	61

Glove sizes are usually the same as in U.K.

SOCKS, STOCKINGS, etc., where different, are measured in cms, not ins.

Answers

Note Some answers and answer hints are in the Teachers' Resource Book.

Chapter 1 Knowing about numbers

1 (a) 17, 19 (b) 16 (c) 20 (d) 18
  (e) 15
2 1 (or 0)
3 4 or 5
4 (a) 23.4 (b) 0.12
5 (a) $-\frac{1}{2}$ ft (b) 1 ft (c) $-1\frac{1}{2}$ ft
6 (a) 5 (b) 300
7 0.009, 0.6501, 0.651, 0.6512,
  3.14, 3.142
8 (a) 10 (b) 2 (c) $2 \times 2 \times 2 \times 3$
  (d) $1\frac{1}{2}$; 1.5; $\frac{2}{3}$; $0.\dot{6}$
9 (a) 24, 119, 15
  (b) 5995 to 6005; 198 to 202
10 (a) 36 (b) 32 (c) 2 (d) 2
  (e) 340 (f) 0.0638
11 (a) 9.85×10^1 (b) 4.6×10^{-1}
  (c) 3.0025×10^2 (d) 6×10^2
  (e) 3.5×10^2
12 (a) 0.4 (b) 60% (c) $\frac{7}{100}$ (d) $\frac{3}{20}$
13 1, 2, 3, 4, 6, 12
14 (a) 2, 3 (b) 1 (c) 1, 2, 4
  (d) 4, 12 (e) 1, 3, 6 (f) 4, 6, 12
15 (a) 1000 (b) 1 000 000
  (c) 200 056
16 (a) 1×10^3 (b) 1×10^6
  (c) $2.000\,56 \times 10^5$
17 (a) 2.8 −04 (b) 3.65 −02
  (c) 8.42 −01
18 (a) 149.9 mm (b) 14.99 cm
19 (a) 20 (b) 126
20 (a) $2 \times 3 \times 3$ (b) 3×13
  (c) $2 \times 3 \times 7$
21 (a) A, $\frac{3}{4}$; B, $1\frac{5}{8}$; C, $2\frac{1}{4}$; D, $3\frac{1}{3}$;
    E, $4\frac{5}{6}$
  (b) A, 0.75; B, 1.625; C, 2.25;
    D, $3.\dot{3}$; E, $4.8\dot{3}$
22 See Figure T1:1.

23 (a) 5, 2 (b) 25, 36 (c) 48, 96
  (d) 42, 56 (e) $\frac{1}{32}$, $\frac{1}{64}$
  (f) 0.01, 0.001 (g) −1, −4
  (h) 0, −0.2 (i) 125, 216 (j) 1, 2
24 12
25 $1 \times 24 \times 4$, $2 \times 12 \times 4$, $3 \times 8 \times 4$,
  $4 \times 6 \times 4$; give your own reason for
  the 'best way'.
26 £3.00 per child, £6.00 per adult
27 $\frac{4}{5}$, $\frac{2}{3}$, 0.6, $\frac{4}{7}$, $\frac{1}{3}$, 15%
28 33 degrees
29 (a) 100 (b) −2
30 11
31 (a) even (b) odd (c) odd
  (d) either (e) odd (f) even
  (g) odd (h) even
32 (a), (c), (e), (f), (h), (i) and (j) are
  rational
33 See question 32(j).
34 112
35 (a) 0.5 (b) 0.143 (c) $\frac{4}{3}$ or $1.\dot{3}$
  (d) $\frac{5}{4}$ or 1.25
36 (a) $\frac{1}{2}$ (b) $\frac{1}{9}$ (c) ± 2 (d) $\pm\frac{1}{2}$ (e) 4
  (f) $\frac{1}{2}$ (g) 1000
37 (a) $2n$ (b) $2n - 1$ (c) n^2 (d) n^3
  (e) $3 + 4(n - 1) = 4n - 1$
  (f) 10^{1-n} (g) $n(n + 1)$
  (h) $(n + 1)(n + 2)$
38 (a) even (b) odd
39 7 frames are possible.
41 (a) 51 200 km²
  (b) about 9 times as many
  (c) Scotland about 66, England about
    358
  (d) 130 500 km²;
    129 500 km²
  (e) 135 000 km²;
    125 000 km²
    Error range increased by
    ±4500 km².

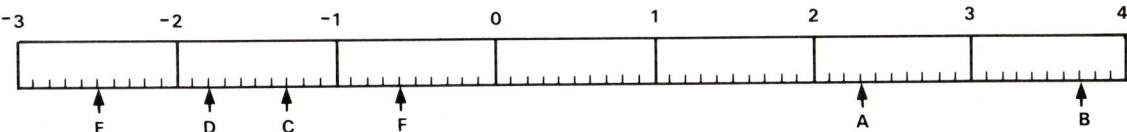

Fig. T1:1

An

Chapter 2 Using numbers

1. (a) 14 688 (b) 208 (c) 29
2. (a) Monday $5 - -1 = 6\,°C$;
   Tuesday $3 - -2 = 5\,°C$;
   Wednesday $4 - -4 = 8\,°C$;
   Thursday $6 - -3 = 9\,°C$;
   Friday $5 - -1 = 6\,°C$
   (b) $3\,°C$ (c) 4
3. $-7\,°C$, $-6\,°C$, $-4\,°C$, $0\,°C$, $3\,°C$, $8\,°C$
4. (a) -4 (b) 7 (c) -9 (d) -7
   (e) 6
5. 3.625
6. (a) $2\frac{3}{20}$ (b) $2\frac{17}{20}$ (c) $1\frac{1}{3}$ (d) $\frac{3}{5}$
7. 8.4
8. (a) e.g. $230 \times 50 = £11\,500$
   (b) e.g. $£240 \div 40 \approx £6$
9. (a) $\frac{2}{9}$ (b) about $107\,m^2$
10. (a) 4320 (b) 35 778 (c) 309
    (d) 84
11. (a) $8\,°C$ (b) $14\,°C$ (c) $5\,°C$
    (d) $44\,°C$
12. $-12\,°C$, $-8\,°C$, $-5\,°C$, $-3\,°C$, $-1\,°C$, $7\,°C$, $9\,°C$, $32\,°C$
13. 10
14. (a) $\frac{17}{20}$ (b) $\frac{1}{21}$ (c) $\frac{1}{3}$ (d) $\frac{3}{8}$
15. (a) $1\frac{7}{8}$ (b) $1\frac{1}{6}$ (c) $\frac{4}{5}$ (d) $\frac{2}{5}$
16. £36 000
17. 256
18. No; 5 min short
19. 54 m.p.h.
20. 9
21. $8\frac{1}{3}$
22. (a) 1 (b) $3\frac{5}{8}$ (c) 5 250 000
23. (a) $-17.8\,°C$ (b) $-23.3\,°C$
24. (a) $\frac{1}{2}$ (b) 45
25. $\frac{3}{4}$
26. (a) about 1.2 million
    (b) about 930 000
    (c) about 493.6 tonnes
    (d) 6 h 55 min (e) 524 m.p.h.
    (f) 3960 ft (g) 8.3p
    (h) (i) 7 h 25 min (ii) 1 h 40 min
    (iii) 9 h 2 min (iv) 25 h 2 min
    (i) about 438 m.p.h.

Chapter 3 Percentages

1. £13.50
2. French 70%; German 80%;
   German best
3. £56
4. 167%
5. (a) $\frac{3}{4}$; 0.75 (b) $\frac{4}{25}$; 0.16
   (c) $\frac{6}{25}$; 0.24 (d) $\frac{7}{100}$; 0.07
   (e) $\frac{1}{3}$; $0.\dot{3}$
6. (a) 65% (b) 42.6% (c) £5.80
7. (a) 4, 16, 256, 65 536, 4.295×10^9, $1.844\,7 \times 10^{19}$
   (b) 12 700%
8. £48
9. £2280
10. £1.14
11. £39.44
12. 20%
13. 60%
14. (a) £42 (b) 25%
15. 21p
16. (a) £2250 (b) £20 250
17. 25%
18. £2160; £2640
19. (a) £162 (b) 720 (c) 48%
20. £1080; £972; £991.44
21. (a) 8.4 million (b) 11.6%
22. £856.09
23. £741
24. 2.5%
25. £6000
26. (a) saved by 47 votes (b) 843
27. (a) 10% (b) 6.974 kg
    (c) 6.5 years
28. 8%

Chapter 4 Ratio

1. (a) direct (b) direct (c) inverse
   (d) direct
2. graph (b)
3. (a) 2:1 (b) 1:50 (c) 15:1
   (d) 13:10
4. (a) 3:1 (b) 2.4:1 (c) 0.75:1
5. 15
6. £240
7. 10 m
8. 5 km
9. 20 litres
10. £2.50
11. (a) $\frac{3}{8}$ (b) 37.5% (c) 3.9 g
12. both the same
13. 96 km/h
14. 100 cm
15. 1:3, 1:1, 3:1

16 9 h 36 min

17 $C = zx/y$

18 10 ft

19 7.2 s

20 No calculated division widths are needed.

21 500 ml, £1.65; 1 litre, £3.15; 2.5 litre, £7.53

22 141 : 1

23 about 0.41

Chapter 5 Using a calculator

3 (a) 265 (b) 500 (c) £2.99
   (d) 143 m

4 (a) 286.84 (b) 493.1 (c) £2.87
   (d) 159.82 m

5 (a) 20 (b) −2 (c) 2 (d) 17
   (e) 23 (f) 6 (g) 0.7 (h) 0.05
   (i) 7 (j) 0.25 (k) 0.25 (l) 1.25

6 (a) 16.92 (b) −1.5 (c) 2.59
   (d) 15.4 (e) 23 (f) 6.4
   (g) 0.725 (h) 0.0452 (i) 6.13
   (j) 0.263 (k) 0.273 (l) 1.21

7 (a) 4.47 (b) 3.46 (c) 0.143
   (d) 7

8 (a) 1.41 (b) 1.82
   (c) 3.70 or −2.7

9 (a) £35.50 (b) £55.10 (c) 26

10 (a) 160 (b) 60 (c) 0.2 (d) 1
   (e) 45

11 (a) 153 (b) 60.4 (c) 0.201
   (d) 0.939 (e) 41.1

12 (a) 4.47 (b) 2.29
   (c) 2.41 or −0.41

13 (a) 2775 (b) Your problem!

14 (a) −3.665 (b) −3.665 (c) 17.48
   (d) −3.75

15 (a) 10.5 (b) 2.71 (c) 0.843
   (d) 0.675 (e) 0.3

18 Finding the exact value of 1 ÷ 17

19 (a) 0.0$\dot{4}$3 478 260 869 565 217 391 $\dot{3}$
   (b) 0.0$\dot{3}$4 482 758 620 689 655 172 413 793 $\dot{1}$

20 (a) $\frac{1}{9}$ (b) $\frac{1}{11}$ (c) $\frac{1}{13}$ (d) $\frac{3}{7}$ (e) $\frac{5}{18}$

21 Write as a fraction with the denominator the same number of 9s as the number of figures in the recurring group. If the first recurring figure is not in the tenths column, multiply by 10^n to move it there.

For question 20(e):
$0.2\dot{7} \times 10 = 2.\dot{7} = 2\frac{7}{9}$
$\frac{25}{9} \div 10 = \frac{25}{90} = \frac{5}{18}$

Chapter 6 Measure

2 (a) 1000 (b) 1000 (c) 1000
   (d) 10 000 (e) 10 000
   (f) 1 000 000

3 (a) 7 oz (b) about 340 g

4 (a) (ii) (b) (iv) (c) (vi) (d) (i)
   (e) (vii) (f) (iii) (g) (v) (h) (viii)

5 (a) 0.015 m (b) 0.012 01 km
   (c) 16 500 g (d) 5.05 m
   (e) 5.007 kg (f) 50 cl
   (g) 0.078 55 m^2

6 (a) 15.5 miles, 14.5 miles
   (b) 12 m.p.h. (c) 22.5 min

7 (a) (i) 120 ft (ii) 0.7 s (iii) 2.1 s
   (b) The thinking time is the same, but the faster your speed, the further you will travel while you are thinking.
   (c) 30 km/h: 5 m, 6 m, 10 m
      60 km/h: 10 m, 25 m, 35 m
      90 km/h: 20 m, 50 m, 70 m

8 14.6 mm; 14.4 mm

9 (a) below (b) 40 kg (c) 32.38 cm

10 No. All different metals have different densities. (Find the story of Archimedes and the king's crown.)

12 (a) $F = 2(C + 15)$ (b) −10 °F
   (c) $C = \frac{F}{2} - 15$ (d) −4 °F
   (e) (i) 65.3 °F, 63.5 °F
      (ii) 64 or 65 °F

13 About 8.5 cm^2; possibly ±0.5 cm^2

14 (a) about 2 hours (b) 600 m.p.h.
   (c) 6.9 miles per second
   (d) about $4\frac{1}{2}$ months

16 (a) 4.12×10^7 m^2 (b) 1.51×10^7
   (c) 9.38 m (d) 6.114×10^9 litres

17 (a) 3.2 V (b) 32 V (c) 6.5 V
   (d) 320 V (e) 0.35 MΩ

18 15.81 m^2, 14.21 m^2

19 (a) 3.925 kg (b) 250 g

20 (a) 32 miles from A at 12 : 32
   (b) 10 min

21 Table should show from 5 rolls to 14 rolls required.

Chapter 7 Finance

1 DM1450; 2592.9 francs
2 £83.99
3 (a) 25p (b) £2.37 (c) £1.97
    (d) £2.37
4 (a) £360 (b) £400 (c) £276
    (d) £124 (e) 182 weeks
5 (a) £150 (b) £75 profit
6 (b) (i) £1323 (ii) £1449 (iii) £487
7 £35 247
8 £9600; £8448; £8025.60
9 £70
10 £30 000
11 £4200
12 £1040
13 (a) (i) £3595 (ii) £1800
    (b) (i) £11 176 (ii) £5400
    (c) £259
    (d) There is a higher risk of him not
        surviving.
    (e) £18; females have a higher life
        expectancy.
14 £4483.75
15 (a) £13 800 (b) £16 500
    (c) £31 200
16 (a) £60 (b) £112.41 (c) £52.50
    (d) £364.91
17 £225
18 (a) A company that lends money
    (b) 23.1%
    (c) (i) £19.11 (ii) £9937.20
    (e) £69.72
19 (a) £2100 (b) 18 (c) 17
20 (a) £108.70 (b) £5.10; £133.69
21 £307.07
22 £6258

Chapter 8 Journey graphs

1 (a) 30 min (b) 0955 (c) 100 km/h
    (d) 50 km/h
2 1115, 50 miles; 1245, 15 miles
3 (a) average speed of 40 km/h for $\frac{1}{2}$ h,
        stationary for $\frac{1}{2}$ h, average speed
        of 20 km/h for 1 h
    (b) average speed of 40 m.p.h. for
        3 h, stationary for 4 h, return
        journey at average speed of
        40 m.p.h.

(c) average speed of 80 km/h for
    15 min, stationary for 15 min,
    average speed of 120 km/h for
    15 min, then average speed of
    20 km/h for 30 min
4 (a) 20 km/h (b) 24 m.p.h.
    (c) 48 km/h
5 1 km/min
6 CD, constant speed of 20 m/s;
    DE, acceleration of 20 m/s^2;
    EF, constant speed of 30 m/s
7 20; 12$\frac{1}{2}$; 45. Total is 112.5 metres.
8 (a) 5 m/s^2 (b) 310 m; 25.8 m/s
9 (b) (i) 0.9 m/s^2 (ii) 0.45 m/s^2
    (c) 1370 m
10 (b) 64 m (c) 134 s
11 2.7 miles

Chapter 9 Practical graphs, tables and flow charts

2 (a) £14 000 (b) 3.6% (c) 38.6%
    (d) £2100
3 (a) apply for mortgage
    (b) consider a cheaper house or find
        someone who will lend you £3000
    (c) consider a cheaper house
4 (a) trapezium (b) isosceles trapezium
    (c) kite (d) parallelogram
    (e) rectangle (f) rhombus
    (g) square
5 CD
6 See Figure T9:1.

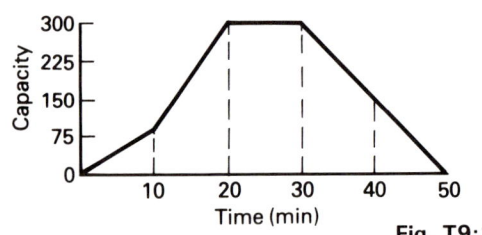

Fig. T9:1

7 A – (b), B – (c)
8 See Figure T9:2 opposite.
9 See Figure T9:3 opposite.
10 About 49 thousand
11 (a) 166 min (b) 26 min
    (c) A, 24; B, 81; C, 60
    (d) 39 m.p.h. (e) 19.44p
    (f) Basingstoke to Winchester
12 See Figure T9:4 opposite.

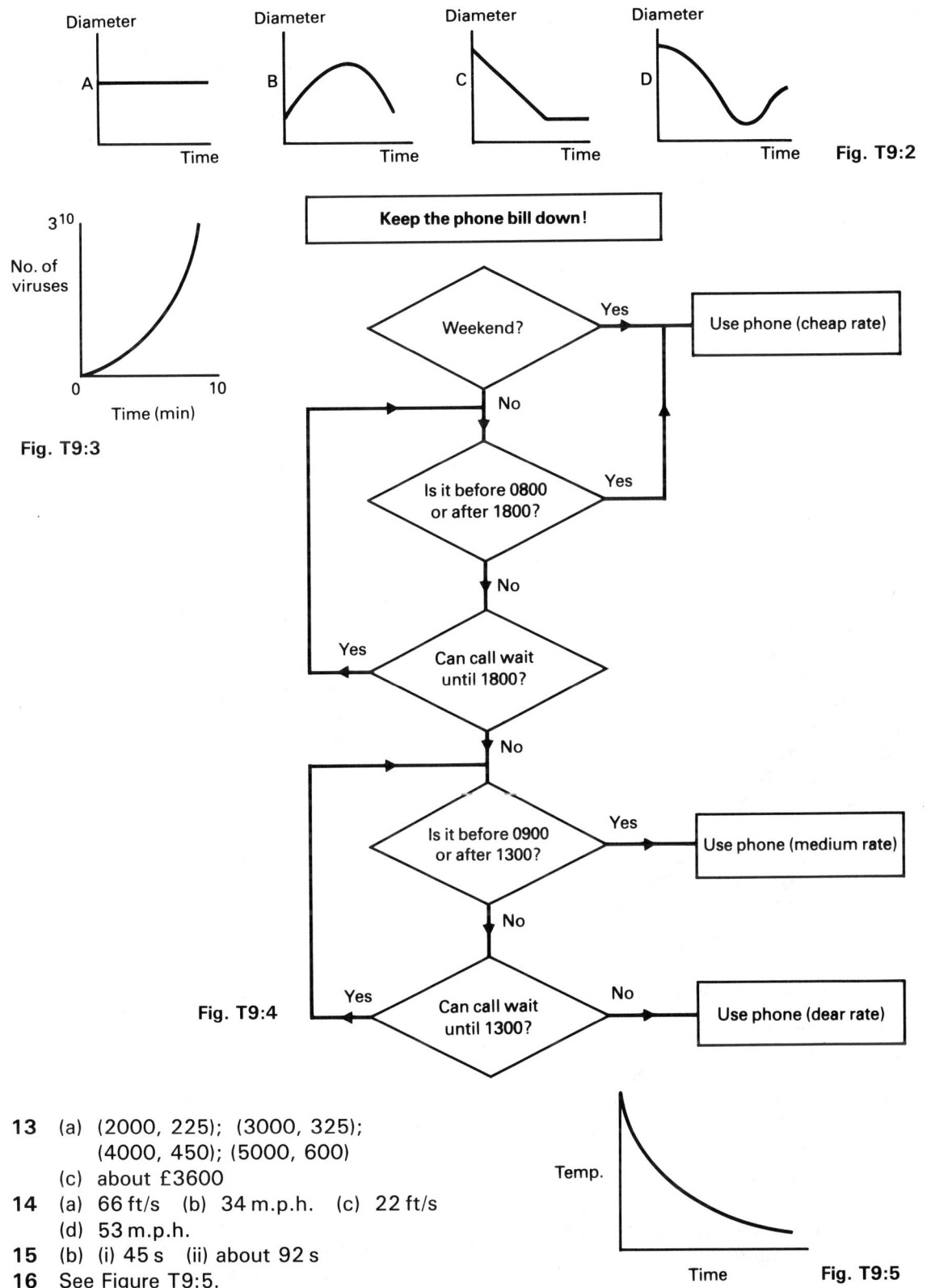

Diameter A — Time

Diameter B — Time

Diameter C — Time

Diameter D — Time

Fig. T9:2

No. of viruses / Time (min)

Fig. T9:3

Keep the phone bill down!

Weekend? — Yes — Use phone (cheap rate)

No

Is it before 0800 or after 1800? — Yes

No

Can call wait until 1800? — Yes

No

Is it before 0900 or after 1300? — Yes — Use phone (medium rate)

No

Can call wait until 1300? — No — Use phone (dear rate)

Yes

Fig. T9:4

13 (a) (2000, 225); (3000, 325);
   (4000, 450); (5000, 600)
   (c) about £3600
14 (a) 66 ft/s (b) 34 m.p.h. (c) 22 ft/s
   (d) 53 m.p.h.
15 (b) (i) 45 s (ii) about 92 s
16 See Figure T9:5.

Temp. / Time

Fig. T9:5

17 (a) e.g. sprint or long jump
  (b) e.g. 10 000 metres or marathon
18 (a) 10p
  (b) No; charge for zero number ordered and supply at zero cost impossible
19 See Figure T9:6.

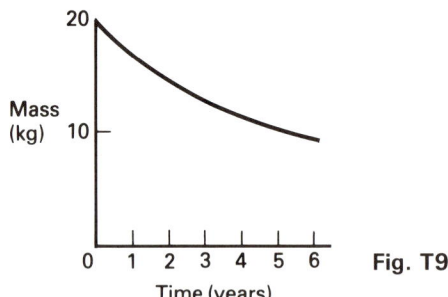

Fig. T9:6

Mass (kg) — Time (years)

20 (b) (i) about 109 beats (ii) about 75 s
  (c) about $\frac{1}{2}$ beat per second
21 (b) Water outflow exceeds inflow for the first 4 min, though the difference is decreasing. Then water inflow exceeds outflow, the difference staying fairly constant.
  (c) 1.4 min
  (d) −7 cm/min, +13 cm/min
22 See Figure T9:7.

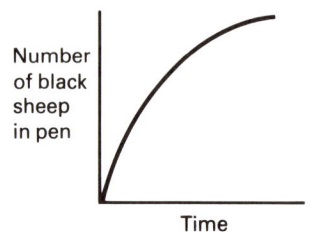

Fig. T9:7

Number of black sheep in pen — Time

24 (a) about 1:24 p.m.
  (b) about 30 km³/h²
  (c) about 163.5 km³
25 (a) 31.5 m.p.g. (b) 53 m.p.h.
27 N(1) 7 1 1 1 1 1
  N(2) 1 7 4 4 4 2
  N(3) 4 4 7 7 2 4
  N(4) 8 8 8 2 7 7
  N(5) 2 2 2 8 8 8

Chapter 10 Formulae and functions

1 (a) 50 °F (b) 30 °F (c) 20 °F

2 6 m²
3 (a) 15 ft (b) 175 ft
4 (a) 3 (b) 9 (c) −1 (d) −1
  (e) 16 (f) $2\frac{2}{3}$
5 Figure 10:1, $4a + 4b + 4c$;
  Figure 10:2, $6x + 3y$;
  Figure 10:3, $4s + 4b$;
  Figure 10:4, $6a$
6 (a) $T = S + L$ (b) $L = T - S$
  (c) $S = T - L$
7 (a) $z = xy$ (b) $x = z \div y$
  (c) $y = z \div x$
8 (a) $h = \dfrac{V}{wb}$ (b) $r = \sqrt{\dfrac{A}{\pi}}$
  (c) $s = \dfrac{v^2 - u^2}{2a}$ (d) $c = 2S - a - b$
  (e) $a = \sqrt{16 - b^2}$ (f) $c = \sqrt{\dfrac{E}{M}}$
  (g) $h = \dfrac{S - \pi r^2}{2\pi r}$
9 (a) $3x$ cm (b) $2x$ cm (c) $24x$ cm
  (d) $22x^2$ cm (e) $6x^3$ cm³
10 (a) 483 (b) 987 (c) 2142
11 Double the number, then write a zero at the end, then subtract the number.
12 10 lb
13 (a) 3p (b) 3p (c) 4p
14 36
15 (a) 10 (b) 13 (c) 130 (d) $1\frac{8}{13}$
16 60 metres
17 $(6 + 6p)$ kg
18 k/v hours
19 (a) 152 cm (b) 22.4 cm
  (c) $f = \dfrac{3h - 256}{10}$ (d) negative length
20 (a) $12s$ cm (b) s^2 cm²
  (c) $5s^2$ cm
21 $6s^3$ cm³
22 $600s^3$ cm³
23 (a) 1 (b) 5 (c) 100 (d) 500
24 (a) 2.407×10^6 (b) 2.9095×10^2
  (c) 8.0245×10^3 (d) 3.494×10^{-2}
  (e) 9.5457×10^0
  (f) -2.1842×10^3
25 (a) $x^2 - x + 41$ (b) 131
  (c) 7, −6 (d) 41
26 $P = 12x + 18y$

22 (a) None (b) −1 (c) None
(d) $1\frac{1}{3}$, 3

23 (a) {−2, −1, 0, 1, 2, 3, 4, 5, 6, 7}
(b) {−4, −3, −2, −1, 0, 1}
(c) $x < -1$, x integral (an infinite set)

24 (a) $x^2 + x - 6 = 0$ (b) $x^2 - 4x = 0$
(c) $2x^2 - x - 10 = 0$

25 −9, 5

26 (b) 1.4583, 1.4789, 1.4688, 1.4738,
1.471; $t \approx 1.47$

27 (a) 3.211 950 3, 3.195 614 2,
3.193 057 2, 3.192 656 8
(b) 3.193; also −2.193 from
$x_{n+1} = -\sqrt{(x_n + 7)}$

28 (a) (i) −1 (ii) 5
(b) One value (at least) between 2
and 3.
(d) (i) 2.2 (ii) 2.227 (e) −0.225

29 8 m by 8 m and 8 m by 5 m

30 0, 100, 48

31 £70

32 (a) −0.28, 1.78 (b) $\frac{3}{4}$
(c) $0 < x < 1\frac{1}{2}$

33 (a) 12.96 (b) 12
(c) 0.34 s, 11.66 s

34 $P = 6x + 10$; $A = 2x^2 + 9x + 4$,
$x = 3.629$, 31.77 cm

35 (a) $\dfrac{240}{x}$ (b) $\dfrac{200}{x-1}$ or $\dfrac{240}{x} + 2$
(c) £5

36 $x = 2\frac{2}{3}$, $y = 1\frac{2}{3}$, $V = 11\frac{23}{27}$ cm^3 or
$x = 2$, $y = 3$, $V = 12$ cm^3

37 (b) 2.2055

38 0.201 64; 2.1284

39 Check your solutions by substitution.

40 2 feet

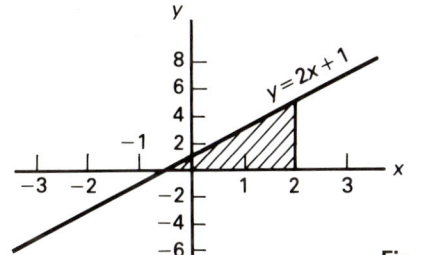

Fig. T13:1

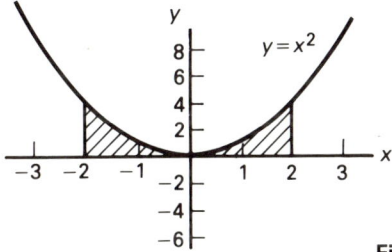

Fig. T13:2

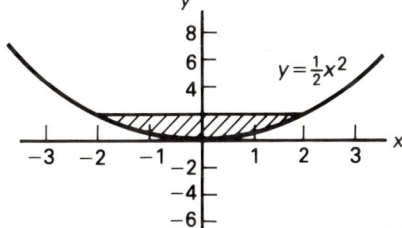

Fig. T13:3

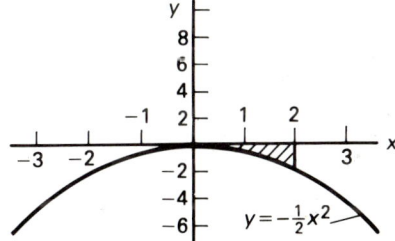

Fig. T13:4

Chapter 13 Function graphs

1 (a) (i) A (−6, −3); B (6, −3)
(ii) I (0, 1); J (−2, 3)
(b) (i) $y = -3$ (ii) $x = -6$
(iii) $y = \frac{1}{2}x$
(c) (i) $\frac{1}{2}$ (ii) $\frac{1}{6}$ (iii) 1
(d) (i) IJ (ii) AH

2 See Figures T13:1 to T13:4.

3 See Figures T13:1 to T13:4.

4 (a) −0.2, −0.22, −0.25, −0.29,
−0.33, −0.4, −0.5, −0.67, −1,
−2, 0.2, 0.22, 0.25, 0.29, 0.33,
0.4, 0.5, 0.67, 1, 2
(b) You cannot divide by zero.

5 (a) $x = 1$, $y = 4$ (b) $x = 3$, $y = 3$
(c) $x = -2$, $y = 0$

6 (a) (i) 3, −1 (ii) 1, 2
(b) $x = 1.5$, $y = 3.5$; see Figure T13:5.
(c) See Figure T13:5.

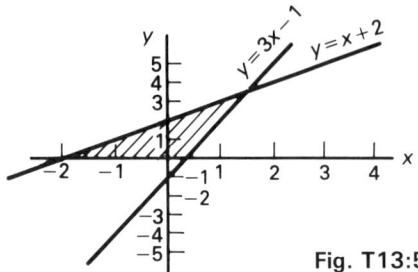

Fig. T13:5

7 (a) $x \geqslant 0$, $y \geqslant 0$, $y \leqslant -\frac{1}{2}x + 4$,
$y \leqslant \frac{1}{2}x + 2$, $y \leqslant -2x + 10$
(b) (i) 6, (4, 2)
(ii) −4; any point on line from (0, 2) to (2, 3)

8 See Figure T13:6. (a) 2 (b) $\frac{1}{2}$
(c) $-1\frac{1}{2}$

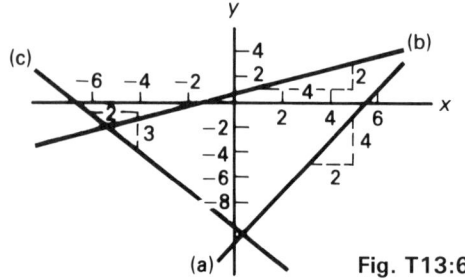

Fig. T13:6

9 (a) $y = 2x - 11$ (b) $y = \frac{1}{2}x + \frac{1}{2}$
(c) $y = -1\frac{1}{2}x - 10$

10 Check your graphs by using a computer graph-plotting program or a graphical calculator.

11 (a) 5 (b) 8 (c) 10 (d) 4 (e) 3
(f) 1

12 See Figure T13:7.

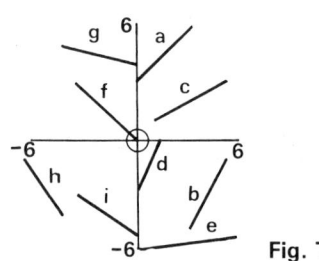

Fig. T13:7

13 (a) 1, $y = x + 3$ (b) 3, $y = 3x - 3$
(c) $\frac{1}{6}$, $y = \frac{1}{6}x - 6$ (d) −1, $y = -x$
(e) $-\frac{1}{4}$, $y = -\frac{1}{4}x + 4$
(f) $\frac{8}{3}$, $y = \frac{8}{3}x + 5$

14 See Figure T13:8.

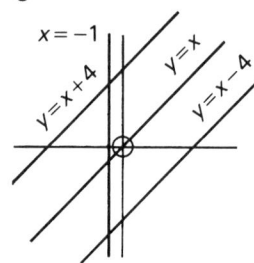

Fig. T13:8

15 See Figure T13:9.

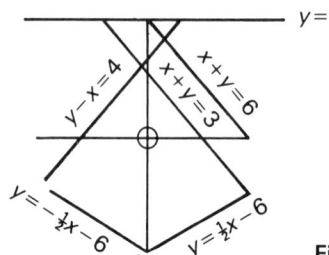

Fig. T13:9

16 See Figure T13:10.

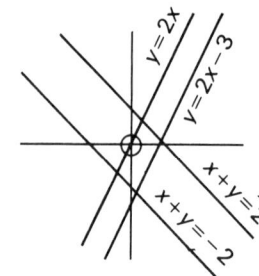

Fig. T13:10

17 (a) See Figure T13:11.
(b) (3, 3); (5, 3); (5, 1); (1, 1)

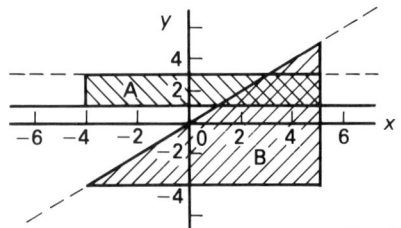

Fig. T13:11

18 Check your graph by using a graph-plotting program.

19 (a) See Figure T13:12.
   (b) (i) about 32.2 cm^2
       (ii) about 2.5 cm

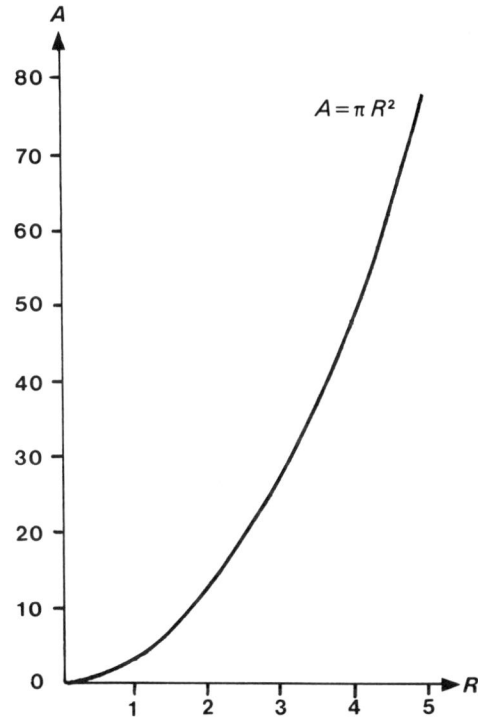

Fig. T13:12

20 See Figure T13:13.

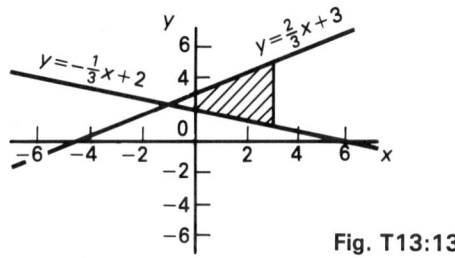

Fig. T13:13

21 See Figure T13:14.
   (b) Implies that given over 15 min he should make no errors
   (c) He cannot know any of the items if given zero time to memorise them
   (d) about -0.8 (e) $y = -0.8x + 13$

22 (a) -2.6, 1.6 (b) -2.3, 1.3

23 (a)

-3	-2	-1	0	1	2
18	8	2	0	2	8
-12	-8	-4	0	4	8
-5	-5	-5	-5	-5	-5
1	-5	-7	-5	1	11

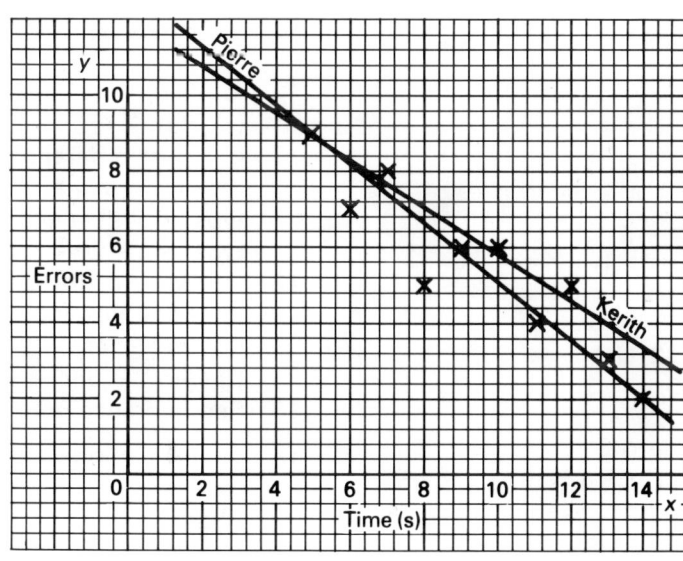

Fig. T13:14

An

(b) See Figure T13:15.

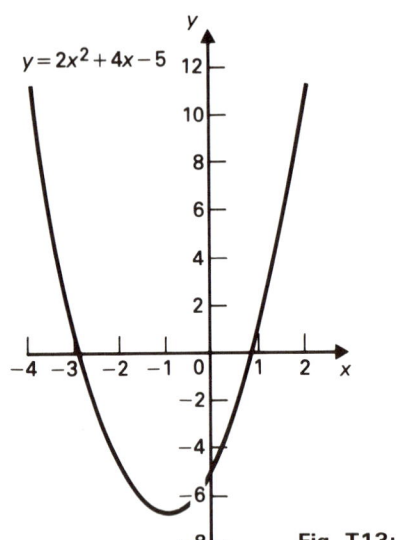

$y = 2x^2 + 4x - 5$

Fig. T13:15

(c) -7 at $x = -1$

(d)   (i) -2, 0, $2x^2 + 4x = 0$
    (ii) -3, 1, $2x^2 + 4x - 6 = 0$
    (iii) -4, 2, $2x^2 + 4x - 16 = 0$
    (iv) 1.4, -3.4, $2x^2 + 4x - 10 = 0$
    (v) 0.7, -2.7, $2x^2 + 4x - 4 = 0$
    (vi) 0.2, -2.2, $2x^2 + 4x - 1 = 0$
    (vii) 1.2, -3.2, $2x^2 + 4x - 8 = 0$

(e) 0.9, -2.9

24 Read the values of x where the parabola crosses the x-axis.

25 (a) $2x^2 - 3x - 4 = 0$
    (b) $2x^2 - 3x = 0$
    (c) $2x^2 - 4x - 1 = 0$
    (d) $2x^2 - 4x - 3 = 0$
    (e) $2x^2 - 2x - 4 = 0$
    (f) $2x^2 - x - 2 = 0$

26 (a) $y = -4x + 4$ (b) $y = -5x + 2$
    (c) $y = x$ (d) $y = -3x - 1$
    (e) $y = -x + 5$

27 (a) 8 (b) -12 (c) -4

28 $x = -2$, $x = 0$, $x = 3$

29 $x = -3$, $x = 0$, $x = 2$;
    $x^3 + x^2 - 6x = 0$

30 (a) 4.8 (b) (i) 1.5 h (iii) 1 h 34 min
    (c) (i) approx 27.5 cm^2
       (ii) 1 cm^2 represents
       $2\frac{1}{2}$ r.p.m. $\times$ 30 min = 75 revs;
       about 83p

31–35 Check your answers with a graphics calculator or a computer graph-drawing program.

36 (a) $-2 \leqslant x \leqslant \frac{1}{3}$ (b) $-1\frac{1}{2} < x < 5$
    (c) $-2 \leqslant x \leqslant 4$

Chapter 14 Sequences

Some answers, all of which you can check for yourself, are not given here.

1 (a) 3, 6, 9, 12, 15, 18
   (b) 2, 3, 4, 5, 6, 7
   (c) 1, 4, 9, 16, 25, 36
   (d) 1, 8, 27, 64, 125, 216
   (e) 2, 6, 12, 20, 30, 42
   (f) 1, 3, 6, 10, 15, 21
   (g) 1, 8, 21, 40, 65, 96
   (h) 3, 14, 39, 84, 155, 258
   (i) 3, 12, 39, 96, 195, 348
   (j) $\frac{1}{2}, \frac{2}{3}, \frac{3}{4}, \frac{4}{5}, \frac{5}{6}, \frac{6}{7}$

3 e.g. 7, prime numbers, or 8, Fibonacci.

4 All answers are of the form $an + b$.

5 (a) 21, 23, 25, 27, 29 (b) 43, 55
   (c) 1   2   3   4   5   6
      1   8   27 64 125 216
      1   4   9   16   25   36
   (d) (i) n^3 (ii) n^2

7 (b) -0.618 and 1.618

9 (b) (i) no (ii) yes (c) 1.3

10 (a) isosceles (b) (i) 13 (ii) $2n - 1$
   (c) (i) 169 (ii) $(2n - 1)^2$
   (d) (i) white (ii) white
   (e) $8(n - 1)$

11 Common difference for:
   type $an + b$ is a,
   type $an^2 + bn + c$ is $2a$,
   type $an^3 + bn^2 + cn + d$ is $6a$.

12 (a) $3n - 1$ (b) $2n^2 + 3$
   (c) $n^2 + 3n$ (d) $3n^2 + 2n - 1$
   (e) $2n^3 + n^2 - 2n$
   (f) $\frac{1}{6}(n^3 + 3n^2 + 2n)$

13 See reference notes on page 240.

14 (a) $\dfrac{n(n + 1)}{2}$; $\dfrac{n^3 + 3n^2 + 2n}{6}$
   (b) $2n + 3$; $n^2 + 4n$
   (c) $4 \times 3^{n-1}$; $2(3^n - 1)$

18 (a) $2t - 1$
   (b) (ii) $1 + p + p^2$ (iii) $\dfrac{p^{n-1} - 1}{p - 1}$

Chapter 15 Variation

1. (a) $C = kd$; $C \propto d$
   (b) $A = kr^2$; $A \propto r^2$
   (c) $y = k/x$; $y \propto 1/x$
   (d) $y = kx/z^2$; $y \propto x/z^2$
   (e) $C = k/n$; $C \propto 1/n$
2. $V \propto AH$
3. $H \propto V/R^2$
4. $A \propto F/M$
5. $y = ks$; $y \propto s$
6. $E = kh$; $E \propto h$
7. (a) $A = kb$ (b) $\frac{1}{3}$ (c) 5 (d) 12
8. (a) $P = k/q$ (b) 30 (c) 1.2
   (d) 3.75
9. (a) $x = k/y^2$ (b) 324 (c) 9
   (d) 5.196
10. 7, direct; 8, inverse; 9, inverse square
11. (a) $t = \sqrt{\dfrac{d}{5}}$ (b) 7.2 metres
12. (a) y varies directly with x^2 and
    inversely with z.
    (b) $\frac{3}{10}$ (c) 10
13. 74
14. (a) 15p (b) 1300
15. $h \propto u^2$
16. (a) $\times \frac{1}{4}$ (b) $\times \frac{1}{2}$ (c) $\times \frac{1}{8}$ (d) $\times 2$
    (e) $\times \frac{1}{2}$
17. (a) $V \propto R^2 H$ (b) $\times 16$ (c) $\times 18$
    (d) $\times 27$
18. Decrease of 20%
19. (a) $H \propto V^2 T/R$ (b) $H = kV^2 T/R$
    (c) $\frac{1}{4}$ (d) $1\frac{1}{4}$
20. 11.9 years

Chapter 16 Lines and shapes

1. (a) rectangle (b) triangle
   (c) octagon
2. (a) 360° (b) 180° (c) 1080°
3. (a) 130° (b) 40°
4. (a) 60° (b) 40° (c) 60°
6. (a) 3 (b) 2 (c) 5
7. (a) no (b) centre of O
8. (a) rectangle, rhombus
   (b) kite, isosceles trapezium
   (c) parallelogram (d) square
9. $x = 55°$, corresponding; $y = 35°$,
   angles in triangle; $z = 35°$, alternate

10. (a) square, rhombus, concave
    dodecagon
    (b) $p = 90°$; $q = 150°$; $r = 30°$;
    $s = 30°$
11. (a) dodecagon; hexagon; square
    (b) 150°
12. rotational symmetry of order 4; point
    symmetry
13. (a) kite (b) 25°
14. (a) 120° (b) 60° or 90°
15. Try drawing it.
16. 162°
17. 60°
18. 100°
19. (a) $2x$ (b) $180 - 2x - 2y$
    (c) AB = BD and AC = CE (isosceles
    triangle)
20. 1440°
22. 18

Chapter 17 Position and movement

1. (a) 315° (b) $022\frac{1}{2}°$ (c) 132°
   (d) 320°

2.
	Cruiser	L1	L2
(a)	2145	2344	2540
(b)	215452	236442	250404

Three-figure bearings

(c)	XXX	342°	293°	323°
	163°	XXX	195°	290°
	113°	015°	XXX	339°
	143°	110°	159°	XXX

Cardinal bearings

XXX	N18°W	N67°W	N37°W
S17°E	XXX	S15°W	N70°W
S67°E	N15°E	XXX	N21°W
S37°E	S70°E	S21°E	XXX

3. (a) 16 km² (b) 8229 (c) 823310
   (d) 800284 (e) N30°W or 330°

4 (a) 240° (b) 070°
  (c) Add 180°; if the answer is more than 360°, subtract 360°.

5 (a) 090°, 330°, 215°
  (b) N 90° E; N 30° W; S 35° W
  (c) About 5 km

6 (a) 12 km (b) JMKL

7 (a) See Figure T17:1.
  (b) Link Bath and Swindon.

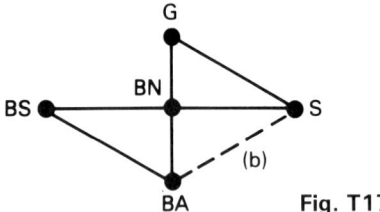

Fig. T17:1

8 (a) 25 miles (b) 15 miles
  (c) See Figure T17:2.

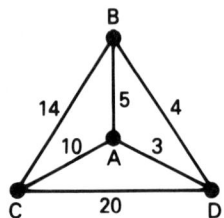

Fig. T17:2

14 (a) See Figure T17:3. (b) 471 mm

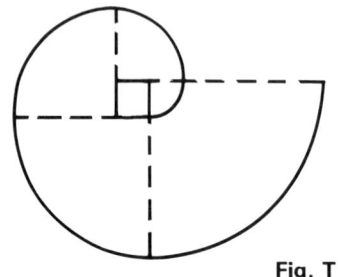

Fig. T17:3

15 See Figure T17:4.

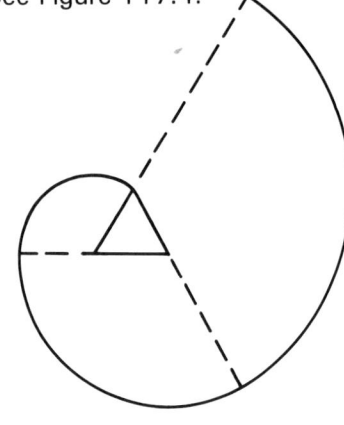

Centres of arcs at corners of triangle

Fig. T17:4

16 (a) and (b) See Figure T17:5
  (half scale).
  (c) 163°, 100 m (d) 900 m

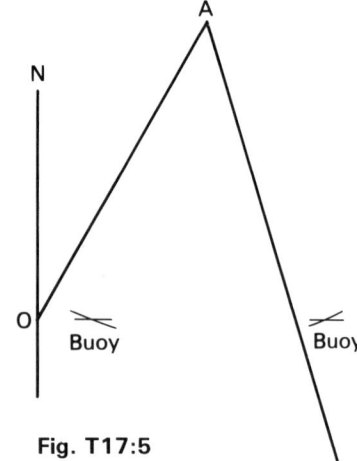

Fig. T17:5

17 See Figure T17:6.

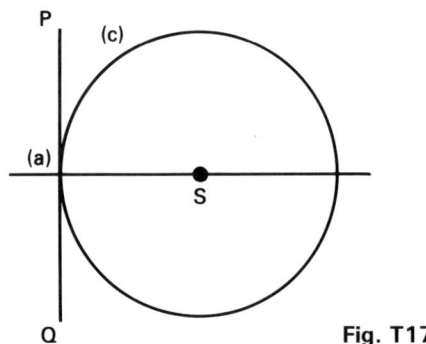

Fig. T17:6

18 See Figure T17:7.

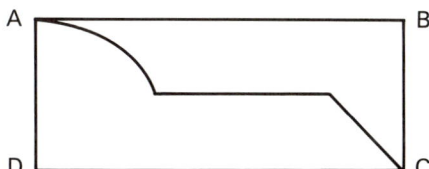

Fig. T17:7

19 (a) and (b) See Figure T17:8.
(c) (ii) 28 hours

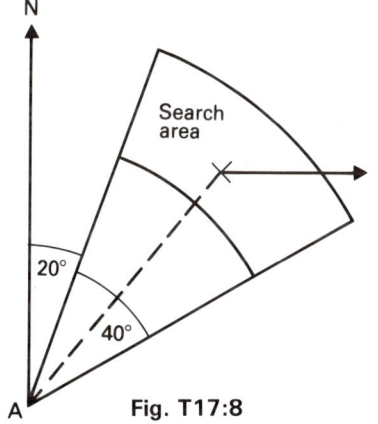

Fig. T17:8

20 (c) 129° (d) 321°
(e) (i) A (ii) 10 min
21 72°
22 See Figure T17:9.

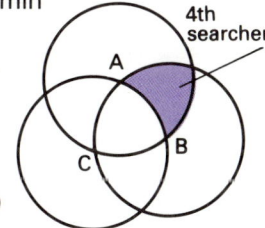

Fig. T17:9

23 (a) Hull and Doncaster
(b) She has a choice but to make her journey shortest she will have to travel from Doncaster to Scunthorpe and Hull to Scunthorpe twice.
(c) York, Leeds, Bradford, Huddersfield, Sheffield, Lincoln, Scunthorpe, Doncaster, Goole, Hull; 216 miles
(d) Same as (c) plus 38 miles back to York
24 (a) and (b) See Figure T17:10. Construct the tangent at A to meet the horizontal line at B. OA produced meets the bisector of angle B at the centre of the circle.

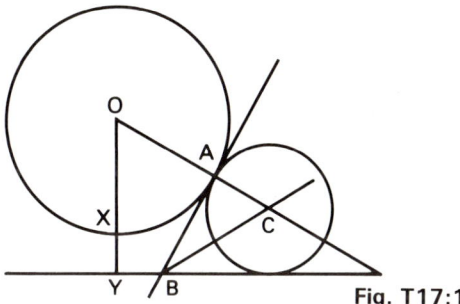

Fig. T17:10

25 The authors cannot get less than a total walk of 1550 metres (not including return journeys for another load). Can you do better?
26 At Q
27 (a) (i) P or Q or S (we think!)
(ii) P (we think!)
(b) P (c) P
(d), (e) Please send us your answers!
28 High. If C turns clockwise, O turns clockwise at about half the speed of C. Normal. If O turns clockwise, C turns clockwise at about twice the speed of O. Low. As for normal. Gear ratios drive: road wheel = High 9:5, Low 5:9
30 (a) On cornering inner wheel turns slower than the outer wheel.
(b) Bevel pinions not moving relative to bevel wheels.
(c) Bevel pinions turn in alternate directions.
(d) As in (c) but they turn faster.
(e) The wheel not 'slipping' does not turn so you cannot drive away unless you can stop the other wheel turning.

Chapter 18 Similarity and congruence

1 (a) and (d)
2 (a), (b), (d) and (f)
3 10 cm (b) 15 cm
4 (a) 3:2 (b) (i) 9 cm (ii) 10 cm
5 4 metres; five doors
6 (a) See the glossary.
(b) Congruent: C, E; F, I; D, J; G, K

7 (a) ABCD; angles in different relative positions
(b) $5\frac{1}{3}$ cm
(c) 12 cm

8 (b) 1 : 4

9 (b) 2 : 3 (c) 4 : 9

10 (a) ∠CAD (b) ∠ADC
(c) (i) DAC (ii) CBA (iii) BAD
(d) (i) DA, AC, AC (ii) AB, CA, BD
(iii) BA, AD, CD

11 $5\frac{1}{3}$ cm

12 (a) RPB (b) QPC, ABC
(c) (i) 12 cm (ii) 3 cm (iii) $\frac{1}{16}$
(d) $3k$ cm²

13 Figure 18:11: not necessarily as first angle is not included
Figure 18:12: yes, three sides
Figure 18:13: yes, two angles, corresponding side
Figure 18:14: yes, two angles, corresponding side
Figure 18:15: yes, two sides, included angle (∠ABG = 90° + ∠CBG and ∠CBE = 90° + ∠CBG)

14 (a) (i) 8 (ii) 27 (iii) 64
(b) (i) 600 cm² (ii) 1350 cm²
(iii) 2400 cm²
(c) (i) 1000 cm³ (ii) 3375 cm³
(iii) 8000 cm³
(d) (i) 1 : 1.1̇6̇ (ii) 1 : 2.5
(iii) 1 : 3.3̇

15 6 cm

16 (a) 4 : 3 (b) 16 : 9 (c) 64 : 27

17 1 : 7

18 0.628

Chapter 19 Solid geometry

1 (a) cube (b) cuboid (c) cylinder
(d) sphere

2 (a) 12 (b) EF (c) A and E
(d) 10 cm (e) 20 cm

3 A

4 (a) Various possibilities.
(b) 14 cm by 10 cm.

5 (a) rectangle, 4 cm by 3.5 cm
(b) triangular prism
(c) See Figure T19:1.
(d) 1.6 cm

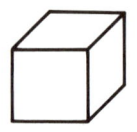

Fig. T19:1

6

	19:5	19:6	19:7	19:8	19:9
(a)	3	1	7	13	7
(b)	3	∞	7	9	6
(c)	4	∞	12	24	12

7 (a) (i) 60 litres (ii) 75 litres
(iii) 90 litres
(b) $52\frac{1}{2}$ litres

8 (a) Figure 19:11, icosahedron; Figure 19:12, tetrahedron; Figure 19:13, cube; Figure 19:14, dodecahedron; Figure 19:5, octahedron
(b)

	19:11	19:12	19:13	19:14	19:15
F	20	4	6	12	8
V	12	4	8	20	6
E	30	6	12	30	12

(c) Check two non-congruent nets for a tetrahedron.

9 A (0, 0, 2); B (0, 6, 2); C (4, 6, 2); D (4, 0, 2); E (0, 0, 0); F (0, 6, 0); G (4, 6, 0); H (4, 0, 0)

10 (a) Figure 19:16, cube; Figure 19:17, square-based pyramid; Figure 19:18, cone; Figure 19:19, cylinder; Figure 19:20, triangular prism; Figure 19:21, tetrahedron; Figure 19:22, probably a triangular prism; Figure 19:23, cube with corner planed to a triangle
(b) See Figure T19:2.

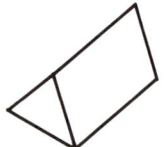

Fig. T19:2

11 (a) e.g. (3, 16, 12); (3, 4, 24); (15, 4, 12)
(b) sphere (c) 13 (d) (0, 0, 24)
(e) 134.8°

12 All right-angles except (d) ∠AHC and (h) ∠ACH

13 8.94 cm (across edge AD)

Chapter 20 Pythagoras' theorem

1 (a) 6.55 cm   (b) 3.66 cm
   (c) 3.54 cm
3 8 m
4 (a) acute   (b) obtuse   (c) acute
   (d) impossible   (e) right-angled
   (f) impossible   (g) acute   (h) acute
5 11.1 cm
6 7 cm, 17 cm
7 (a) 5   (b) $\sqrt{2}$ (c) 5   (d) $\sqrt{26}$
   (e) $\sqrt{74}$ (f) $\sqrt{(X-x)^2 + (Y-y)^2}$
8 (a) 6.32 cm   (b) 7 cm
9 20.8 cm
11 13.42 cm
12 6 cm
13 $x = 7.5$ or 5

Chapter 21 Mensuration

1 (a) 257 m   (b) 6.12 m/s
   (c) 1363.5 m^2 (d) 25.1 m
2 57.3 cm
3 97
4 (a) 12.5 kg   (b) 4091 cm^2
   (c) 398.2 cm^3
   (d) 26 (If you can get more please
       write to us.)
5 593.4 g
6 Consider whether the formula involves
   length × constant, length × length, or
   length × length × length. Other
   combinations, e.g.
   length + (length × length), are
   impossible.
7 (a) length   (b) area   (c) volume
   (d) volume   (e) area   (f) impossible
   (g) impossible   (h) length
   (i) volume
8 £129.60;   yes
9 £1125
10 50
11 (a) £26   (b) 5 m^2 (c) 0.5 m^3
   (d) £13   (e) £16.25
12 (a) 25 cm   (b) 4710 cm^3
   (c) 94.2 cm   (d) 25 cm
   (e) 94.2 cm   (f) 216°
   (g) 1204.76 cm^2
13 (a) 457 cm^3 (b) 7.18 cm
14 $\frac{1}{6}\pi d^3 + \frac{1}{4}\pi d^2 h$, $\pi d^2 + \pi dh$; volume
   from product of three lengths; area
   from product of two lengths

15 (a) 163.62 mm^3 (b) 60%
   (c) Yes. Lengths cancel out
       completely in % change formula.
16 (a) $7\frac{1}{4}$ m   (b) 87.2°   (c) 220.7 m^2
   (d) 275 m^3 (e) Please tell us.
17 (a) 5.66 cm   (b) 9 cm
   (c) 74.$\dot{6}$ cm^2 (d) 50.9 cm^2
   (e) 4.40 cm
18 (a) 6.2 cm   (b) 14.1 cm^3
   (c) 96.7%
19 (a) 12 m
   (b) (i) 1018 m^3
       (ii) 2413 m^3; 1395 m^3
   (c) 1.39 × 10^6 litres
   (d) (i) 20 m   (ii) 5 m
   (e) (i) 330 m^2 (ii) 254 m^2
       (iii) 98 litres
20 3183 cm/min

Chapter 22 Trigonometry

1 (a) 4.66   (b) 4.23   (c) 26.6°
   (d) 30°   (e) 8.39   (f) 6.43
2 (i) 9.42 cm; 10.2 cm
   (ii) 1.7 cm; 4.35 cm
   (iii) 3.68 cm; 1.56 cm
3 (i) 59°; 31°   (ii) 36.9°; 53.1°
4 About $11\frac{1}{2}$ metres
5 About 27 metres!
6 63.4°
7 11.6°
8 44°
9 2.9°
10 $6\frac{1}{4}$ km
11 15 n.m. south; 10 n.m. east
12 (a) 2.66   (b) 2.6   (c) 1.32
   (d) 5.49   (e) 56.3°   (f) 41.8°
   (g) 31°   (h) 48.6°   (i) 6.06
   (j) 1.46   (k) 38.7°   (l) 60.3°
   (m) 44.4°   (n) 53.1°   (o) 34.8°
   (p) 33.7°   (q) 8.66   (r) 32°
   (s) 10.3
13 (a) 7.99 cm   (b) 7.54 cm   (c) 48.9°
14 45°
15 (a) 4.87 cm   (b) 10.2 cm
   (c) 17.5 cm
16 See Figure S57 on page 258.
17 Check your graphs using a graph-
   plotting program or a graphical
   calculator.

18 (a) 15 m (b) 19.1 m (c) 46.4°
19 (a) 545 m (b) 204 km/h (c) 17°
20 (b) 200 sin x° − x = 60; 25; 117.5
21 (a)

	sin	cos	tan
30°	0.5	0.866	0.5774
150°	0.5	−0.866	−0.5774
210°	−0.5	−0.866	0.5774
330°	−0.5	0.866	−0.5774

(b)

	sin	cos	tan
45°	0.7071	0.7071	1
135°	0.7071	−0.7071	−1
225°	−0.7071	−0.7071	1
315°	−0.7071	0.7071	−1

23 (a) 0°, 180° (b) 90° (c) 270°
   (d) 90°, 270° (e) 0°, 360°
   (f) 180° (g) 26.57°, 206.57°
   (h) 153.43°, 333.43°
   (i) 108.43°, 288.43°
   (j) 48.59°, 131.41°
   (k) 113.58°, 246.42°
   (l) 56.31°, 236.31°

25 (a) 4.46 cm; 16.5°; 138.5°
   (b) 4.60 cm; 41.8°; 88.2°
   (c) 75.5°; 46.6°; 57.9°
   (d) 78.5°; 57.1°; 44.4°
   (e) 3.40 cm; 48.9°; 59.8°
   (f) 85.7°; 57.7°; 36.6°
   (g) 4.6 cm; 9.2 cm; 11.5 cm;
       49.5°; 22.3°; 108.2°

26 3.66 km; 4.48 km; 3.17 km
27 21.8 m
28 66.8°; 113.2°
29 (a) **Hint:** Use cosine rule.
   (b) $x = 5$
   (c) See Figure T22:1.

30 6.14 cm^2
31 17.3 cm^2
32 (a) 35.4 n.m. (b) 21.7 n.m.
   (c) 11.8 n.m. (d) 16.3 n.m.
33 (b) 33.7 m
   (c) $B_1T_1 = 33.7 + (x − 40) \sin 61°$
   (d) $x \simeq 58.6$ m
34 (b) $\sqrt{128}$; $\sqrt{128}$ (c) 60° (d) 45°
35 16.8 cm
36 (a) 16 hours; 8 hours
   (b) Daylight greater than 24 hours
37 Between 1700 and 2200

Chapter 23 Vectors

1 (a) $\begin{pmatrix} 0 \\ 2 \end{pmatrix}$ (b) $\begin{pmatrix} 1 \\ 0 \end{pmatrix}$ (c) $\begin{pmatrix} 1 \\ -1 \end{pmatrix}$

   (d) $\begin{pmatrix} 2 \\ 2 \end{pmatrix}$ (e) $\begin{pmatrix} 0 \\ -4 \end{pmatrix}$

2 (a) $2\underset{\sim}{a}$ (b) $-2\underset{\sim}{b}$ (c) $\underset{\sim}{c}$ (d) $\frac{1}{2}\underset{\sim}{b}$
   (e) $3\underset{\sim}{a}$ (f) $\underset{\sim}{d}$ (g) $-1\frac{1}{2}\underset{\sim}{b}$ (h) $-2\underset{\sim}{d}$
3 (a) 2 (b) $\sqrt{8}$ (c) $\sqrt{18}$
4 (a) 5 (b) $\sqrt{29}$
5 (a) $\begin{pmatrix} 5 \\ 0 \end{pmatrix}$ (b) $\begin{pmatrix} 1 \\ 1 \end{pmatrix}$ (c) $\begin{pmatrix} 1 \\ -1 \end{pmatrix}$

   (d) $\begin{pmatrix} -3 \\ -3 \end{pmatrix}$

6 $\begin{pmatrix} -4 \\ 4 \end{pmatrix}$

7 (c) $\underset{\sim}{a}$; $2\underset{\sim}{b}$; $-\underset{\sim}{d}$; $-2\underset{\sim}{d}$; $\underset{\sim}{a}$; $2\underset{\sim}{d}$; $\underset{\sim}{a}$; $1\frac{1}{2}\underset{\sim}{c}$
   $1\frac{1}{2}\underset{\sim}{a}$
8 See Figure T23:1 (reduced scale).

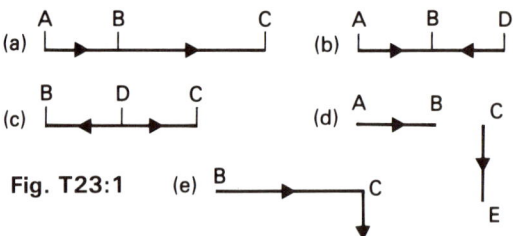

Fig. T23:1

9 (a) $\begin{pmatrix} 7 \\ 2 \end{pmatrix}$ (b) $\begin{pmatrix} 9 \\ 1 \end{pmatrix}$

10 (a) (i) $\underset{\sim}{b}$ (ii) $-\underset{\sim}{b}$ (iii) $\frac{1}{2}\underset{\sim}{b}$ (iv) $-\frac{1}{2}\underset{\sim}{b}$
       (v) $\frac{1}{2}\underset{\sim}{b}$ (vi) $-\frac{1}{2}\underset{\sim}{b}$
   (b) (i) $-\underset{\sim}{b} - \underset{\sim}{a}$ (ii) $-\underset{\sim}{b} + \underset{\sim}{a}$
       (iii) $\underset{\sim}{a} + \frac{1}{2}\underset{\sim}{b}$ (iv) $-\frac{1}{2}\underset{\sim}{b} - \underset{\sim}{a}$
       (v) $\frac{1}{2}\underset{\sim}{b} - \underset{\sim}{a}$ (vi) $\underset{\sim}{a} - \frac{1}{2}\underset{\sim}{b}$
11 (a) $-2\underset{\sim}{a}$
   (b) $2\underset{\sim}{a} + 2\underset{\sim}{d}$ (c) $\underset{\sim}{a}$ (d) $\underset{\sim}{d} - \underset{\sim}{a}$
12 354.29° or 354°17'; 402 km
13 7.779 kg; 5.077 kg
14 14.168 knots, 154°
15 (a) (i) $\underset{\sim}{x} - \underset{\sim}{y}$ (ii) $4\underset{\sim}{x} - 4\underset{\sim}{y}$
   (c) (i) $3\underset{\sim}{x}$ (ii) $3\underset{\sim}{x} - 3\underset{\sim}{y}$
   (d) (i) 9 (ii) 6
16 (a) (i) $\underset{\sim}{p} + \underset{\sim}{q}$ (ii) $\frac{1}{3}\underset{\sim}{q} - \underset{\sim}{p}$
   (b) (i) $k(\frac{1}{3}\underset{\sim}{q} - \underset{\sim}{p})$
       (iii) $\frac{3}{4}$ (as $1 - k = \frac{1}{3}k$)

Fig. T22:1 (B triangle figure): 8 cm, 7 cm, 60°, 98.2°, 3 cm, A, C

17 (a) (i) $2q + p$ (ii) $2p + q$ (iii) $-q + p$

(b) $p - q$

(c) Parallel and same length

(d) $2p - 2q$

(e) Parallel, with RT = 2DB

(f) $\begin{pmatrix} \sqrt{3} \\ 1 \end{pmatrix}$, $\begin{pmatrix} -\sqrt{3} \\ 1 \end{pmatrix}$

(g) $(2\sqrt{3}, 2)$; $(\sqrt{3}, 3)$; $(-\sqrt{3}, 3)$; $(-2\sqrt{3}, 2)$

(h) $4\sqrt{3}$

18 (a) $-a + b$; $2a$; $-2a + 2b$

(b) $-4a + 4b$; $-2a + 4b$; $2a - 3b$

(c) ka

Chapter 24 Transformations

1 (a) E3 (b) D4 (c) C1 (d) D4, E3

(e) B4, C5 (f) D1

(g) C3, C4, D3, D4

2 (b) (i) (6, 6), (6, 12), (12, 6)

(ii) $(-2, -2)$, $(-2, -4)$, $(-4, -2)$

(iii) (1, 1), (1, 2), (2, 1)

(iv) $(-4, -4)$, $(-4, -8)$, $(-8, -4)$

3 (a) (2, 2), (2, 6), (6, 2)

(b) (4, 0), (2, 0), (4, -2)

(c) (5, 5), (4, 5), (5, 4)

4 (b) (i) $13\frac{1}{2}$ cm², $2\frac{1}{4}$ times

(ii) $1\frac{1}{2}$ cm², $\frac{1}{4}$ times

5 (a) See Figure T24:1.

(b) Reflection $y = x$

6 (a) $\begin{pmatrix} 5 \\ 4 \end{pmatrix}$ (b) (i) $+90°$ (ii) $(-3, 4)$

(c) (i) $y = -x$ (ii) $(-7, -2)$

7 (b) $(-3, 4)$, $(-4, -3)$, $(0, -5)$, $(5, 0)$; isosceles trapezium

8 (a) 40° (b) 100°

(c) AB = A'C, AD = BC, A'B = BC - A'C, BD = AD - AB, therefore A'B = BD

9 See Figure T24:2. P is invariant under DTM(Q), therefore rotation $+90°$ about (1, 3).

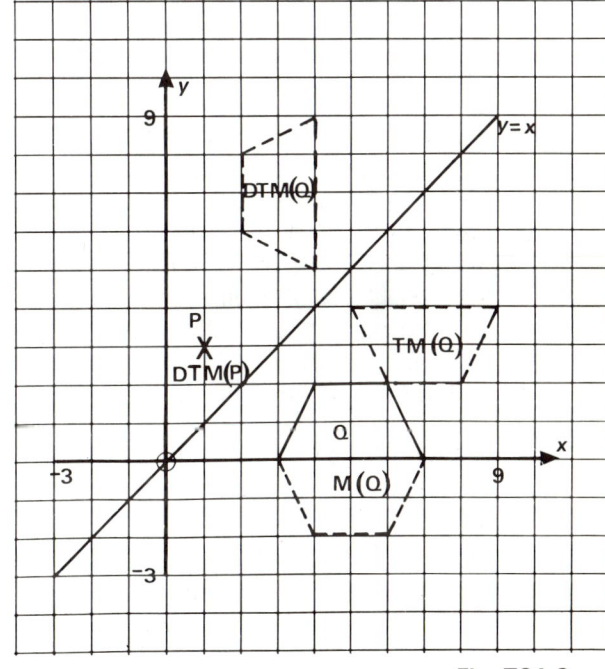

Fig. T24:2

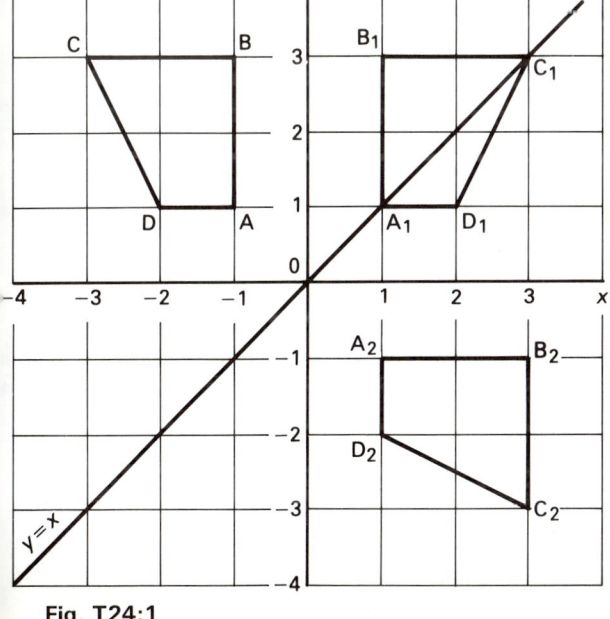

Fig. T24:1

Chapter 25 Representation and interpretation of data

2 (a) 36.9 °C

(b) beats/min; breaths/min

(c) 71 beats/min

(d) 18 breaths/min

(e) Yes: high temperatures; high pulse rate; low respiration rate

(f) 28th February

3 See Figure T25:1.
  (b) Test too hard, pupils did not revise, etc.

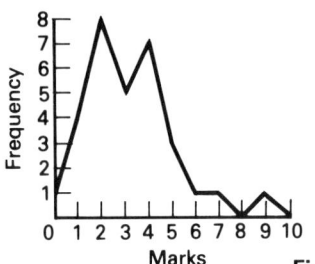

Fig. T25:1

4 (a)

$150 \leqslant h < 155$	JHT I	6
$155 \leqslant h < 160$	JHT III	8
$160 \leqslant h < 165$	IIII	4
$165 \leqslant h < 170$	III	3
$170 \leqslant h < 175$	JHT II	7
$175 \leqslant h < 180$	II	2

  (b)

$40 \leqslant w < 45$	II	2
$45 \leqslant w < 50$	JHT II	7
$50 \leqslant w < 55$	JHT II	7
$55 \leqslant w < 60$	JHT	5
$60 \leqslant w < 65$	II	2
$65 \leqslant w < 70$	III	3
$70 \leqslant w < 75$	I	1
$75 \leqslant w < 80$	I	1
$80 \leqslant w < 85$	II	2

  (c) See Figures T25:2 and T25:3.

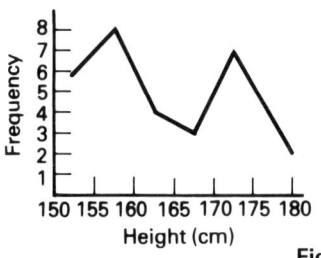

Fig. T25:2

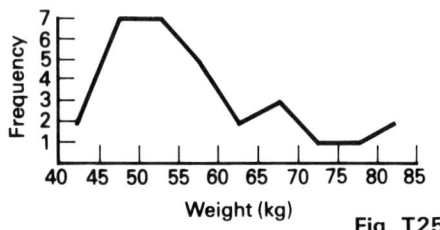

Fig. T25:3

5 Low female literacy level correlates with high birth rate.

6 (a) 7% raw material production; 48% manufacturing; 45% services
  (b) See Figure T25:4.

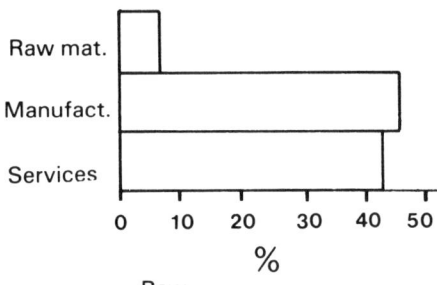

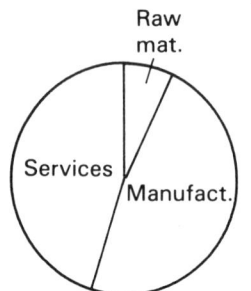

Fig. T25:4

7 (a) See Figure T25:5.

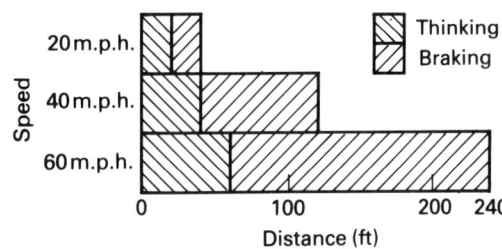

Fig. T25:5

  (b) See Figure T25:6.

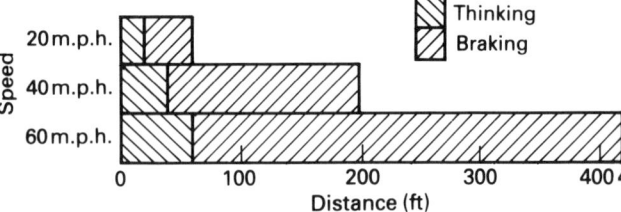

Fig. T25:6

8 (a) and (b) See Figure T25:7.

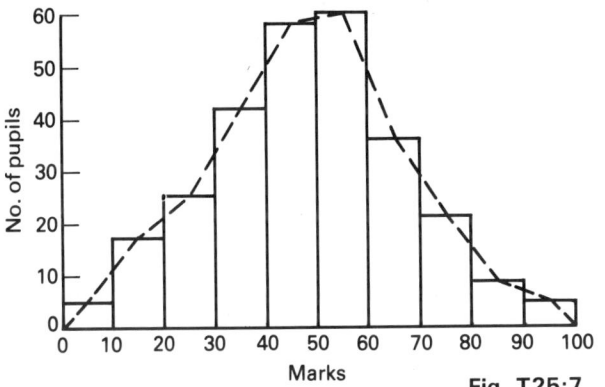

Fig. T25:7

9 (a) See Figure T25:8.

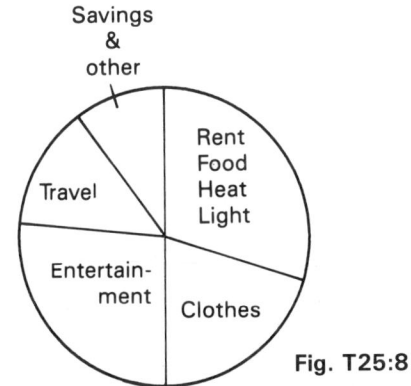

Fig. T25:8

(b) (i) £1980 (ii) 27° (iii) $6\frac{2}{3}$%

10 (a) positive (b) uncorrelated
(c) positive (d) negative

11 Figure 25:6 positive, Figure 25:7 negative, Figure 25:8 uncorrelated

12 See Figures T25:9 to T25:12.

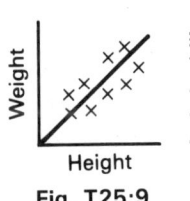

Fig. T25:9 Fig. T25:10 Fig. T25:11

Fig. T25:12

13 (a) and (b) See Figure T25:13.

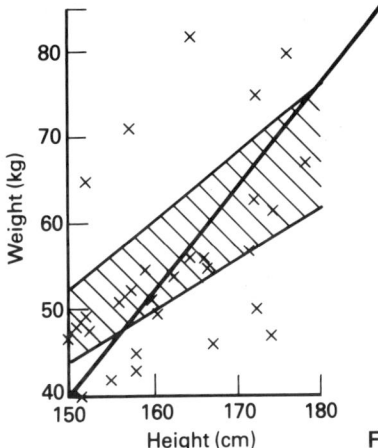

Fig. T25:13

14 As speed increases, so does fuel consumption. An average of 58 m.p.g. is unlikely as the speed is not a constant 45 m.p.h.

15 (a) men: £300, £200; women £175, £120
(b) Most women are lower paid workers. Men's pay is more spread out.

16 (a) 20 15 12 15 13 10 8 4 2 1
7 9 12 14 15 15 12 8 5 4
(b) 20 35 47 62 75 85 93 97 99 100
7 16 28 42 57 72 84 92 97 101
(d) Longevity increased, birth-rate decreased, total population much larger, average age higher.

17 (a) 39, 33, 46, 13
(b) about 135; about 42
(c) New curve to the left of old curve; vehicles travelling more slowly

18 (b) City A has a wider spread of prices and more dearer ones.
(c) The data fits a normal (bell-shaped) curve so is probably true.

19 Most viewers are over 20, with a peak around age 60. Female viewers are more evenly distributed over the age range.

20 1.25 0.556 0.417 0.278
2.00 2.889 1.555 1.555
Country A. High number of children, shorter life expectancy, e.g. Jamaica.

21 (b) 108 g (c) 112 (d) 25
22 (a) 79 (b) 58 (c) 40 (d) 12
　(f) (i) about 22 years
　　　(ii) about 28
23 Brides 5½, grooms 6. The ranges do not show a big difference, but the median for males is higher, showing brides tend to be younger than their grooms.
24 $a = 5$, $b = 10$, $c = 125$, $d = 25$
25 See Figures T25:14 and T25:15.

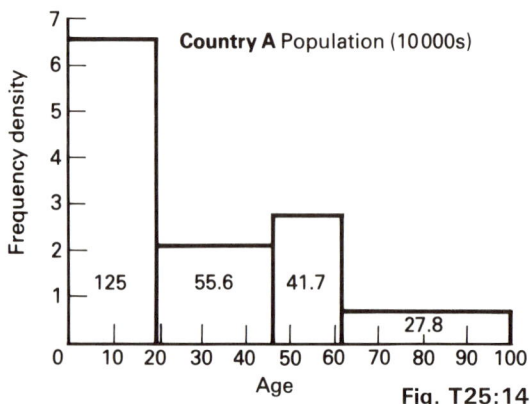

Fig. T25:14

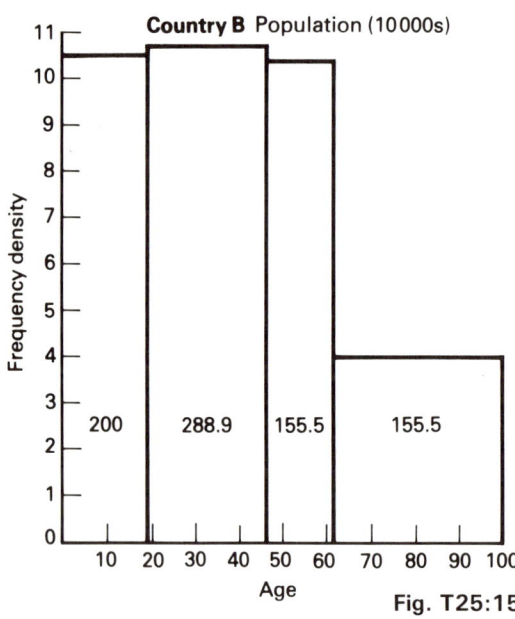

Fig. T25:15

26 (a) 170 age 11; 200 age 14
　(b) 60 age 11; 130 age 14
　(c) They have the same range, but 14-year-olds tend to be paid more.

27 Comments should include: almost the same number watching (probably the same sample), but the second day far more watched for a long time (more than 6 hours). Maybe first is summer, second is winter, or second has a big event happening.
28 (a) 1 152 495; 1 147 500
　(b) 219; 54.8 cm, 9.6 cm, 1.4 cm, 0.5 cm, 0.1 cm
　(c) Comments should include: the graph is very large, but because the data has such a disparity between 16–19 and 45–60 it is difficult to reduce it any further.

Chapter 26 Averages and dispersion

1 All the averages are 3, and all ranges are 6. It is pointless to calculate averages for small amounts of data, especially when they do not form a continuous rising and falling normal (or skewed normal) curve. See also question 20 to see how the measure of standard deviation helps indicate non-normal-curve data.
2 Non-numerical, e.g. car colours.
3 When people use the term 'average' in everyday life they often give a figure around which experience shows them the data lies. This might well be the centre number in the modal class for grouped data.
4

	mean	mode	median
Ann	£1000	£0	£0
Brenda	£1000	—	£999.50
Camille	£1000	£0	£50
Davian	£1000	£2000	£500
Erica	£1000	£100 & £1000	£600

5 (a) mean (b) mode (c) median
6 Fifty people, with such a wide range of earnings, cannot give statistically significant results.
7 Some possible answers: 1, 3, 4, 4, 4, 5, 7, or 2, 3, 3, 4, 4, 4, 8 or 0, 2, 4, 4, 5, 6, 7

8 120, 220, 90, 72, 28, 8; mean 3.547, mode 4, median 4, range 7

9 (a) 4, 8, 5, 4, 0, 0, 2, 1
 (b) (i) 24 (ii) 49 (iii) 1 (iv) 2.04
 (v) 1.5

10 (a) no effect (b) increases to 2.12
 (c) increases to 2

11 (a)

0–9	I	1	4.5	4.5
10–19	JHf I	6	14.5	87
20–29	JHf II	7	24.5	171.5
30–39	JHf IIII	9	34.5	310.5
40–49	JHf JHf II	12	44.5	534
50–59	JHf JHf	10	54.5	545
60–69	JHf III	8	64.5	516
70–79	IIII	4	74.5	298
80–89	II	2	84.5	169
90–99	I	1	94.5	94.5
Totals		60		2730

 (b) (i) 40–49 (ii) 40–49 (iii) 45.5
 (iv) 86 (v) 59 − 31 = 28
 (c) No; too many pupils getting under 50

12 (a)

13:30–13:59	5
14:00–14:29	3
14:30–14:59	5
15:00–15:29	12
15:30–15:59	14
16:00–16:29	12
16:30–16:59	15
17:00–17:29	9
17:30–17:59	5
18:00–18:29	3
18:30–18:59	4
19:00–19:29	6
19:30–19:59	3
20:00–20:29	2
20:30–20:59	1
21:00–21:29	1

 (b) modal class, 16:30–16:59;
 median class 16:00–16:29;
 mean 16:39

13 (a)

	Mean	Median	Mode	Range
A	£100	£100	£100	0
B	£100	£100	£105 £100 £95	£10
C	£3400	£100	£100	£9900
D	£366.83	£100	£1000 £100 £0.50	£999.50
E	£100	£50	£50	£150

 (b) Yes, at least one average is £100 for each.
 (c) C, D and E

14 (a) 191 (b) 5–7 (c) 5–7 (d) 10

15 £3.04

16 $23\frac{1}{3}$

17 1, 3, 5, 13, 13

18 2, 10, 10, 22

19 See Figure T26:1.

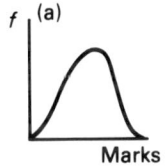
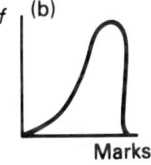
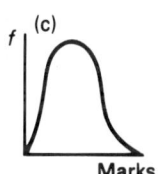

Fig. T26:1

20 Question 4: 3000, 12.38, 1386, 836.66, 1189
 Question 9: 1.881
 Normal: Brenda and question 9

21 20.06, 40%, 99%, 100%
 Data near normal, although a bit 'short' of very high and very low marks.

22 $\sigma = 0.707$
 (a) mean increases by 2, σ unchanged
 (b) mean doubles, σ doubles

23 (a) 545.7, 66.12
 (b) Snack foods; it has a larger standard deviation.

24 (a) 36, 15.35
   (b) Overall less consistency. Advise her to return to her previous aim.

25 Only the mode makes any sense in this context as the data has no continuous property.

26 No set of data has everyone above average, although it is possible to have no-one below average (when?). The maximum is 99% above average, though this depends on which average is used:
   Median: 50% will be above average
   Mode: 89% could be above average (11% at level 1, 10% each at eight other levels, and 9% at the tenth.)
   Mean: 99% could be above average; for example, with 1% at level 1 and 99% at level 2, the mean level would then be 1.99

27 A All employees are paid £200
   B e.g. 20% earn £20, 20% earn £30, 20% earn £50, 40% earn £450
   C e.g. 20% earn £150, 60% earn £200, 20% earn £250
   D e.g. 20% earn £4, 20% earn £6, 20% earn £10, 40% earn £490
   E e.g. 80% earn £5, 20% earn £980
   F e.g. 40% earn £100, 20% earn £150, 20% earn £300, 20% earn £350
   G e.g. 1% earn £0, 24% earn £150, 50% earn £200, 24% earn £250, 1% earn £400

28 B, D, E, F negative skew; C, G normal; A is flat.

Chapter 28 Probability theory

1 (a) $\frac{1}{2}$ (b) $\frac{1}{4}$ (c) $\frac{1}{13}$ (d) $\frac{1}{52}$
2 $\frac{3}{4}$
3 Discuss
4 A and C
5 Sample, say picking ten pens at random from every thousand made.

6 1 and 4 are equally as likely as 2, 3, 5 and 6, so $P(1 \text{ or } 4) = \frac{2}{6} = \frac{1}{3}$. Winning, losing, drawing are almost certainly not equally likely; for example, a very good team matched against a poor one might have $P(\text{win}) = 0.9$, $P(\text{draw}) = 0.08$ and $P(\text{lose}) = 0.02$.

7 The sample is really too small to be sure of anything, but if it was truly random then you might reasonably expect more people to say red than another colour. Otherwise you cannot tell.

8 (a) 15% (b) 30% (c) 80%
   (d) 6%

9 (a) No. The pattern is random, and 17H/13T is quite likely for a few tosses where $P(H) = P(T)$.
   (b) About 50. The greater the number of trials, the lower the percentage difference between actual outcome and the calculated probability of a half of each.
   (d) All are equally likely. However, two heads and two tails *in any order* is ten times as likely as four heads.

10 (a) 0.5 (b) 0.4

11 If a cat is white, it is more likely to be deaf than if it is another colour; i.e. the colour a cat is alters the probability that it is deaf.

12 Because the possible outcomes are HH, HT, TH and TT, you are twice as likely to score H and T than both the same. $P(HH) = P(TT) = \frac{1}{4}$ and $P(H \text{ and } T) = \frac{1}{2}$. Note however that $P(H \text{ then } T)$ is $\frac{1}{4}$.

13 (a) 37, 38, 73, 78, 83, 87
   (b) (i) $\frac{1}{6}$ (ii) $\frac{4}{6} = \frac{2}{3}$

14 These dots should **not** be ringed:
   (0, 0), (5, 1), (6, 1), (3, 2), (4, 2), (5, 2), (6, 2). (a) $\frac{3}{14}$ (b) $\frac{3}{7}$ (c) $\frac{9}{14}$

15 HTHH, HTHT, HTTH, HTTT, THHH, THHT, THTH, THTT, TTHH, TTHT, TTTH, TTTT. Six of the sixteen outcomes have equal heads and tails, so the probability is $\frac{6}{16} = \frac{3}{8}$.

16 Boy/Girl probability is 'two state' like heads/tails, so you can use the tree you drew in question 15 to give:
(a) $\frac{1}{16}$ (b) $\frac{3}{8}$ (c) $\frac{15}{16}$

17 (b) (i) $\frac{1}{18}$ (ii) $\frac{1}{6}$ (iii) 0 (iv) $\frac{5}{18}$ (v) $\frac{5}{6}$

18 (a) SL + SR, SL + MR, SL + ML, SR + MR, SR + ML, MR + ML
(b) $\frac{1}{3}$ (c) 3

19 (a) If $P(1) = x$,
then $P(2) = P(4) = P(5) = x$,
$P(3) = 2x$, $P(6) = 4x$.

Total $10x$, so $P(1) = \dfrac{x}{10x} = 0.1$.

(b) $P(2) = P(4) = P(5) = 0.1$,
$P(3) = 0.2$, $P(6) = 0.4$

(c) They cover all possible outcomes, and $P(\text{certain}) = 1$.

20 (a) 32 (c) (i) $\frac{1}{6}$ (ii) $\frac{7}{30}$

21 You should change, as you then have a $\frac{2}{3}$ chance of winning the £1. Not everyone agrees! Experiment!

Chapter 29 Probability calculations

1 (a) $\frac{1}{120}$ (b) $\frac{1}{80}$ (c) $\frac{1}{960}$ (d) $\frac{11}{12}$
(e) $\frac{89}{320}$

2 (a) 0.512 (b) 0.008 (c) 0.488

3 (a) $\frac{8}{27}$ (b) $\frac{2}{27}$

4 (a) (i) 0.75 (ii) 0.7
(b) See Figure T29:1. (c) $\frac{9}{40}$ (d) $\frac{31}{40}$

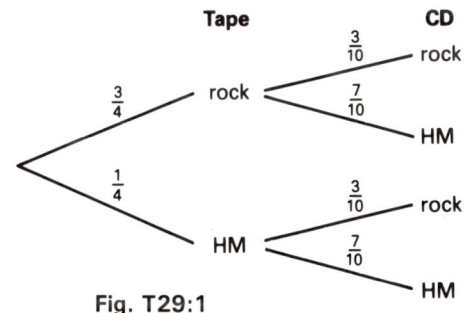

Tape   **CD**

rock $\frac{3}{10}$ rock
$\frac{3}{4}$ rock $\frac{7}{10}$ HM
$\frac{1}{4}$ HM $\frac{3}{10}$ rock
HM $\frac{7}{10}$ HM

Fig. T29:1

5 (a) $\frac{1}{4}$ (b) $\frac{1}{3}$ (c) $\frac{3}{4}$ (d) $\frac{1}{64}$ (e) $\frac{5}{36}$

6 (a) $\frac{1}{13}$ (b) $\frac{4}{663}$ (c) $\frac{13}{204}$ (d) $\frac{1}{11\,050}$
(e) $\frac{2}{17}$ (f) $\frac{28}{1105}$

7 (a) (i) $\frac{27}{125}$ (ii) $\frac{3}{1000}$ (iii) $\frac{27}{250}$ (iv) $\frac{27}{500}$
(b) $\frac{27}{1000}$ (c) $\frac{81}{625}$

8 (a) 0.6, 0.3, 0.7
(b) (i) 0.28 (ii) 0.224

9 Because by the fourth failure $P(\text{passing}) = 100\%$, or certain.

10 (a) $\frac{1}{150}$ (b) $\frac{8}{125}$ (c) $\frac{17}{100}$ (d) $\frac{71}{1500}$

11 (a) See Figure T29:2. (b) $\frac{3}{5}$ (c) 4

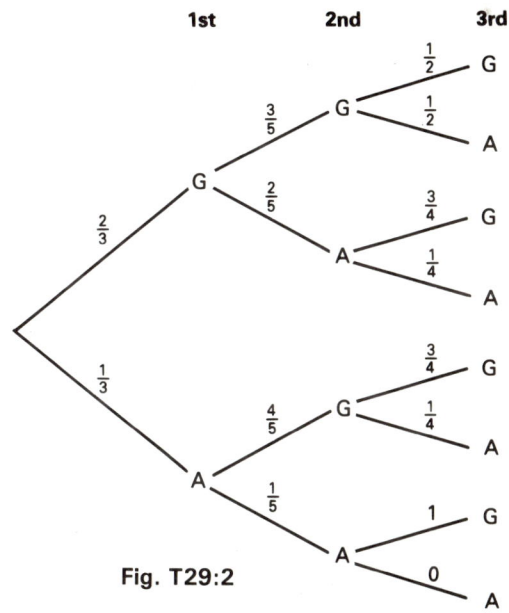

1st   **2nd**   **3rd**

$\frac{2}{3}$ G
  $\frac{3}{5}$ G
    $\frac{1}{2}$ G
    $\frac{1}{2}$ A
  $\frac{2}{5}$ A
    $\frac{3}{4}$ G
    $\frac{1}{4}$ A
$\frac{1}{3}$ A
  $\frac{4}{5}$ G
    $\frac{3}{4}$ G
    $\frac{1}{4}$ A
  $\frac{1}{5}$ A
    1 G
    0 A

Fig. T29:2

12 (a) $\frac{1}{45}$ (b) $\frac{2}{9}$ (c) $\frac{1}{6}$ (d) $\frac{1}{120}$ (e) $\frac{4}{15}$
(f) $\frac{11}{20}$

13 (a) $\frac{4}{51}$, as after the first pick only 51 cards are left; ×2 as it can happen two ways (Ace 5 and 5 Ace).

(b) They are not-independent events; $P(\text{second ace}) < P(\text{first ace})$ because one ace has been removed.

(c) Because the events are not mutually exclusive. They can both happen in one pick (i.e. a red ace). When the ace is red the other card could be anything. The events can also happen either way round, but $P(\text{ace then red}) = P(\text{red then ace})$, so we double.

$P(\text{red ace and any}) = \frac{2}{52} \times 1 \times 2$

$P(\text{black ace and red}) = \frac{2}{52} \times \frac{26}{51} \times 2$

$P(\text{ace and red}) = \frac{1}{13} + \frac{2}{51} = \frac{77}{663}$

14 (a) $\frac{1}{17}$ (b) $\frac{1}{26}$ (c) $\frac{1}{26}$ (d) $\frac{13}{34}$
15 (a) (i) $\frac{1}{15}$ (ii) $\frac{8}{15}$ (iii) $\frac{2}{5}$
   (b) (i) 91 (ii) 130
16 (a) $\frac{1}{6}$ (b) $\frac{11}{15}$ (c) $\frac{2}{87}$ (d) $\frac{1}{10}$
   (e) $\frac{3}{290}$
17 (a) e.g. draw names from a hat.
   (b) (ii) is more likely. (c) $\frac{1}{9}$ (d) $\frac{1}{9}$
18 (a) P(shapes) totals 1
   (b) P(colour) ≠ 1 (c) 0.6
   (d) No cone is red. (e) Total > 1
   (f) Not mutually exclusive, i.e. being a sphere and being white can happen in one pick.
   (g) 7 cubes, 8 spheres, 5 cones
   (h) $\frac{7}{38}$ (i) $\frac{35}{76}$
   (j) (i) A (ii) C (iii) D (iv) B (v) D (vi) D and A
   (k) two
19 (a) $\dfrac{3}{10}$ (b) (i) $\dfrac{N}{2}$ (ii) $\dfrac{3N}{20}$ (iii) $\dfrac{93N}{220}$
   (c) $\dfrac{11}{31}$
20 (a) (i) $\frac{1}{300}$ (ii) $\frac{253}{300}$ (iii) $\frac{23}{150}$ (b) 18
21 (b) (i) $\frac{1}{64}$ (ii) $\frac{20}{64}$ (iii) $\frac{1}{64}$ (iv) $\frac{22}{64}$
22 (a) $\frac{16}{81}$ (b) $\frac{1}{81}$ (c) $\frac{90}{243}$ (d) $\frac{160}{729}$
   (e) 5:3 can only follow from 3:3, so $\frac{160}{729} \times \frac{2}{3} \times \frac{2}{3} = \frac{640}{6561}$.
23 (a) 6 (b) 24
24 (a) (i) 4 (ii) 8 (iii) 16
   (b) 2^n numbers with n digits
   (c) (i) 32 (ii) 64 (iii) 128

Chapter 30 Circle geometry

1 A, 30:3 B, 30:2 C, 30:1 D, 30:5 E, 30:4
2 Perpendicularly bisect chord to give diameter; perpendicularly bisect diameter to give centre. Or use the perpendicular bisectors of two chords.
3 6 cm
4 (a) 40° (b) 48° (c) 48° (d) 80° (e) 40°
5 115°
6 (a) OT, OS (b) ATX, ASY
   (c) (i) 90° (ii) 90° (iii) 120° (iv) 240°
7 (a) ∠S = 70°, ∠Q = 110°, ∠R = 110°, ∠P = 70°
   (b) 90°

8 ∠s in same segment; vertically opposite ∠s
9 ∠ sum of right-angled △
10 ∠s in same segment
11 100° (Join AC; ∠s in same segment; ∠ sum of isosceles △)
12 105°
13 (a) ∠s in same segment
   (b) opp. ∠s of cyclic quad.
   (c) AYXB; AZXY
14 (a) (i) 90° (ii) 90°
   (b) (i) 20° (ii) 70° (iii) 70°
15 (a) $v = s$ (b) $r = s$ (c) $u = q$
22 4 cm

Chapter 31 Matrices

1 (a) $\begin{pmatrix} 2 \\ -1 \end{pmatrix}$ (b) $(2 \quad 2)$
   (c) not possible (d) $\begin{pmatrix} 9 \\ 6 \end{pmatrix}$
   (e) $(-1 \quad \frac{1}{2})$ (f) $\begin{pmatrix} -2 & 6 \\ 0 & -4 \end{pmatrix}$
   (g) (-7) (h) not possible
   (i) $(0 \quad -2)$ (j) $(2 \quad -8)$
   (k) $\begin{pmatrix} -3 \\ -2 \end{pmatrix}$ (l) $\begin{pmatrix} 6 & 1 \\ 8 & 2 \end{pmatrix}$ (m) $\begin{pmatrix} 2 & 4 \\ 2 & 6 \end{pmatrix}$
   (n) $\begin{pmatrix} 0 & -2 \\ -2 & 6 \end{pmatrix}$ (o) $\begin{pmatrix} 6 & -1 \\ -4 & 0 \end{pmatrix}$
2 (c) $\begin{pmatrix} 1 & 0 \\ 0 & 1 \end{pmatrix}$ identity
   $\begin{pmatrix} 0 & -1 \\ 1 & 0 \end{pmatrix}$ rotation +90°
   $\begin{pmatrix} -1 & 0 \\ 0 & -1 \end{pmatrix}$ rotation 180°
   $\begin{pmatrix} 0 & 1 \\ -1 & 0 \end{pmatrix}$ rotation +270°
   $\begin{pmatrix} -1 & 0 \\ 0 & 1 \end{pmatrix}$ reflection in y-axis
   $\begin{pmatrix} 0 & 1 \\ 1 & 0 \end{pmatrix}$ reflection in y = x
   $\begin{pmatrix} 0 & -1 \\ -1 & 0 \end{pmatrix}$ reflection in y = -x
   $\begin{pmatrix} 1 & 0 \\ 0 & -1 \end{pmatrix}$ reflection in x-axis

3 (a) shear, *x*-axis invariant,
  (1, 1) → (3, 1)
  (b) shear, *y*-axis invariant,
  (1, 1) → (1, −1)
  (c) enlargement, scale factor 2,
  centre (0, 0)
  (d) enlargement, scale factor −2,
  centre (0, 0)
  (e) stretch from *y*-axis, stretch factor 2
  (f) stretch from *x*-axis, stretch factor 2

5 (a) no change
  (b) enlargement by scale factor 5
  (c) enlargement by scale factor 2
  (d) no change

6 (a) shear, *y* = 0 invariant,
  (2, 1) → (4, 1)
  (b) enlargement, centre (0, 0), scale
  factor 2
  (c) $\begin{pmatrix} 1 & 2 \\ 0 & 1 \end{pmatrix}$ (d) $\begin{pmatrix} 2 & 0 \\ 0 & 2 \end{pmatrix}$
  (e) 4 square units; $\begin{pmatrix} 2 & 4 \\ 0 & 2 \end{pmatrix}$

7 (a) $y = \frac{1}{2}x$
  (b) Inverse of $\frac{1}{5}\begin{pmatrix} 3 & 4 \\ -1 & 7 \end{pmatrix}$
  $= \frac{1}{5}\begin{pmatrix} 7 & -4 \\ 1 & 3 \end{pmatrix}$, hence

  final matrix is $\frac{1}{5} \times \frac{1}{5}\begin{pmatrix} 7 & -4 \\ 1 & 3 \end{pmatrix}$

  $= \frac{1}{25}\begin{pmatrix} 7 & -4 \\ 1 & 3 \end{pmatrix}$.

8 (a) (0, 0), (2, 2), (1, 3) and (0, 0),
  (0, 4), (−2, 4)
  (c) rotation of +45°; enlargement,
  centre (0, 0), scale factor $\sqrt{2}$
  (d) *n* = 8, (0, 0), (32, 0), (32, 16),
  $\begin{pmatrix} 16 & 0 \\ 0 & 16 \end{pmatrix}$

9 (b) (i) −45° (ii) $\sqrt{2}$
  · (c) *n* = 8, *k* = 16

10 Multiply by $\begin{pmatrix} -1 & 0 \\ 0 & -1 \end{pmatrix}$ then
  add $\begin{pmatrix} -2 & -2 & -2 & -2 \\ 0 & 0 & 0 & 0 \end{pmatrix}$

Chapter 32 Linear programming

1 (a) See Figure T32:1.
  (b) (1, 2); (1, 3); (2, 2); (3, 2)
  (c) (1, $3\frac{1}{3}$)

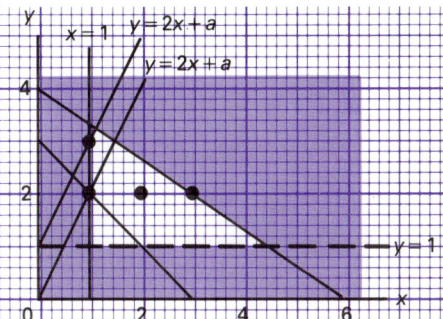

Fig. T32:1

2 (a) See Figure T32:2.
  (b) *x* = 4, *y* = 5

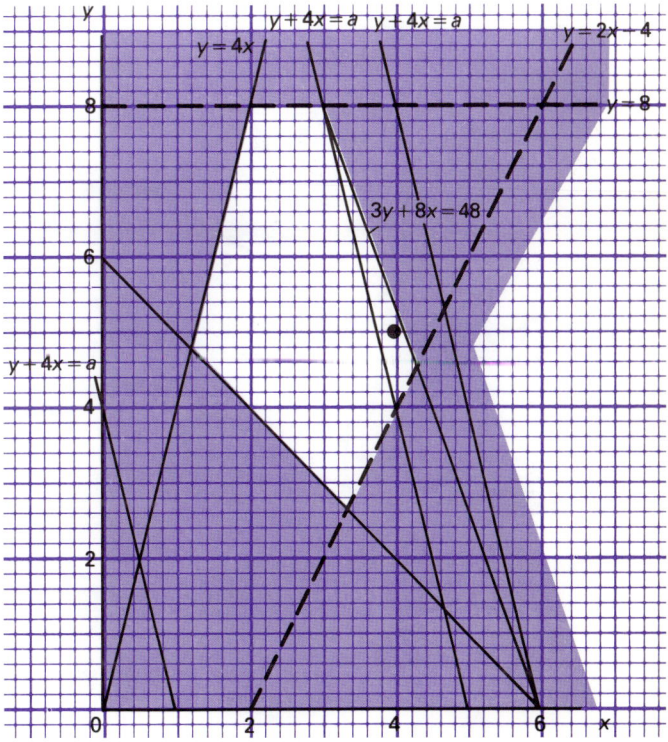

Fig. T32:2

5   Articles a cm², pictures p cm², then
    $a + p \leqslant 700$; $a \geqslant 400$; $p \geqslant 150$;
    $a \geqslant 2p$
    (a) 550 cm²
    (b) Just under 470 cm² ($466\frac{2}{3}$ cm²)
6   four 48-seaters; three 64-seaters
7   0.25 g of A; 0.5 g of B
8   (a) £1400 on dresses; £1000 on
         blouses
    (b) £1600 on dresses; £800 on
         blouses
9   (a) $x + y \leqslant 22$; $x \leqslant 10$; $y > x$;
         $2x + y > 26$
    (c) 9 dogs; 10 cats
10 (a) 11   (b) 25

Chapter 33 Critical-path analysis

1   (a) DE, 13 hours   (b) 1600
    (c) See Figures T33:1 and T33:2;
         ABE; 15 hours

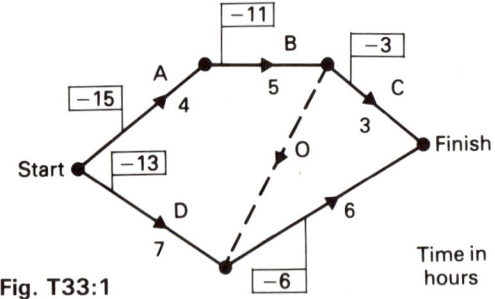

Fig. T33:1

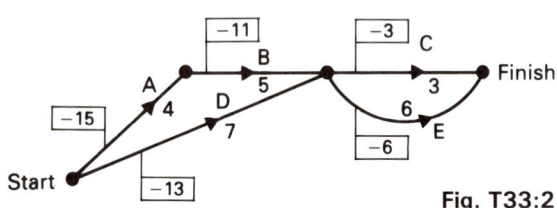

Fig. T33:2

2   (a) 11:25 p.m.   (b) A and E
    (c) H, 12:40; G, 12:20; F, 12:10;
         E, 11:30; D, 12:55; C, 12:50;
         B, 11:55
    (d) A, E, B, F, G, H, C, D
    (e) 12:35 p.m.; A, B, F, G
    (f) See Figure T33:3. H, 12:40;
         G, 12:20; F, 12:10; E, 12:00;
         D, 12:55; C, 12:35;
         B_2, 12:25; B_1, 12:05;
         A, 11:55

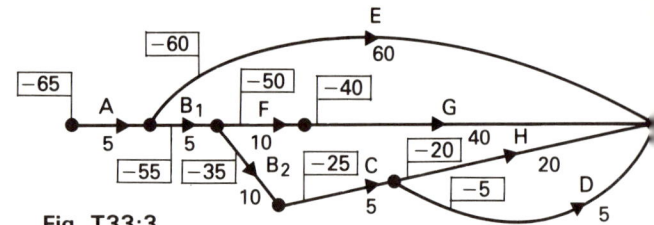

Fig. T33:3

    (g) A, E, B_1, F, G, B_2, C, H, D
3   See Figure T33:4. One worker A, G, H.
    One worker B, C, D, E, F (others
    possible).   Critical path B, C, D, E, F,
    H;   65 minutes

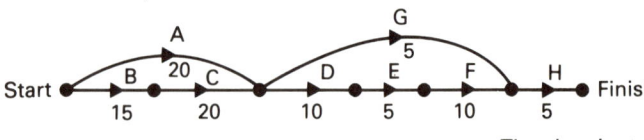

Fig. T33:4
                                    Time in minutes

4   (a) See Figure T33:5.

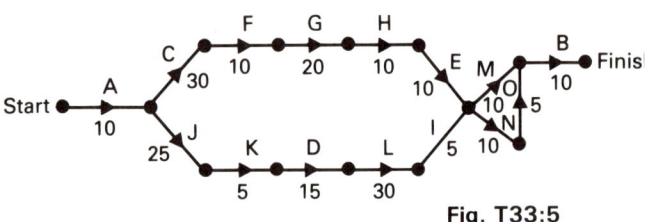

Fig. T33:5

    (b) A, C, F, G, H, E, N, O, B or A, J,
         K, D, L, I, N, O, B;   115 min.
         e.g. Val A, C, F, G, H, E, M,
              B;   110 min
              Winston A, J, K, D, L, I, N,
              O, B;   115 min
    (c) From Figure 33:6 the critical path
         for three workers is A, J, K, L, N,
         B;   90 min. (J and K must be
         done before L, so it is not possible
         to do all three in less than 60
         min.) Time saved 25 min.
    Now, what if they could work
    together on some of the jobs?

Glossary and index

Words in *italics* but without page numbers may themselves be looked up in this glossary.

A

Acre An area of 4840 square *yards* or about 0.4 *hectares*.

Acute An angle between 0° and 90°. See *Basic geometry 1* (page 243).

Adjacent Next to each other. See *Basic geometry 1* (page 243).

Allied Joined together. See *Basic geometry 1* (page 243).

Allowances Tax allowances are amounts you may earn free of income tax.

Alternate On opposite sides. See *Basic geometry 1* (page 243).

Altitude Height measured from a chosen base line; a triangle has three possible heights.

Amount In *compound* interest, the *principal* plus the *interest*.

Ante Before.

Apex The top point, especially of a triangle and a *cone*.

Appreciation Increase in value.

Approximate About the same as, represented by the symbol ≃ or ≈.

Arc Part of a curve; usually part of the *circumference* of a circle. Also a line joining two *nodes* in a *network*; see *Movement 5* (page 246).

Are (say 'air') A metric unit of the same area as a square 10 m by 10 m (i.e. 100 m²).

Ascending Rising; e.g. 1, 3, 6, 10 are numbers writting in ascending order of *magnitude*.

Associative An *operation* in arithmetic which is unaffected by brackets; e.g. addition, because $3 + 4 + 5 = (3 + 4) + 5 = 3 + (4 + 5)$.

Average An attempt to represent a whole range of data, as in 'Julian is of average height for his age.' Statistical averages, see *Averages 1* (page 273).

Axis (pl. axes) A line about which a body rotates. In mathematics we call the numbered lines on a graph the x and y axes. See also *symmetry*.

B

Bar chart See *Charts 3* (page 269).

Bar-line graph A chart showing *frequency* by lines rather than bars as in a *bar chart*.

Basic hours The hours at work for which you receive the *basic wage*.

Basic wage The wage you agree to work for, usually given 'per hour'.

Bearing See *cardinal points* and *three-figure bearings*.

Biased Not being fairly balanced, affecting the impartial outcome of a *trial*.

Billion A thousand million. The Earth is about 5 billion years old. In Britain a billion used to be a million million.

Binary	Based on 2. Binary codes like 10010 are used in computer graphics character definitions.
Bisect	To divide into two equal parts.
Block graph	Similar to a *bar chart* with the bar divided up to look like bricks piled up on each other.
BODMAS	An aid to remembering the order of *operations*: work out what is in **B**rackets first, then **O**f (times), **D**ivide and **M**ultiply, then **A**dd and **S**ubtract. See *Calculators 1* (page 211).
Bonus	An extra payment made to an employee.

C

Cancelling	To cross out, as in cancelling fractions to make them simpler; e.g. $\dfrac{{}^2\!\!\not4}{{}_3\!\not6} = \dfrac{2}{3}$.
Capacity	The amount of space inside a container.
Cardinal point	One of the directions north, south, east and west.
Cent	To do with a hundred, as in centimetre, century and *percentage* (but not centre!).
Centi	A hundredth.
Chain	The *imperial* measurement of 22 *yards* (just over 20 *metres*).
Change of subject	To alter a formula. See *Basic algebra 3* (page 218).
Chord	A straight line joining two points on the *circumference* of a circle. The *perpendicular bisector* of a chord is a *diameter*. See *Circle geometry 1* (page 265).
Circumcircle	See *Movement 4* (page 245).
Circumference	The boundary line of a circle. Also the length of this line.
Class	A statistical grouping. See *Averages 3* (page 273).
Coefficient	The *constant* multiplying an algebraic term; e.g. the 3 in $3x^2$.
Column matrix	A *matrix* with one column of figures; e.g. the vector $\begin{pmatrix} 2 \\ 3 \end{pmatrix}$.
Commission	Pay that depends on the value of a salesperson's sales.
Common fraction	A fraction less than a whole, written like $\frac{3}{4}$.
Commutative	An arithmetical *operation* which is unaffected by the order it is worked; e.g. multiplication, because $3 \times 4 \times 5 = 12 \times 5 = 3 \times 20 = 4 \times 15$.
Complement	One of two parts that make up a whole. Complementary angles add up to $90°$.

Compound	Made up of a number of parts, as in compound *interest*, where each new interest is based on the original amount plus all previous interest. See *Percentages 8* (page 208).
Concave	Curving or pointing inwards.
Concurrent	Meeting at the same point.
Cone	A *pyramid* with a circular base. Volume of, see *Mensuration 5* (page 255).
Congruent	Exactly the same, in size and shape. See *Similarity 5* (page 249).
Conjecture	A guess or forecast without proof.
Consecutive	In *rank order* with no member of the set omitted; e.g. 3, 4, 5, 6 or 1, 4, 9, 16.
Constant	Not changing. In algebra, a value which is the same for all values of the *variables*.
Construct	To draw accurately.
Converge	To move towards, or meet at, one point. The opposite is 'diverge'.
Convert	To change. Conversion graph, see *Practical graphs 3* (page 217).
Convex	Curving or pointing outwards.
Co-ordinates	The two numbers which fix the position of a point on a grid, in the order (how far *horizontally*, how far *vertically*).
Corresponding	In the same kind of position. See *Basic geometry 1* (page 243) and *Similarity 2* (page 248).
Cosine (rule)	See *Trigonometry 6* (page 260).
Credit	A bank credit is money paid into your account. Credit also means time to pay, as in a *credit card*.
Credit card	'Plastic money'. A card agreeing that you can pay for the goods you are buying at a later date, or over a long time (paying quite a lot extra for the *interest*!).
Critical-path analysis	See *Critical-path analysis* (page 281).
Cross-section	The *plane* surface made by cutting a solid.
Cube/cuboid	A cuboid is a solid with six rectangular faces, like a plank of wood. When all the rectangles are squares it is called a cube.
Cubic	In the shape of a *cube*. Also an algebraic expression involving a term of third power, like $5x^3$.
Cumulative	Collected together. Cumulative frequency curve, see *Charts 9* (page 272).
Cyclic quadrilateral	One with its four corners on the *circumference* of a circle.
Cylinder	A *prism* with a circular *cross-section*.

D

Data	Facts from which other facts may be found.
Debit	A debit to a bank account is money paid out of your account.
Decade	Ten years.
Decagon	A ten-sided *polygon*.
Decimal	Based on ten. Often short for a decimal fraction, like 0.5.
Decimal places	The number of figures to the right of the *decimal point*.
Decimal point	The marker to show where the whole numbers (units) end and the fractions (tenths) begin.
Deduce	To reach a conclusion by reasoning.
Deduction	Something taken away.
Denary	Based on ten, as is our usual number system.
Denominator	The bottom number in a *common fraction*, telling you into how many pieces the whole was divided.
Density	The heaviness of a material, measured in grams per cubic centimetre. The density of water is 1 g/cm^3.
Deposit	To pay an amount into a bank or savings account. Also a part payment to guarantee a contract.
Depreciation	Loss of value.
Depression	Angle of, see *Trigonometry 3* (page 257).
Diagonal	A straight line joining two *vertices* of a *convex polygon*.
Diameter	The *chord* of a circle that passes through the centre.
Die	The correct singular of dice.
Difference	The result of subtracting a smaller number from a larger.
Digit	One of the figures 0, 1, 2, 3, 4, 5, 6, 7, 8, 9.
Digit sum	The result of repeatedly adding the digits of a number until the result is a single *digit*. The digit sum of 49 is 4.
Directed number	A number that has a sign to indicate whether it is above or below zero. (See *positive* and *negative*.)
Direct debit	Permission to your bank to *debit* your account at the request of named people or firms.
Direct proportion	Two amounts are in direct proportion when they change at the same rate; see *Variation 2* (page 241).
Discount	The amount taken off a price.
Discrete	Able to be counted, like sheep.
Dividend	The number being divided; e.g. the 8 in 8 ÷ 4. Also the payment made to shareholders in a company.

Divisor	The number you are dividing by; e.g. the 4 in 8 ÷ 4.
Domain	The set of numbers, or the *region* on which *operations* are performed.
Double time	Overtime paid at twice the *basic wage*.
Dozen	Twelve. (A 'baker's dozen' is 13.)

E

Elevation	View, see *Solid geometry 2* (page 251). Angle of, see *Trigonometry 3* (page 257).
Empirical	Found by experiment or past experience.
Equilateral	Equal sided.
Equivalent	Having the same value; e.g. $\frac{3}{6} = \frac{1}{2} = 0.5 = 50\%$.
Estimate	To give an *approximate* value.
Evaluate	Find the value of.
Event	See *Probability theory 1* (page 276).
Exchange rate	The amount of currency you can buy using another currency.
Exclusive	For 'mutually exclusive', see *Probability calculations 1* (page 277).
Expand	To multiply out brackets; e.g. $5(x + 6)$ expands to $5x + 30$.
Express	To write in a certain way, e.g. express the answer in metres.
Expression	A collection of terms with no equality, like $4x - 2y^2 + 3$.
Exterior angle	See *Basic geometry 3* (page 244).

F

Face	The flat side of a *polyhedron*.
Factorise	To split a number or an algebraic expression into *factors*; e.g. 12 factorises to 3×4 or 2×6 or $2 \times 2 \times 3$, and $3x - 9$ factorises to $3(x - 3)$.
Factor	An *integer* or algebraic term which multiplies another integer or term to make the original number or term. See *factorise* and *highest common factor*.
Fibonacci sequence	A *sequence* where each new term is made by adding the previous two terms, often starting 1, 1, 2, 3, 5.
Foot (pl. feet, abbrev. ft)	An *imperial* length of 30.48 cm.
Formula	A recipe to make something; e.g. the formula for the area of a rectangle is multiply the base by the height, or $A = bh$.
Formulate	To express in a clear or definite form.
Frequency	The number of times something occurs.

Frustum	The part of a *pyramid* or *cone* between the base and a *plane* parallel to the base.
Function	An algebraic 'event'; e.g. a function of x such that x becomes x^2, often written f(x): $x \rightarrow x^2$. Then f(2) = 4. See page 50.

G

Gallon	An *imperial* measure of liquid volume, about 4.55 litres.
Generalise	Express in general terms, usually using an algebraic *formula*.
Googol	A large number, 1×10^{100}, or 1 followed by 100 zeros.
Gradient	The slope of a line; see *Algebraic graphs 6* (page 238) and *Trigonometry 3* (page 257).
Gram (abbrev. g)	A metric measure of weight. Two drawing pins weight about 1 gram.
Grid reference	See *six-figure reference*.
Gross	Without any *deduction*. Also 144 items.
Growth and decay	See *Charts 7* (page 271).

H

Hatch	To shade an area by drawing parallel sloping lines on it.
Hectare	100 *ares*.
Hecto	A hundred.
Heptagon	A seven-sided *polygon*.
Hexadecimal	A number system of base 16, used in computer programming. The digits are 0 to F, so F6 is 15 sixteens and 6 units = 246 in *denary*. A BBC computer uses '&' before a hexadecimal number.
Hexagon	A six-sided *polygon*.
Highest common factor (abbrev. HCF)	The highest number which is a *factor* of each number in a set of numbers. See *Basic arithmetic 5* (page 203).
Hire purchase (abbrev. HP)	Payment for an article by *instalments*. Largely replaced by *credit cards*.
Histogram	See *Charts 8* (page 271).
Horizontal	Across the page. Remember it by 'horizon'.
Hourly rate	The *basic wage* paid per hour.
Hypotenuse	The longest side of a right-angled triangle. It is always opposite the right-angle.

I

Image	The result of the *transformation* of an object.
Imperial	The system of measures which were used in the British Empire.

Improper fraction	A fraction, more than a whole one, where the *numerator* is more than the *denominator*.
Inch	An *imperial* measure of length, equal to 2.54 cm. There are 12 inches in a *foot*.
Incircle	See *Movement 4* (page 245).
Inclusive	Including both ends.
Independent	Not relying on anything else. Independent events, see *Probability calculations 2* (page 278).
Index	The raised figure used in a power, e.g. the 2 in 5^2 (meaning multiply two fives together) and the -2 in 10^{-2} (which is another way of writing $\frac{1}{100}$).
Income tax	Money taken from your earnings to finance government spending.
Increment	An increase.

Inequality	An expression showing that two things are not equal.
	Symbols: $-5 > -7$ meaning -5 is more than -7.
	           $x \geqslant 8$ or $x \not< 8$ meaning the value of x is 8 or more.
	           $3 < 6$ meaning 3 is less than 6.
	           $x \leqslant 4$ or $x \not> 4$ meaning x is not more than 4.

Infinite	Without end.
Inflation	In finance, the decrease in the value of money; e.g. a loaf costing 12p in 1970 cost 70p in 1990.
Instalment	One of a series of payments.
Integer	A whole number, like 4 and -4. Zero is usually taken to be an even integer.
Intercept	Part of a line between two crossing points.
Interest	An amount paid by the borrower of money to the lender.
Interior angle	An inside angle; see *Basic geometry 3* (page 244).
Intersection	The crossing point.
Inverse	Opposite or upside down. The inverse of $+$ is $-$. The inverse of $\frac{5}{6}$ is $\frac{6}{5}$.
Inverse proportion	See *Variation 1* (page 240).
Invoice	A bill, setting out the payment required.
Irrational	Unable to be written as an exact fraction. All *square roots* of *prime numbers* are irrational, as is π.
Irregular	Not *regular*.
Isometric	A two-dimensional view of a solid. See *Solid geometry 2* (page 251).
Isosceles	With two equal sides. (*Iso* – 'equal', *sceles* – 'legs')
Iteration	Repetition. See *Equations 8* (page 233).

K

Kilo	A thousand, as in kilometre, a thousand metres. Sometimes kilo is taken as short for kilogram.
Kite	A *quadrilateral* with two pairs of equal *adjacent* sides.

L

Like fractions	Fractions with the same *denominator*.
Like terms	Terms of the same kind; e.g. 3x and 4x; 2a^2b and 5a^2b.
Line of symmetry	The fold line or mirror line.
Linear equation	One that has no *power* above 1; e.g. 3x + 4 = 9.
Linear graph	A straight line. Its equation can always be simplified to the form $y = mx + c$.
Linear programming	See *Linear programming* (page 280).
Litre	A metric unit of liquid volume. A litre is about 1.76 pints.
Locus (pl. loci)	The path made by a moving point. See *Movement 3* (page 245).
Lowest common multiple (abbrev. LCM)	The smallest number which is in the multiplication table of each of a set of numbers. See *Basic arithmetic 6* (page 204).

M

Magnitude	Size. Magnitude of a vector, see *Vectors 8* (page 263).
Mapping	The action of a *function*. Also a *transformation* in geometry.
Mean	See *Averages 1* (page 273).
Median	See *Averages 1* (page 273).
Mega	A *million*.
Metre	The unit of length after which the metric system is named, originally taken as 2.5×10^{-8} of the equator.
Micro	A millionth.
Milli	A thousandth.
Million	A thousand thousand.
Mixed number	An *integer* and a fraction, as $1\frac{1}{2}$.
Mode	The most frequent *score(s)* in a set of *data*.
Modulus	The size of a vector. See *Vectors 8* (page 263).
Mortgage	A loan secured on a property. If the loan is not paid back as agreed, the property can be sold by the lender.
Multiple	A number made by multiplying another number, as in multiplication tables.

Mutually exclusive	See *exclusive*.

N

National Insurance	The state system by which people pay to support the social services.
Natural numbers	Numbers used for counting.
Nautical mile	The distance round the Earth which *subtends* an angle of 1′ (1 minute) at the centre. About 1.15 land miles.
Negative number	A number below zero, indicated with a − sign.
Net	The *plane* shape that can be folded to make a solid. An amount after *deductions*.
Network	A system of *intersecting* lines. See *Movement 5* (page 246).
NI	*National Insurance*.
Node	A junction in a *network*. See *Movement 5* (page 246).
Nonagon	A nine-sided *polygon*.
Not-independent	See *Probability calculations 3* (page 279).
Notation	The symbols used in a mathematical problem.
***n*th term**	See *Equations 1* (page 225).
Numerator	The top number in a *common fraction*.

O

Object	The shape that is to be *transformed* to give the *image*.
Obtuse	An angle between 90° and 180°.
Octagon	An eight-sided *polygon*.
Odds	A way of stating probability. Odds of 5 to 1 against means 1 chance that the *outcome* will be favourable, 5 chances that it will not, equivalent to probability $\frac{1}{6}$.
Operation	In arithmetic, one of +, −, × and ÷, or something else defined in a particular case.
Ordered pair	Another name for the *co-ordinates* of a point.
Order of rotational symmetry	The number of times that a shape fits into a tracing of itself in one full rotation. See *Plane symmetry 2* (page 244).
Orthographic	Two-dimensional views of a solid. See *Solid geometry 2* (page 251).
Outcome	In probability, the result of a trial. See *Probability theory 1* (page 276).
Overtime	Extra work beyond your agreed *basic hours*.

P

Parabola	The curve obtained when a *cone* is cut parallel to the *slant height*. Also the curve taken by, e.g., a golf ball. See *Algebraic graphs 4* (page 236).

Parallelogram	A *quadrilateral* with both pairs of opposite sides parallel.
PAYE	Pay As You Earn; a method of collecting *income tax* from your pay.
Pension	An amount paid to you after you have retired from full-time work.
Pentagon	A five-sided *polygon*.
Per annum	Each year.
Percentage	For every hundred.
Percentage error	Error over true amount times 100%.
Percentile	See *Charts 9* (page 272).
Perfect square	An algebraic expression of the form $(ax + b)^2$.
Perimeter	The boundary of an area.
Perpendicular	At right angles. Perpendicular bisector, see *Movement 4* (page 245).
Personal allowance	An amount the government allows people to earn free of *income tax*.
Pictogram	See *Charts 1* (page 269).
Pie chart	See *Charts 5* (page 270).
Piecework	Pay based on how many pieces of work you do.
Pint	An *imperial* measure of liquid volume, about 0.57 litres.
Place value	The value of a *digit* according to the column it is in.
Plan	A view looking from above. See *Solid geometry 2* (page 251).
Plane	Having only two dimensions; 'flat'.
Point symmetry	See *Plane symmetry 3* (page 244).
Polygon	A *plane* figure bounded by straight lines. A circle can be thought of as a polygon with an *infinite* number of straight sides.
Polyhedron	Any solid made up of *plane* surfaces.
Position vector	See *Vectors 2* (page 262).
Positive number	A number above zero, having a + sign, or no − sign.
Post	After.
Pound (abbrev. lb)	An *imperial* measure of weight, about 0.45 kg.
Power	The result of multiplying a number by itself a number of times. It can be shown using an *index* (raised figure) as in 5^3.
Prime	A number which has only two different *factors*. See *Basic arithmetic 4* (page 203).
Principal	The amount on which *interest* is calculated.
Prism	A *three-dimensional* shape with a *constant cross-section*. Volume of, see *Mensuration 5* (page 255).
Produce	To make a line longer.

Product	The result of a multiplication.
Proportion	See *Variation 1* (page 240).
Proportional division	Dividing an amount in a given *ratio*. See *Ratio 5* (page 210).
Proportionate bar chart	See *Charts 4* (page 270).
Pyramid	A three-dimensional shape with any shape base, and triangular sides meeting at a point.
Pythagoras' theorem	See *Mensuration 1* (page 253).

Q

Quadrilateral	A four-sided *polygon*.
Quadratic	An equation involving a square term. It may have two, one, or no solutions.
Quadrant	A quarter of a circle.
Quotient	The result of a division.

R

Radius (pl. radii)	The distance(s) from the centre to the *circumference* of a circle or *arc* of a circle.
Range	In statistics, the difference between the highest and lowest number in the *data*.
Range of error	See *Basic arithmetic 2* (page 202).
Rank order	In order, usually from smallest to biggest.
Ratio	The relation of two amounts to each other; see *Ratio* (page 209).
Rational	Able to be expressed as an exact fraction. *Recurring* decimals are rational. See *irrational* and *Calculators 7* (page 214).
Raw data	The result of a survey before any work is done on it.
Real number	A number that exists, unlike $\sqrt{-1}$ which is unreal or imaginary.
Ream	Originally 480, but a printer's ream was 516, and now a ream is often 500!
Reciprocal	One divided by the number, e.g. $\frac{1}{2}$ is the reciprocal of 2. Calculators often have a $\boxed{1/x}$ reciprocal key. See *Calculators 6* (page 213).
Rectangular number	Any number which is not *prime* (though 1 is doubtful).
Recurring	Repeating. See *Calculators 7* (page 214).
Redundancy	Losing your job because your employer no longer has work for you. If you have been employed for some while you receive a cash sum from your employer.
Reflection	A *transformation* when the *object* and *image* are *symmetrical* about a mirror line.

Reflex	An angle between 180° and 360°.
Regular	Regular *polygons* have all sides and all angles equal.
Representative fraction	The scale of a map; e.g. 1:1000 or $\frac{1}{1000}$, meaning 1 unit on the map is 1000 units on the ground.
Resultant	In vectors, the single vector that gives the same final position as all the others in a set. See *Vectors 6* (page 263).
Rhombus	A *parallelogram* with four equal sides. Sometimes called a 'diamond'.
Root	Short for 'square root'. See *Calculators 5* (page 213) and *Basic algebra 6* (page 220). Also a solution of an equation, especially a non-linear one.
Rotational symmetry	See *Plane symmetry 2* (page 244).

S

Salary	The amount a worker is paid for a year's work, usually paid monthly.
Scale drawing	A drawing that is mathematically *similar* to the original shape.
Scale factor	The *factor* by which a *similar* shape is made bigger or smaller.
Scalene	Having no two sides the same length.
Scattergram	Also scatter graph. See *Charts 6* (page 270).
Scientific notation	Another name for *standard form*.
Score	Twenty. Also the total made.
Secant	A straight line cutting a circle. A *chord* is part of a secant.
Sector	Part of a circle bounded by two *radii* and an *arc*.
Segment	Part of a circle bounded by a *chord* and an *arc*.
Sequence	A set of numbers that are linked by a rule.
Shear	A *transformation* in which all points move parallel to a fixed line. The distance they move is proportional to their distance from the fixed line.
Shift vector	See *Vectors 3* (page 262).
SI	Système International d'Unités; the correct name for the metric system.
Simple interest	The interest is not added to the amount loaned. See *compound interest*.
Significant figures	An *approximation*. All figures, except a string of zeros at the beginning, count as significant.
Similar	See *Similarity* (page 247).
Simultaneous	At the same time. See *Equations 3* (page 227).
Sine rule	See *Trigonometry 5* (page 259).

Six-figure grid reference	This fixes the position of a point on an Ordnance Survey map. See Chapter 17.
Sphere	The mathematical name for a ball whose *cross-section* is always a circle. Volume of, see *Mensuration 5* (page 255).
Square numbers	Numbers made by multiplying an *integer* by itself.
Square root	See *root*.
Standard form	See *Basic arithmetic 3* (page 202).
Standing order	A request to a bank to make regular payments to someone else.
Statement (bank)	A notice from a bank showing the *credits* and *debits* made on your account over a period of time.
Stick graph	Another name for a *bar-line graph*.
Subtend	Angle APB is subtended at a point P by the lines AP and BP.
Sum	The result of an addition.
Superannuation	Payments made through your employer towards a *pension*.
Supplementary angles	Angles which together make 180°.
Surd	An *irrational* number.
Symmetry	See *rotational symmetry* and *line of symmetry*.
Symmetrical	Having *line symmetry*.
Symbol	Using a letter or a sign to represent something else. In mathematics and science a letter usually represents a number, but sometimes it represents a measurement, like h for height.
T	
Tally	Counting by using marks, like ///. Often every five are connected, so 5 is 卌.
Tangent	Trigonometrical, see *Trigonometry 1* (page 256). Geometrical, see *Circle geometry 6* (page 267).
Tariff	A table of prices.
Tax code	A code to show the income a person may have free of income tax. For reasons best known to the Inland Revenue, the final figure is omitted, so a code of 356 means you can earn £3560 free of tax.
Taxable pay	The amount of earnings on which *income tax* has to be paid.
Tessellation	A pattern of shapes which entirely covers a surface.
Tetrahedron	A *pyramid* on a triangular base. A regular tetrahedron has four equilateral triangles as its sides.
Three-figure bearings	Directions given by the amount of turn clockwise from north.

Time and a half	Overtime (more than your *basic hours*) when you are paid half as much again as your *basic wage*.
Tonne	A metric unit of weight, equal to 1000 *kilograms*.
Top-heavy fraction	Another name for an *improper fraction*.
Transformation	A change of position or shape according to a rule. In this course we have met *reflection, rotation, enlargement* and *translation*.
Translate	To slide to a new position in one direction; the action of a vector.
Transpose	See *change of subject*.
Trapezium	A *quadrilateral* with one pair of parallel sides.
Trapezoidal rule	See *Practical graphs 2* (page 216) and *Algebraic graphs 7* (page 238).
Travel graph	Also called a journey graph or a time/distance graph. See *Practical graphs 1* (page 216).
Tree diagram	Used in probability to represent choices. See *Probability calculations 2* and *3* (page 278).
Trial	In probability the carrying out of the experiment, e.g. picking two cards from a pack.
Triangular numbers	Numbers that can be represented by an *equilateral* triangle of dots, starting 1, 3, 6, 10.
Trundle wheel	A wheel of standard *circumference*, usually a metre, that can be rolled along the ground to measure distance.

U

Unbiased	Free from prejudice. See *biased*.
Unitary method	Finding the cost or value of one item to help find the cost or value of several.

V

Variable	Liable to change, like the letters in algebra.
VAT	Value added tax. A tax put on goods and services by the government.
Vertex (pl. vertices)	The corner(s) of a shape.
Vertical	Upright, or going towards the top of a page.
Vertically opposite angles	See *Basic geometry 1* (page 243).
Vulgar fraction	Another name for a *common fraction*.

d	An *imperial* length, 0.9144 metres.